Counterman's Guide to Parts and Service Management

by Gary A. Molinaro

WE ENCOURAGE
PROFESSIONALISM

THROUGH TECHNICIAN
CERTIFICATION

Delmar Publishers Inc.®

NOTICE TO THE READER

Cover Photo
Electronic Catalog™, Triad Systems Corporation

Text, Design, and Production by
Scharff Associates, Ltd.
RD 1   Box 276
New Ringgold, PA 17960

Scharff Staff
Editor: Lois Breiner
Contributing Writers: Drew Corinchock, Darla Lotz, Richard J. Paquette
Cover Design: Eric Schreader

Delmar Staff
Editor-in-Chief: Mark W. Huth
Associate Editor: Joan Gill

For information address Delmar Publishers, Inc.
2 Computer Drive West, Box 15-015
Albany, New York 12212

Printed in the United States of America
Published simultaneously in Canada
by Nelson Canada,
a division of International Thomson Limited

10 9 8 7 6 5 4 3 2 1

Library of Congress Cataloging-in-Publication Data

Molinaro, Gary.
    Counterman's guide to parts and service management.
        p.   cm.
    Includes index.
    ISBN 0-8273-3629-2.—ISBN 0-8273-3630-6 (instructor's guide)
    1. Automobile supplies industry—Management.   2. Automobile
supplies industry—Inventory control.   I. Title.
HD9710.3.A2S33   1989
629.2'068'8—dc19   89-1078
                                   CIP

518-53

CONTENTS

PREFACE

There was a time when buying auto parts was a difficult task. The shops were hidden away with little in the way of signage, front display windows or even a listing in the phone book.

For the most part, these were wholesale operations only, catering to professional garage mechanics and some hot-rodding do-it-yourselfers. These people knew what they wanted to buy when they called, and the man behind the counter—there were no women in the parts store then—took the order, looked it up in a catalog, wrote up the part number and the price, and took the part from the shelf in the back room. These order-takers worked behind a counter that was located close to the front door, with little or no floor space between the counter and the door. The parts store was a dingy place, not much cleaner than the service bays they served, and saw little need to cater to anyone outside of the inner circle of the automotive replacement parts.

Knowledge of cars was optional for these order-takers, ability to serve a customer was not necessary, and sales skills were never a prerequisite for the position of counterman. Obviously, a lot has changed since then.

In those days, there were a limited number of places to buy auto parts, and the motoring public and the mechanics who serviced their vehicles were held to a limited choice in service and parts. All that changed in the 1960s, '70s, and '80s, and now everyone is trying to get a piece of the auto parts trade. Drug stores, grocery stores, even department stores now sell parts, and there are hundreds of specialized service facilities available to the driving public.

In this highly competitive environment, parts stores became more retail-oriented and custom service-oriented. Both professional mechanics and walk-in customers have become vital to a profitable parts store. And the order-taker of the past, that so-called "counterman," has become a vital link in the automotive aftermarket's sales chain.

The men and women selling parts today, especially those selling in tomorrow's replacement parts market, must be knowledgeable of automotive systems, know how to sell a customer the right parts the first time, and do it all with an eye on the bottom line.

Today's and tomorrow's countermen have to be customer service oriented and trained to sell the parts that their predecessors just took orders for.

The title "counterman" carries with it a lot of out-dated notions, but it is a title worn proudly by the many parts professionals in modern auto parts stores—today's countermen.

ACKNOWLEDGMENTS

Grateful acknowledgment is made to the following companies, which provided reference material and information used in the preparation of this book.

A-C Delco, Division of General Motors Corporation
Allied Aftermarket, Division of Allied Signal Corporation
Car Quest
Dana Corporation
Fel-Pro Inc.
Master Products Manufacturing
Moog Automotive, Inc.

National Automotive Parts Association (NAPA)
Niehoff, Inc.
Triad Systems Corporation

Appreciation and thank you are extended to the following businesses for their cooperation in taking photographs for this book:

Moyer's Car Care Center, Inc.
St. Clair Auto Supply

I would like to thank the following for reviewing the manuscript and for their helpful comments:

Mr. Allen Copland
Keyaho College
Ft. McMurray
Alberta, Canada

Mr. Steve Jones
Clover Park Vocational Technical Institute
Tacoma, Washington

Mr. Jerry Laverty
Lincoln Technical Institute
Indianapolis, Indiana

Mr. Edward O'Riley
Chemeketa Community College
Salem, Oregon

Special thanks to Tom B. Babcox, Daniel J. Cook, and Vicky Poulsen of *Counterman* magazine, a Babcox publication.

CHAPTER ONE

INTRODUCTION TO AUTO PARTS MANAGEMENT

Objectives

After reading this chapter, you should be able to:
- Explain the function and importance of the automotive aftermarket.
- Explain the organization and various levels of the automotive aftermarket parts supply network.
- Explain the function and importance of jobber stores in the aftermarket parts supply dealership network.
- Explain the important role played by jobber store parts managers and counterpeople in servicing aftermarket customers.
- Explain the various types of accounts and customers serviced on the jobber store level and have a basic understanding of how these accounts differ in their needs
- Explain the basic duties and job tasks of counterpeople and managers at the jobber store level.
- Identify the qualities and personality traits required for a successful career as a counterperson or jobber store manager.

The automotive industry is second only to the food industry in its contribution to America's gross national product. Servicing and maintaining vehicles is an incredibly large, diverse, and exciting business. Consider the following facts:

- 140 million cars and 42 million trucks and buses are registered and operating on American roads. In Canada vehicle registration is 14.6 million.
- Americans drive their vehicles nearly 1.0 trillion miles per year, or the equivalent of more than 10,000 round-trips to the sun.
- The average life span of American vehicles is 12 years.
- The average vehicle age is 6.1 years, with 80 percent of all cars on the road three years or older, which is the prime age for maintenance and repair dollars.
- Owners will spend approximately $7,000 to repair and maintain a vehicle during its life span.
- In the United States and Canada, the sales volume of automotive parts and supplies at the wholesale (jobber) level is nearing $30 billion per year.

THE AUTOMOTIVE AFTERMARKET

Few people are aware of the complexity and operation of what is generally called the Automotive Service Industry or Automotive Aftermarket. *Automotive aftermarket* refers to the network of businesses that supplies replacement parts and services to vehicle service stations, independent garages, specialty repair shops, car and truck dealerships, fleet and industrial operations, and the general buying public—consumers interested in maintaining and servicing their own vehicles.

Necessities include replacement items such as spark plugs and ignition parts, engine parts, brake system components, chassis, steering, suspension parts, filters and cartridges, belts and hoses, exhaust system parts, heating and cooling system components, paint and body repair products, batteries, tires, oil, lights, wipers, and hundreds of other parts so readily available for virtually every make and model of car or truck on the road that most people do not realize the vastness of the distribution network needed to move the parts from the manufacturer to the final user.

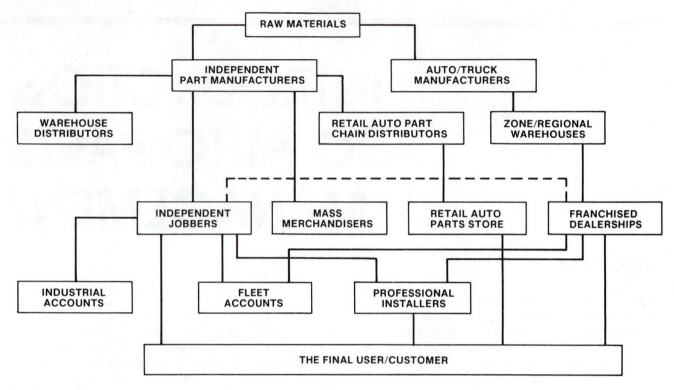

FIGURE 1-1 Parts supply network

PARTS SUPPLY NETWORK

The vast distribution of thousands of auto parts and items through jobber outlets is made possible through an extensive parts supply network that operates on various levels as illustrated in Figure 1-1.

The traditional automotive aftermarket distribution system begins with the manufacturer of the parts. Vehicle manufacturers and independent parts manufacturers sell and supply parts to approximately 1,000 warehouse distributors throughout the United States. These warehouse distributors, or WDs, as they are often called, carry substantial inventories of many part lines.

Warehouse distributors serve as large distribution centers or parts hubs. In turn, the WDs sell and supply parts to auto parts wholesalers. In the aftermarket industry, parts wholesalers are commonly known as *jobbers*.

THE ROLE OF THE JOBBER

Auto part wholesalers or jobbers sell parts to a variety of markets. The jobber's traditional wholesale customers are the operators of service stations and independent garages who install the parts and products in their customer's vehicles. The parts sales to these two types of installers account for an average of 55 to 60 percent of a jobber's business.

Jobbers also supply parts to fleet operators, such as taxi or delivery truck companies, farmers and agricultural accounts, and industrial and institutional clients. They might also sell a number of products to car and truck dealerships, franchised repair shops, and tire dealers, although this is usually a smaller percentage of a jobber's traditional market.

Retail or Walk-In Trade

Walk-in customers are the second largest segment of jobber customers. These retail customers include both experienced and novice do-it-yourselfers interested in repairing and servicing their own vehicles. They also include the general buying public interested in buying auto accessories and basic maintenance and upkeep items such as oil, antifreeze, waxes, wipers, etc. Walk-in business represents a large percentage of sales. Small jobbers might rely on walk-in trade for up to 40 percent of all business. Larger jobbers can sell 20 to 25 percent of all parts and products to these retail customers.

JOBBER DIVERSITY

More than 28,000 jobbers are operating in the United States. These jobbers include a number of very large, well-financed organizations, many of which operate numerous branch stores.

About one-third of the largest jobbers in the United States supply about 83 percent of the parts, equipment, and services sold through wholesalers.

Many of the larger jobber stores also operate machine shops that offer important services to the auto repair industry. Professional installers and serious amateur mechanics rely on jobber machine shops for block boring, cylinder head reconditioning, rod rebuilding, complete engine rebuilding, crankshaft grinding, flywheel machining, brake lathe work, and other machining operations.

About two-thirds of all jobbers are small to medium sized operations that supply about 17 percent of the professional and DIY trade. Small, independent "Mom and Pop" parts stores are frequently located in small towns or rural areas. They generally do not have the financial strength or floor space to carry anything but parts that can be sold quickly.

THE SYSTEM'S ADVANTAGES

Why does a replacement part move through so many hands before it becomes part of the motorist's vehicle?

The answer is that over the years this parts supply network has proved to be the most economical method of distribution for the automotive aftermarket industry. It is also the most efficient—virtually any auto part is available to a customer in less than 24 hours.

Before warehouse distributors became part of the system in the 1940s, manufacturers operated their own distribution systems with factory warehouses and sales forces across the nation. This was not only costly for the manufacturers, it also forced them into a type of business operation they preferred not to handle. WDs, by providing their own factory warehousing and sales forces for large numbers of products, permitted manufacturers to share the costs of warehousing, sales, and distribution.

CHANGES IN THE TRADITIONAL PARTS SUPPLY SYSTEM

For many years, the manufacturer to warehouse to jobber wholesaler to installer/final user distribution system was the standard method of moving parts through the aftermarket. The highly structured system worked well for a number of reasons.

- All but a small percentage of wholesale trade involved professional installers and mechanics. Jobbers could ignore the retail or walk-in trade and still be successful.

- Jobber stores were often the sole source of parts in the marketplace. Competition for customers remained on the jobber level. And because most jobbers worked through the WD system, pricing remained more or less constant, with store service being the key to landing and holding accounts.

- The number of makes and models of vehicles on the road, although growing, was still at a manageable level. The number of parts needed to sustain an acceptable fill-rate and the size of the cataloging system needed to locate and track parts were still manageable.

- The basic design and technology of vehicles remained relatively constant. The introduction of new technology with the accompanying new parts and installation skills was at a much slower rate than today. The "mechanics" of vehicle repair were more clearly understood by jobbers and their customers.

But by the mid 1970s, the automotive aftermarket began to experience rapid and dramatic change that has affected the way manufacturer's WDs and jobbers do business.

MASS MERCHANDISERS AND RETAILERS

In the past twenty years, do-it-yourselfers, eager to save maintenance dollars by doing more and more of their own repairs and vehicle upkeep work, have accounted for much of the growth in the aftermarket industry. Beyond performing periodic oil changes, the DIY, through self-training, trial-and-error, and general automotive classes, has advanced in skill and knowledge. Unfortunately, many of the traditional jobber stores were not ready or willing to deal with this growing segment of the parts market. Mass merchandisers, such as department store chains and discount houses, had played a very small role in the parts supply network for several decades. These merchants quickly recognized this DIY trend and acted. They began stocking larger and larger inventories of popular, fast-moving items such as spark plugs, filters, wipers, and oil. Other types of businesses, such as drug, hardware, and even food stores, also began selling these items at or below cost to help generate customer traffic and offer true one-stop shopping.

Failure to forecast and prepare for this growth in retail trade hurt many traditional jobbers and professional installers. Fortunately, successful jobber stores are now aggressively pursuing this lucrative walk-in trade through changes in marketing strategy, additional services offered, and increased merchandising efforts.

AUTO PARTS RETAIL CHAINS

National or regional auto parts retail chain stores now make up a significant segment of the independent jobber's competition. Like the mass merchandisers these retail chains compete largely for the DIY, shade-tree mechanic, and regular retail customer, but they will sell to all types of accounts.

The auto parts retailers sell only auto parts and closely related items such as tools and accessories. Unlike general mass merchandisers, auto retail chains deal in many hard parts lines such as alternators, starters, master cylinders and brake items, exhaust system and suspension parts.

Their stock might not be as complete as that of independent jobbers, but they concentrate on popular and fast-moving items.

The organization of retail auto parts stores also differs from jobber stores in a number of ways. In the traditional jobber distribution system, both the warehouse distributorships and the jobber stores are independently owned and operated companies. The jobber might have a headquarters store and branch locations, but the owner or manager is free to do business with the WDs and manufacturers of his or her choosing. The same is true of independent warehouse distributors. And while independent jobbers and WDs might join co-op buying groups or programmed distribution groups to increase their buying power and marketing strength (see Chapter 9), they are not obligated to do so.

In contrast the retail auto chains operate through their own distribution system. Chain A auto retail stores source parts through Chain A warehouse distributors, and Chain A warehouse distributors serve only Chain A retail stores.

The actual auto retail stores may be owned and operated by one or several owners, but they act more as franchises in the retail chain. So rather than a series of independent companies working together, the operators of retail chain franchises buy into a preset program and distribution chain that is controlled by top management in the mother company. So managers at the retail level may have very little input as to how parts are sourced or company policies are formulated.

But the size and central organization of auto retail chains give them some distinct advantages. One is tremendous purchasing power. Another is an aggressive and systematic approach to marketing and merchandising. Extensive help with store layout, merchandise planographs and displays, and advertising help are just some of the services the chain stores offer their associated stores. Their services also include help in running, managing, and controlling all phases of the business.

These factors have made retail auto parts chains a strong force in the auto aftermarket and have altered the way traditional independent jobbers go about their business.

AUTO FRANCHISED DEALERSHIPS

The competition has also increased for the professional trade. Vehicle manufacturers are now competing for an ever-increasing share of the service aftermarket through the service departments of their franchised dealerships.

To promote their service departments, vehicle manufacturers have set up their own part supply networks complete with zone/regional warehouses that operate like traditional WDs for the dealerships. The parts department in a large dealership can be as large as many independent jobber stores.

The dealerships hurt traditional jobbers in that they take away business from independent garages and service stations, traditionally two of the jobber's best customers. Extended warranties, service contracts, and factory-trained service technicians are other ways manufacturers attempt to increase traffic in their dealership service bays.

Jobbers can help their independent garage accounts compete by helping them keep abreast of the latest service bulletins and techniques offered through parts manufacturers.

Franchised dealerships also compete with the jobber for some types of fleet accounts. But as Figure 1-1 illustrates there is often some friendly interaction between independent jobbers and vehicle dealerships. A dealership will often purchase from a jobber to fill rush orders, to stock up on other automotive supplies not carried by the zone warehouse, or simply when it is more convenient to deal with the jobber down the street than the corporate warehouse two states away. The jobber can also turn to friendly franchised dealerships as a source of emergency or hard-to-access parts.

SPECIALTY REPAIR FRANCHISES

The growth of fast repair franchises, offering quick, economical replacement of brakes, mufflers, shocks, filters, tune-up parts, etc., has also cut into the traditional jobber-installer market. Franchised shops often have the buying power to bypass the jobber or even the WD in the distribution chain. They have also contributed to losses in business for service stations and garages, traditionally the jobber's strongest customers.

Within their local market area, the average jobber store competes against six franchised installation specialists, five to six retail auto stores, three

Manufacturers	Group 1 (Small)	Group 2 (Medium)	Group 3 (Large)	All Groups
Percent respondents buying direct from manufacturers	46.9	58.3	76.2	65.2
Average number bought from	4.3	5.7	32.3	21.5
Warehouse Distributors				
Percent respondents buying from WDs	96.9	95.9	88.1	92.1
Average number bought from	6.9	6.2	13.1	9.8
Redistributing Jobbers				
Percent respondents buying from redistributing jobbers	37.5	20.8	26.2	26.8
Average number bought from	2.6	4.1	10.3	6.7

FIGURE 1-2 Number of suppliers from which jobbers buy (*Chart courtesy of Jobber Topics Magazine*)

to four discount houses, and three to four mass merchandisers. This intense market competition is one reason the aftermarket industry is no longer the clearly defined traditionally structured industry it was originally. Although the basic distribution machinery is still in place, there are so many deviations in the way products get to the marketplace today that generalizations can no longer be made.

For example, many manufacturers bypass the WD and jobber and sell directly to mass merchandisers, large retail auto chains, and franchised repair facilities.

Likewise, jobbers have begun sourcing more and more of their parts lines directly from the manufacturer. As shown in Figure 1-2, more than 65 percent of all jobbers buy some of their part lines directly from the manufacturer. And while smaller jobbers deal directly with four to six manufacturers, the larger jobbers deal with dozens of manufacturers.

Figure 1-2 also indicates that warehouse distributors still play a key role in the distribution chain. More than 92 percent of all jobbers still rely on an average of almost 10 WDs to supply parts. And roughly 27 percent of all jobbers buy some of their lines from redistributing jobbers, jobbers who buy certain products in large quantities and resell to other jobbers.

THE AUTOMOTIVE TECHNOLOGY EXPLOSION

In the past ten years the automobile has been going through a period of change that has been more profound than any since its inception. Due to various federally mandated changes concerning fuel efficiency, emission controls, and safety concerns, the vehicles of today run better, get better fuel mileage, are more crashworthy, and last longer than any automobiles ever built.

Because of these changes, today's vehicles are also more complex. Fuel injection, distributor-less (electronic) ignition, four-wheel steering, driver adjustable suspension, and computerized controls have revolutionized the way vehicles operate and are serviced. Additionally, MacPherson strut suspensions, CV joint-driven front-and four-wheel drives, electronically controlled rack-and-pinion steering, antilock braking systems, overdrive transmissions, and unitized body construction have made the modern vehicle a complex array of wiring, vacuum hoses, and sensors.

THE GROWING IMPORT MARKET

In addition to rapid technological change, the aftermarket industry is faced with servicing a greater variety of makes and models of vehicles than ever before, many of which are foreign made. Since 1980, more than fifty foreign car manufacturers have sold more than 22 million passenger cars in the United States. Imports accounted for 28 percent of all U.S. car sales in 1986. By 1990 this figure might reach 40 percent.

At present, the average jobber devotes 19.1 percent of a store's inventory to import parts, up from 14.5 percent in 1983, and this trend will only increase as imports continue to play an important role in the world car market.

The rapid growth in the number of makes and models of vehicles on the road is not only due to the

Product Lines	Group 1 (Small)	Group 2 (Medium)	Group 3 (Large)	All Groups
Average	71.3	105.1	132.4	113.8
Median	59.0	95.0	103.0	90.0
Part Numbers				
Average	15,077	17,603	29,434	23,977
Median	12,500	15,500	24,000	19,261

FIGURE 1-3   Average number of product lines stocked *(Chart courtesy of Jobber Topics Magazine)*

influx of import models. In 1987, U.S. auto makers (Chrysler, Ford, and General Motors) offered more than seventy different model lines.

PARTS PROLIFERATION

Advances in technology, coupled with the dramatic increase in vehicle makes and models, have resulted in thousands of new parts and components flooding the aftermarket.

Figure 1-3 lists the average number of product lines and part numbers stocked by small, medium, and large volume jobbers. On the average, a jobber with annual sales of more than $750,000 stocks 132 different product lines and carries nearly 30,000 different parts numbers in inventory. Smaller operations, with annual sales less than $350,000, carry 71 lines and more than 15,000 part numbers. Jobbers with a medium size business ($350,000 to $750,000 in annual sales) carry 105 lines and almost 18,000 part numbers.

This increase in parts numbers has made careful inventory selection and control a prime concern for both jobbers and their suppliers. To assist in inventory control, billing, and other business functions, more and more jobbers are using computer systems. The percentage of small, medium, and large jobbers using computers is listed in Figure 1-4.

A ranking of a computer's chief benefits is given in Figure 1-5.

As both parts numbers and inventory have increased, catalogs for tracking part numbers have grown in size and number. Efficient, accurate cataloging is one of the prime skills of counter personnel.

	Percent respondents having computers		
	Most Helpful	Second Most Helpful	Third Most Helpful
Inventory	75.4	19.7	4.1
Point-of-sale	36.9	22.1	9.0
Accounts receivable	20.5	18.9	23.8
Customer billing	22.1	12.3	23.8
Order entry	16.4	10.7	16.4
Gross profit analysis	13.9	4.9	9.0
Machine shop management	4.1	1.6	0.8
	*		

*Totals do not equal 100 percent due to multiple mentions.

FIGURE 1-5   Ranking of the chief benefits of a computer *(Chart courtesy of Jobber Topics Magazine)*

To streamline catalog work, jobbers are now adding electronic cataloging to their computer systems. These electronic cataloging programs quickly locate part numbers, suggest related sales items, and perform inventory control and invoicing functions. More information on electronic cataloging can be found in the disk included with this book.

PROGRAMMED DISTRIBUTION

To help jobbers and WDs remain competitive on price, control inventory and business functions and successfully market and promote product lines, program distribution groups have become increasingly popular within the aftermarket industry. The first of these groups was formed in 1925 and their growth has rapidly increased during the last fifteen to twenty years.

These groups offer WDs and jobbers private brand labels on numerous product lines. Other services offered to group members include:

Year	Small	Medium	Large
1977	5.3	10.2	35.8
1983	19.2	42.0	71.0
1988	40.6	64.6	90.5

FIGURE 1-4   Percentage of jobbers using computer systems

- Tie-ins to computer systems that handle point-of-sale, inventory control, sales analysis by part and customer, lost sales, core tracking, price updating, customer billing, accounts receivable and accounts payable, general payroll, machine shop sales, gross profit analysis, financial spreadsheets, branch store operations, and cataloging
- Co-op advertising programs including TV, radio, and newspaper
- Clinics for jobber dealers and DIY
- Display units and point-of-sale materials
- Inventory control and management assistance
- Obsolescence protection of fading part numbers
- Signage inside and outside
- Store layout assistance
- Sales training

Roughly one-half of all jobbers are now members of programmed distribution groups; jobbers with small and medium sized businesses account for the bulk of group memberships. Independence from groups is more successfully handled by large-volume jobbers, many of whom possess sufficient buying power to stay competitive on pricing.

A listing of services offered through both independent and programmed warehouse distributors is summarized in Figure 1-6

THE JOBBER'S ROLE IN TODAY'S AFTERMARKET

The advances in auto technology, the dramatic increase in aftermarket parts, and changes in the way parts are brought to market and merchandised are not all bad news for today's jobbers. On the contrary, foresighted jobbers and parts management teams can benefit from the state of today's market.

Rapid technological change has created an information void in the auto service industry. Both professional installers and DIY customers often lack the information and skills to service the latest generation of vehicles. Jobbers with well-informed counter personnel and parts managers can fill this void and perform a valuable service to their customers. In

	Program Group Members %	Non-Members %	All %
Co-op advertising/promotion programs	92.1	46.2	58.3
(Wholesale)	73.7	34.0	44.4
(Retail)	63.2	21.7	32.6
Clinics	89.5	50.9	61.1
(For jobber dealer customers)	89.5	48.1	59.0
(For jobber DIY customers)	21.1	1.9	6.9
Computerized ordering	60.5	28.3	36.8
Delivery service	94.7	75.5	80.6
(Daily)	84.2	56.6	63.9
(Weekly)	10.5	20.8	18.1
(Other)	13.2	1.9	11.1
Display units, gondolas	73.7	28.3	40.3
Financial assistance	63.2	26.4	36.1
Inventory control assistance	79.0	52.8	59.7
Management assistance/counseling	76.3	34.0	45.1
Obsolescence protection	94.7	67.9	75.0
Point-of-sale materials	81.6	57.6	63.9
Private branding	84.2	20.8	37.5
Repair manuals	60.5	19.8	30.6
Signage, inside/outside	81.6	24.5	39.6
Store layout assistance	65.8	20.8	32.6
Training programs	71.1	34.0	43.8
(For jobber outside sales personnel)	44.7	19.8	26.4
(For jobber counter personnel)	63.2	30.2	38.9

FIGURE 1-6 Services warehouse distributors provide for their jobbers (*Chart courtesy of Jobber Topics Magazine*)

fact, helping customers accurately locate the parts for their needs and offering the technical assistance and advice needed to perform the repair correctly are the most valuable assets of today's jobber stores.

For, unlike twenty years ago, when many jobbers could successfully operate as middle men and order filling houses for the aftermarket distribution network, today's jobbers must operate a service-intensive business. For the jobber store owner and staff, providing thorough, knowledgeable service is the key to maintaining the traditional professional trade and winning a large percentage of the DIY market.

JOBBER PERSONNEL

This book will concentrate on successful parts management and store operations on the jobber level. In any successful business, employees must function as a team working toward a common goal. Depending on the jobber's size, staff personnel might include:

1. An overall parts manager to oversee general store operations.
2. A lead counterperson and several counterpeople to service customer orders.
3. An inventory control clerk to monitor inventory levels and records.
4. Receiving and shipping personnel to handle incoming WD shipments and outgoing customer orders.
5. Stockroom personnel to assist in organizing inventory and pulling orders from stock.
6. An accountant/bookkeeper to monitor billing, accounts receivable, expenses, payroll, and all financial matters.
7. Cashiers to register and ring out customer transactions once orders have been filled.
8. A machinist to work in the store's machine shop.
9. Dispatchers and delivery personnel to provide the store's delivery service.
10. Buyers to work through warehouse distributors and manufacturers in procuring parts lines.
11. Outside sales personnel to call on professional dealers, fleet owners, industrial clients, etc.

Large jobber stores with a staff of twenty or more might have employees that fit precisely into these job descriptions. However, in most small wholesale stores, a single employee often performs the functions of several job categories. This multi-

skilled role is most often filled by the key ingredient in any successful jobber operation—the counter staff.

COUNTER PERSONNEL— THE KEY INGREDIENT

The men and women who work behind the counters of jobber stores form the final vital link to the aftermarket parts distribution chain. Face to face or over the telephone, counter personnel deal one-on-one with customer accounts of all types. The performance and attitude of the counter staff is pivotal to the jobber's success. Profit and continued growth at the jobber level can only be assured by staffing with the most knowledgeable, skillful, and productive counterpeople. And these counter professionals are not born, they are trained and developed.

The counterperson's primary job is to sell to fill the customer's needs and to initiate the sale of related items. On the average, a counterperson performs forty-five sales transactions per day. In fact, every hour, counter professionals in the United States perform about 360,000 selling transactions.

Surveys of jobber store owners and professional counter personnel confirm the importance placed on selling and servicing customers.

When asked what is the most important aspect of a counter staff's job, jobber owners overwhelmingly rated selling to the professional mechanic as number one. Handling complaints was rated second and selling to the do-it-yourselfer was third. It is interesting to note that all the answers—even in the complaint area—are directly related to selling.

When counter professionals were asked the same question their responses closely mirrored those of their bosses. According to counterpeople, selling to the professional mechanic is their most important job aspect followed by selling to the do-it-yourselfer and then handling customer complaints.

The important role played by counter personnel affects all aspects of store operations. Seventy percent of all jobber owners stated that input from their counterpeople directly influenced the line of parts and inventory carried by their stores. Counterperson influence also included important decisions on catalog rack arrangement, sales policies, pricing, in-store displays, and store advertising and marketing.

THE MEANING OF TRUE SELLING

At one time or another everyone has dealt with a hard sales pitch. Many have purchased items they did not want or need when pressured by a relentless salesperson. These are the types of sales experiences most people remember, and they help create a negative image of sales and sales personnel.

The reason for this is simple. People tend to forget successful sales experiences, because when a salesperson does the job correctly, the customer does not feel as if he or she is being sold; the person feels as though he is being given the assistance and opportunity to make the right choice.

This is the attitude professional parts managers and counter personnel must have when approaching the selling aspects of their jobs. Consistently matching products and parts to customer needs is the sales goal of every counter staff. Building and sustaining a large customer base is essential for success. Jobbers need repeat business to survive. Hard-sell techniques and overselling by counter personnel will result in one-time sales opportunities with little or no return business.

JOBBER STORE CUSTOMERS AND ACCOUNTS

The types of customers and accounts serviced by jobbers vary greatly, but can be categorized into five basic groups:

1. Professional installers
2. Retail or walk-in trade
3. Agricultural accounts
4. Fleet accounts
5. Industrial and institutional accounts

Although many jobber stores might tailor their marketing and merchandising efforts toward one or several of these major market segments, most parts stores will eagerly service any type of account. Yet each of these account types requires special attention. A store's pricing and discount structure is often keyed to account classifications. The needs of each account will differ. Counter personnel must be aware of these distinctions and approach each type of account from the proper perspective to ensure each receives the best possible service.

PROFESSIONAL INSTALLERS

Service stations that perform repair and maintenance work and independent garages and specialty repair shops are the jobber's main source of professional accounts. Building a large client list of professional mechanics and installers is essential to a jobber's success. Professional accounts still form the customer foundation in many jobber stores and

are a very valuable portion of the market to capture. They offer steady, repeat business with high volume.

A store's professional accounts might also include franchised repair outlets and car and truck dealerships. Although these businesses might deal directly with manufacturers as a source for most of their aftermarket needs, many will rely on jobbers to fill a portion of their orders.

Many of the services offered by jobber stores, such as professional price discounts, credit accounts, delivery, outside sales personnel, and machine shop work, are designed to meet the needs of the professional trade.

Installers must have their parts orders filled quickly, efficiently, and accurately. To professional mechanics, time is truly money. They expect to deal with professional counterpeople and parts managers who can help them run their business smoothly.

The automotive technology explosion and parts proliferation in recent years have increased the importance of the counterperson's role in dealing with professional accounts. In the past, professional mechanics could more precisely determine their parts needs for many repairs. Today, the location of parts and related items needed to complete the repair might not be as easily determined.

In this regard, a knowledgeable, competent counterperson becomes a valuable asset to the professional mechanic. He or she can confirm the parts needed for repair, suggest related items to avoid future trouble, and mention any service tips offered in the parts catalogs or other literature that might not be available to the mechanic.

RETAIL OR WALK-IN TRADE

Retail trade includes auto do-it-yourselfers, more serious amateur or "shade-tree" mechanics, and the general public. Retail trade accounts for 20 to 25 percent of an average jobber's sales volume.

Retail customers are more likely to purchase items that relate to the appearance and general upkeep of vehicles such as waxes, cleaners, splash guards, seat covers, wheel covers, oil, antifreeze/coolants, windshield washer fluids, and so on.

For most retail customers, actual repairs involve items and parts that are not too complex or difficult to install. Examples would be filters, spark plugs, wipers, headlights, etc. Figure 1–7 lists the most often used product lines sold to the retail market.

A certain segment of the walk-in trade includes serious amateur mechanics ready to perform most of their own repair work. But the advances in auto technology have hurt the serious amateur much more than the professional. The end result is that

Percent total respondents	
	All Groups
Filters (oil, air, fuel)	61.5
Ignition parts	36.7
Chemicals	35.9
Brake parts (brake shoes, pads, drums, rotors, cylinders)	30.8
Spark plugs	26.5
Oils and lubricants	26.5
Belts, hoses	21.4
Exhaust system parts (mufflers & pipes)	17.9
Electrical	16.2
Tools	11.1
Batteries	9.4
Paint/supplies	8.5
Starters/parts	6.8
Wiper blades	6.8
Accessories	6.8
Engine parts	5.9
Front end parts	5.9
Pumps (water, fuel)	5.1
Shocks	4.3
Remanufactured units	4.3
Other	23.1
No answer	10.3
	*

Other
Gaskets, thermostats, antifreeze, high-performance parts, carburetors/parts, headlights, tires, air conditioning parts, gas caps, jacks, bearings, bolts, gauges.

*Totals over 100 percent due to multiple mentions.

FIGURE 1-7 Products most often purchased by walk-in customers *(Chart courtesy of Jobber Topics Magazine)*

while more people than ever want to work on their vehicles, the number of people who are truly qualified to do so is rapidly declining.

That is why the technical advice of knowledgeable counter personnel is extremely helpful in servicing the DIY trade. Most DIY customers will need assistance in understanding the parts, related items, and service techniques needed to perform their repair. Assessing the skill level of DIY customers and providing the help needed to ensure a successful sale and repair are some of the most important aspects of today's counter professionals' job.

AGRICULTURAL ACCOUNTS

In some areas, agricultural accounts, such as farmers and ranchers, can make up a sizable and profitable segment of a jobber store's client base.

Some personal farm maintenance shops would shame the service facilities of many big-city garages.

Farmers are serious about maintenance and repair. The farmer is the ultimate do-it-yourselfer. If a hydraulic hose bursts, the farm mechanic must fix it. If a blade brakes, he or she does the welding repair with his or her own equipment.

The potential to sell to the farm market is great. Farm equipment functions under the worst conditions. Filters operate in dust-filled environments. Axles are buried in mud. Silt and water damage brake assemblies and king bolts. Shocks and bearings are pounded over dirt roads.

Jobber stores who take the time to assess and stock for this market will sell everything from a tiny chain saw carburetor to piston and cylinder assemblies for big diesels.

Other advantages in selling to the farm market include:

1. *Knowledgeable Customers.* Farmers know their machinery. They often know the exact part and supplies they need. They also research related items to make certain they buy everything needed on one trip to the jobber store.
2. *Value-Minded Customers.* As a group, farmers are less price conscious. They require quality parts that last and are wary of bargains that could compromise quality.
3. *Volume Buying.* Farmers often buy in large quantities. They do not have the time for repeat visits to the store. A purchase of a high-use item, such as penetrating oil or grease, can involve gallons and cases, not merely a 12-ounce can.

Farm accounts should be viewed as professional clients. Many jobbers servicing rural areas conduct surveys to determine the parts and supplies needed by their agricultural accounts. Catalog listings for farm equipment are often sparse or nonexistent. Successful stores often generate their own files on part numbers and stocking recommendations. Research into identifying which manufacturers have the best parts coverage and catalogs for farm equipment might also be needed.

FLEETS

Fleet accounts are any customers that use several vehicles (or other pieces of equipment) for which a jobber can supply parts in their operations. From the small cabinetmaker with a few pickup trucks, to a huge government garage, fleets offer the jobber a large and profitable market. Examples of fleet accounts include

- Appliance and furniture stores
- Beverage distributors
- Construction firms
- Dairy product distributors
- Delivery companies
- Emergency services
- Government agencies
- Industrial laundry services
- Schools
- Taxi companies
- Trucking firms
- Utility companies

Many fleets have their own shops and mechanics to maintain their equipment, although some rely on outside garages.

Free-lance mechanics are another valuable source of fleet trade. Generally working out of a service truck, they perform repairs on the fleet's premises, or in the field, without being on the fleet payroll. These independents can offer jobbers a large volume of business because they often work for several accounts that cannot afford to employ a full-time mechanic. The effort to locate these fleet free-lancers and acquire their loyalty is extremely worthwhile.

Fleet accounts are attractive for a number of reasons. The first and most obvious is the prospect of high-dollar volume. Fleets use large quantities of parts and supplies. Most fleets operating in the United States will spend several thousand dollars a month with parts stores. Some fleets spend tens of thousands of dollars.

Dollar volume is not the only advantage of fleet accounts. Most fleets include large groups of similar or identical vehicles, making it possible for the jobber to become very familiar with his or her customer's equipment. This reduces research time in locating parts numbers and allows the jobber to tailor inventory to closely match customer needs. With several fleet accounts a jobber can move a relatively large amount of parts through a comparatively small inventory.

Servicing fleets is not easy. Competition for fleet accounts is fierce within the aftermarket industry. Large fleet accounts have considerable purchasing power and often buy directly from specialty distributors and WDs. Because they might not be able to compete directly on price, jobbers must reply on completely dependable service to capture a segment of the fleet market.

Fleet accounts that take 60 or 90 days to pay can also cause problems with a jobber's cash flow. A policy of offering discounts for prompt payment could solve this problem.

INDUSTRIAL AND INSTITUTIONAL ACCOUNTS

Industrial accounts are those whose primary use of jobber's products is nonautomotive. An industrial account can purchase belts, filters, and oil for air compressors in their plant, extension cords and worklights, hand tools, air hoses and couplings, and many other products.

Industrial accounts will seldom use jobber products as raw materials in their products, but they can come to rely heavily on a jobber for their maintenance needs. In most industrial operations there is a steady need for screw extractors, drills, taps, dies, thread repair kits, and thread locking compounds. Gasket material, sealants, and adhesives are also products that can be sold quickly.

Hydraulics are such an important part of many industrial operations that becoming an expert in hydraulics will generate interest at the industrial level. Jobbers pursuing industrial accounts should be knowledgeable in regard to the various types of hoses and fittings, the proper way to assemble hoses, and the difference between fittings such as JIC and SAE.

Other areas of potential sales include

- Equipment cleaners such as steam cleaning soaps, descalers, high-pressure hoses, and degreasers
- Hose, tubing, and fitting for plumbing and liquid transfer
- Belts for motor-driven equipment
- Paint supplies such as spray guns, cups, strainers, paint, solvent, masking tape, paper, abrasives, and materials for surface cleaning and preparation
- Ignition parts, batteries, starters, alternators, hydraulics, and switches for lift trucks and other plant vehicles
- Machine shop services

Industrial accounts often begin as an extension of an existing fleet account. Jobbers who service a company's fleet should be aware of the potential of selling supplies inside the plant as well.

Landing industrial accounts involves outside selling by the jobber owner or sales staff. Making the correct contacts within the company and giving a clear sales presentation of the store's products and services are the keys to success in this market.

Processing and handling industrial accounts usually involves a substantial amount of formal paperwork. Most industrial accounts will have strict requirements about how merchandise is to be ordered and how the jobber's paperwork must be submitted for payment.

Some accounts will issue a blanket purchase order that will cover all sales for a given period of time. Others will require a separate purchase order for each transaction. Some will allow their buyers to issue an oral purchase order, others will refuse to pay an invoice if a signed copy of their purchase order is not attached. Government agencies might issue a contract for certain merchandise and require that the contract number appears on all invoices.

In addition to giving a buyer a copy of the jobber invoice when the merchandise is delivered, the jobber might be required to submit a second copy to the industrial accounts home office before payment can be authorized.

When dealing with industrial accounts everyone on the jobber's staff—parts managers, counterpeople, sales personnel, and delivery drivers—must be aware of the importance of properly handling paperwork. Many larger industries pay their bills on a 30-, 60-, or even 90-day schedule. Mix-ups in paperwork can delay payment even longer and cause serious cash flow problems at the jobber level.

To eliminate this problem, everyone on the jobber staff should understand all the details of doing business with each of the store's industrial accounts. A binder with a section for each account outlining policies, listing the authorized buyers, giving the appropriate purchase order or contract number, and listing special prices or discounts should be within reach of every counterperson.

Matching automotive parts numbers to industrial parts numbers will also require work on the part of counterpersonnel and inventory control managers. As with farm and fleet accounts, conducting an on-site parts survey of the clients' needs is the best method of establishing a basis for maintaining a correct parts inventory.

CAREERS IN PARTS MANAGEMENT

The need for well-trained, dedicated counter and parts management personnel eager to pursue careers in the automotive aftermarket has never been greater. Parts management is not simply filling part orders or "pushing parts." It is helping run a dynamic, growing, and ever-changing business. To succeed, skills must be mastered in many areas. And the need to add to this knowledge will never stop.

PRODUCT KNOWLEDGE

A key ingredient in the makeup of any professional counterperson is a keen interest in automobile technology. Many of the finest counterpeople began their careers as service mechanics or serious amateur auto buffs. Having a clear understanding of the systems that make up today's vehicles allows for better communication with professional installers, especially the DIY walk-in trade. Counter personnel who lack product knowledge and do not keep abreast of changes in the auto industry will be at a serious, if not fatal, disadvantage. Professional counter workers have an obligation to jobber owners and their customers to know as much as practically possibly about the products they offer and their applications.

SALES AND SERVICE SKILLS

Parts managers and countermen and women deal with people. They deal with customers, suppliers, manufacturers, and numerous other segments of the aftermarket industry.

The ability to communicate is essential, whether it be explaining the features and benefits of a product to a customer, placing an order with a warehouse distributor, or routing the proper paperwork and data through the various departments in the jobber store.

And all will not always go smoothly. By nature, interacting with other people is not always easy. Personalities can clash, misunderstandings will occur. A counterperson or manager must be a problem-solver, an astute negotiator, and a person who gains satisfaction through serving and helping others.

PROFESSIONAL DISCIPLINE

With the very best countermen and women, all of the activities that occur at the counter are supported by an entire operational discipline that is taking place at a deeper level. This textbook outlines many of the skills and attitudes that go into making up this discipline.

There are many less obvious aspects of a counterman's job that require attention.

Business Sense

Basic information on how the jobber store fits into the distribution system can help new counterpeople relate to customers and suppliers. The general distribution network was outlined earlier in this chapter, but the situation of every jobber store is unique.

Counterpeople should know their store's suppliers. They must understand what market forces make it easy for their store to compete in some areas and difficult to compete in others. Counter personnel must also have the information needed to order parts intelligently and to give customers credible explanations concerning store pricing and policies.

Inventory control is another area that every counterperson needs to understand. While it is true that much of what goes into making inventory decisions is philosophical and that most final decisions are the responsibility of store owners or managers, counter personnel must understand their store's philosophy and the basic guidelines for sound inventory management.

The modern wholesaler can rely on a computer to control inventory, but this is merely using technology to execute the store's philosophy of buying, stocking, and pricing. That philosophy must be transmitted to each person who buys and sells parts. For example, under what circumstances is a lost sale recorded? What parts are considered important enough to record by buyouts? How well should a part be moving before it is given a spot in the store's regular inventory? How poorly should it sell before it is eliminated from stock?

Pricing is also based on an interplay between market realities and philosophy. As counter personnel source parts, quote prices, and defend pricing, they need to be aware of that interplay.

For example, parts of varying quality are available at various prices, and, for any part sold, the jobber needs a given markup to cover operating expenses. But competition puts a ceiling on selling prices. Now within those constraints, the jobber owner might want a certain minimum profit or perhaps the maximum possible profit. To achieve these goals, counter personnel must understand the store's philosophy.

Mentioning inventory control and pricing strategies logically leads to one of their results, the return a jobber receives on his or her investment (ROI). Counter personnel must be aware of what factors determine ROI and the steps they can take to keep in line with expectations.

Return on investment is not just a management concern. Each counterperson manages ROI whenever he or she engages in a transaction that affects inventory. An example would be when a counterperson makes a decision to place a part returned by a customer into stock rather than sending it back to the warehouse distributor.

Cash flow is a related concern that counter personnel can affect, but only when they understand it and its importance. An account that charges several hundred dollars a month and pays a little slowly might not be a major concern. But when this same account orders a $2,000 engine, a business-wise counterperson will insist that this particular invoice

be paid on time, perhaps even tying the quoted price into timely payment.

Productivity

Productivity encompasses all the little attitudes and work habits that make the difference between working hard and working smart.

With the dozens of different tasks performed by today's counter personnel, time management has become essential. For example, eliminate special trips to the stockroom or display areas of the store to restock items. Develop the habit of returning restock items whenever regular duties require a trip to the stockroom or display shelves.

When working on the telephone, make effective use of the "hold time" by inserting new price sheets into the catalog rack, pricing out an invoice, or organizing the counter area.

The business of doing more than one activity at a time does not come naturally to many people, but it can be taught and developed with the correct work attitude.

Even the simple task of answering incoming telephone calls can be made more efficient. For example, suppose two lines are ringing. The counterperson first answers line one, puts it on hold, then answers line two and puts it on hold. Finally, he or she returns to help the party on line one.

Consider the efficiency of answering line two first, putting it on hold, and then picking up and helping the party on line one. With this approach, line one is still the first served, but the counterman saves punching two extra buttons and a little extra mental shifting around.

Taken by itself, this little example of an efficient work strategy might appear to be trivial and save only a minute or two each day. But the most powerful thing about time management techniques such as these is that once a worker begins to use and appreciate one or two problem-solving methods, he or she tends to develop more of them. The result can be a substantial increase in productivity.

QUALITIES FOR SUCCESS

Men and women interested in a career in parts management must be skilled in many areas. They must be sound in their knowledge of automotive technology. They must be service- and sales-oriented people interested in dealing with and helping customers of all types. They must be effective time managers, disciplined in their work habits. They must be able to perform duties in an organized, accurate manner, and realize the importance of the small details that affect efficient parts management. And they must be business-oriented thinkers, understanding how their store functions and how all policies and actions affect the store on a dollar and cents level.

Most important, a successful counterperson must be willing to learn, change, and grow as the aftermarket industry changes. For example, computerized parts cataloging and inventory control is becoming increasingly popular at the jobber level. Successful counterpeople are learning the skills needed to use this valuable time-saving technology.

Even the most experienced counterperson is learning on the job every day. Many intangible characteristics are the keys to the proper work attitude in today's aftermarket service industry.

ENTHUSIASM

Enthusiasm is a combination of interest and belief, of energy and activity. Every counterperson needs a broad curiosity about the products, customers, and self. Enthusiasm leads a counterperson to become acquainted with his or her work from all possible angles to expand his or her knowledge and to better assist customers.

OPTIMISM

Optimism in a counterperson instills faith in customers. A counterperson must be optimistic about the job and his or her abilities, about the products sold, and about the store and the staff worked with. If the counterperson offers maximum assistance and service, if the products offered are what they are purported to be, and if the store is everything it should be, then optimism will be a natural result.

CONFIDENCE

Confidence is the result of knowledge plus experience, and for this reason, it can be developed and is more easily acquired than optimism and enthusiasm. To the counterperson, confidence is a healthy respect for one's capabilities. To a customer, a confident counterperson is an able person who instills trust.

SINCERITY

Sincerity is vital to the continued trust and confidence of the customer. A counterperson must mean what he or she says. Customers can easily detect the false ring of half-truths or an insincere

presentation. If a mistake is made, admitting it, not covering it up, is the best method for saving the sale and customer. Honesty and loyalty are closely linked to sincerity. Sincerity is a particularly valuable asset for the long run because it attracts new customers and holds current ones.

DETERMINATION

Determination is the will to succeed. Patience and perseverance, parts of determination, keep the counterperson working until a customer's need is fulfilled.

DEPENDABILITY

Customers are always in search of certainty in an area which, by and large, is one of change. One such certainty in the changeable aftermarket is a dependable counterperson. A dependable counterperson can be relied on to keep all promises and secrets and to remember requests and instructions.

 SALES TALK _____

Although this text might appear to be slanted toward the independent parts jobber or store, it is not. The techniques and procedures given apply to all types of counterperson operation, including those working for mass merchandisers and chain store retail concerns, specialty repair franchises, and automotive dealerships. With today's auto manufacturer warranties ranging from three to seven years, the factory automotive dealer has a major advantage and a need of counter-trained personnel. Parts under warranty generally must be ordered as OE (original equipment) parts from the vehicle manufacturer or recommended by him or her. Though this market is almost an exclusive area of the automotive dealership counterperson, the techniques of selling, stocking, and managing are the same as the jobbership.

INITIATIVE

A counterperson with initiative combines ambition with a healthy amount of industry. He or she works efficiently without close supervision. By understanding the overall goals of the store, a counterperson with initiative understands the correct course of action without being told.

MENTAL AGILITY

Working the counter in a busy jobber store requires a high level of mental agility. Every action taken or every comment made to a customer affects the store's operation and the customer's perception of the store and its ability to meet his or her needs.

Dealing with parts numbers, catalogs, invoicing, inventory, price structures, and customer concerns requires quick, accurate thinking and a logical approach. Counter personnel are thinking professionals. They must understand the entire jobber store and aftermarket parts industry to perform their duties competently. This ability to "see the entire picture" will lead to success on the counterperson level and to career advancement to lead countermen and women to parts managerial positions.

REVIEW QUESTIONS

1. What is the automotive aftermarket?
   a. the automotive service industry
   b. the network of businesses that supply automotive replacement parts and services
   c. service stations, repair shops, vehicle dealerships, and so on
   d. all of the above

2. The traditional automotive aftermarket distribution system begins with the _____ .
   a. parts store
   b. parts manufacturer
   c. consumer
   d. mechanic

3. In the aftermarket industry, parts wholesalers are commonly known as _____ .
   a. jobbers
   b. WDs
   c. reps
   d. distributors

4. An average of 55 to 60 percent of a jobber's business involves sales to _____ .
   a. service stations
   b. independent garages
   c. both a and b
   d. the general public

5. What factor has caused changes in the way manufacturer's WDs and jobbers do business in the past twenty years?
   a. increase of the number of do-it-yourselfers
   b. involvement of department store chains and discount stores
   c. increase in the number of repair specialists
   d. all of the above

6. Which of the following is not a cause of the thousands of new parts and components flooding the aftermarket?
   a. advances in technology
   b. increase in domestic vehicle makes and models
   c. increase in number of jobber stores
   d. growing import market

7. Which of the following is not a membership benefit of a programmed distribution group?
   a. free delivery
   b. sales training
   c. computer tie-ins
   d. display units and point-of-sale materials

8. What is the counterperson's primary job?
   a. handle telephone calls
   b. sell to fill the customer's needs
   c. handle customer complaints
   d. keep the counter clean

9. Which of the following customers or accounts are serviced by jobbers?
   a. retail or walk-in trade
   b. fleet accounts
   c. professional installers
   d. all of the above

10. Which of the following statements about selling to the farm market is not true?
    a. Farmers often buy in large quantities.
    b. As a group, farmers are more price conscious.
    c. Farmers are knowledgeable customers.
    d. Farm equipment requires quality, durable parts.

11. What are fleet accounts?
    a. customers that use several vehicles or pieces of equipment in their operations
    b. customers with boats
    c. farm accounts
    d. all of the above

12. What type of accounts require nonautomotive products?
    a. fleet
    b. walk-in retail
    c. industrial
    d. farm

13. Which of the following is an important skill of successful counter personnel?
    a. product knowledge
    b. sales skills
    c. efficient use of time
    d. all of the above

14. Which of the following personality traits concerning work attitude is an important quality for success?
    a. enthusiasm
    b. sincerity
    c. both a and b
    d. neither a nor b

15. Counterperson A is concerned with cash flow and ROI. Counterperson B is concerned with inventory control. Which counterperson is concerned with something that should not be important to him or her?
    a. Counterperson A
    b. Counterperson B
    c. Both A and B
    d. Neither A nor B

16. Counterperson A attempts to sell as many items as possible, even unrelated items, to every customer that enters the store in an effort to boost sales. Counterperson B attempts to sell only what would clearly fill each customer's present and related needs even if sales are not high that day. Who is right?
    a. Counterperson A
    b. Counterperson B
    c. Both A and B
    d. Neither A nor B

CHAPTER TWO

STORE OPERATIONS AND PERSONNEL DUTIES

Objectives

After reading this chapter, you should be able to:
- Describe differences in personnel organization and duties between small and large jobber stores.
- Describe the layout and function of the three distinct areas in a typical jobber store: back of store operations, counter area operations, and front of store operations.
- Explain the advantages of developing a planograph.
- Explain the steps for checking in and stocking shipments from WDs and manufacturers.
- Describe advantages and disadvantages of a jobber's machine shop, including increased sales and handling hazardous wastes.
- Explain the procedure for filling a part order using catalogs.
- Define and describe the following paperwork used by counter personnel: price sheets, invoices, accounts receivable report, lost sales report, damage claims, credit memorandums, warranty claims, and work tickets.
- Describe the procedure for handling telephone orders.
- Explain how to organize deliveries to customers.
- Explain the function and operation of wholesale distributors and why their organizational structures differ.
- Explain the difference between a manufacturer's representative and a manufacturer's agent and the guidelines for dealing with either.

Chapter 1 explained the role played by independent jobber stores in today's changing aftermarket industry. This chapter details the organization and structure of jobber stores, including a breakdown of the store's main departments and how the personnel in each department work together to meet customer demands. Many of the tasks are the same for chain and dealership operations as for an independent jobber.

STORE ORGANIZATION

In the traditional manufacturer to warehouse distributor to jobber distribution chain, both the warehouse distributor and jobber are independently owned and operated businesses. (In contrast, many of the larger retail auto parts chain stores are corporate run with their own system of distribution.)

Independent ownership at the jobber store level makes most jobber stores a reflection of the owner's business philosophy and managerial style. Individual job responsibilities and policies are established by the owner or the owner's selected manager(s). So the job responsibilities of an inventory clerk at Smith's Auto Parts might not be the same as those at Jones' Auto Parts. In fact, Jones might not even have a designated inventory clerk. The responsibility of the inventory clerk's position might be filled by the counter staff and parts manager.

Job classification in jobber stores is largely a function of store size. In larger stores, workers can specialize in one area of operation, such as stockroom operations or dispatch and delivery. In smaller stores, a single employee might perform key operations in several phases of the store's operation. For example, a small jobber store might be owned by a husband and wife team who also employ two other workers. All four people take shifts working at the counter. Everyone understands the inventory system and can check in and stock incoming parts orders in the proper bins and shelves.

After conferring with the staff, the husband makes all purchasing decisions and usually meets with manufacturer's representatives and WDs. He also does a little outside selling, periodically calling on several key professional and industrial accounts. His wife maintains the store's financial records, tracks lost sales inventory, and bills accounts. She also keeps payroll and personnel records.

Their workers are versatile counter professionals. They can efficiently work the counter and deal with both professional and retail trade. They write up and ring out their own sales tickets and do a good job of maintaining accurate records and accounts. They understand the store's client base and know most customers on a first-name basis. Working closely with the store owners has also given them a good insight into the owners' business philosophy and goals. Everyone understands the store's pricing and discounting system and how an inflated inventory or past-due accounts can severely hamper the store's financial stability.

The small staff is very team oriented. Everyone pitches in to keep the store clean, neat, and organized. They brainstorm concerning merchandise displays and how to best use the store's limited advertising and promotional budget.

While the store is too small to offer regular or "hot-shot" delivery service, it will occasionally, in a pinch, see that an important order is run across town to a good customer. The store is also too small to maintain its own machine shop, but it works closely with a number of independent shops to meet its customers' machining needs.

In contrast, a large jobber store might have twenty or more employees working in the store and a separate staff for machine shop operations. Their inventory and stockroom clerks might check in shipments and organize stock but never work at the counter and wait on customers. The counter personnel might write up or key in sales to the store's computer system, but ringing up the final sale is handled by a separate cashier. The store might have several delivery vehicles, drivers, and its own dispatcher. Several outside salespeople might work in the store's market territory, helping the store's professional installers establish their own small in-garage inventories and calling on industrial and fleet accounts. These sales people might spend very little time in the actual store.

The store might have its own management team consisting of a parts manager, sales manager, lead counterperson and service manager. There might be a separate bookkeeping and accounting department. The store might have a buyer to deal with its numerous WD and manufacturer suppliers.

A typical organization chart for a large jobber operation is shown in Figure 2-1.

While the size and staff of the jobber store might vary greatly, the functions of the store do not change greatly. With the exception of delivery and machine shop services, the four-person staff at the "Mom and Pop" store in the example performed the same tasks and services as the large store with a staff of two dozen people. The main difference is sales volume, not services offered.

The operations of a typical jobber store can be broken down into three distinct areas:

1. Front of store
2. Back of store
3. Counter area

This is largely due to the physical design and layout of the typical jobber store. The focal point of the store is the counter. As shown in Figure 2-2, the counter usually divides the store into two discernible halves. The area in front of the counter is re-

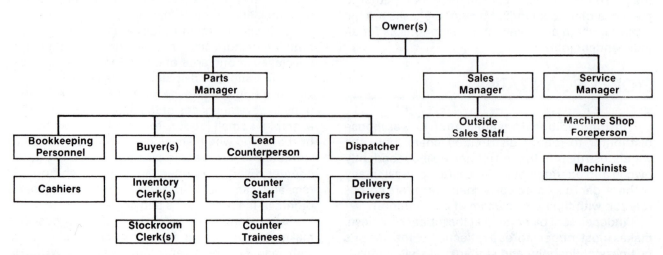

FIGURE 2-1 Management and personnel classification and organization in a typical large jobber operation

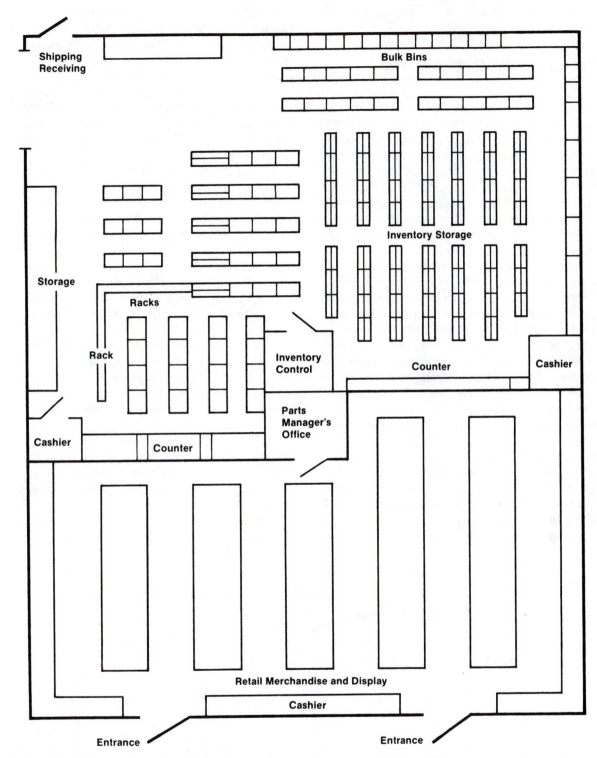

FIGURE 2-2 Typical layout of a jobber store showing retail merchandising, counter area, and back-of-store operations

served for the display and stocking of merchandise devoted to the retail or walk-in trade. That is not to say that professional customers never purchase items from this area of the store. They do so quite frequently. But the merchandise stocked in the retail area is normally an item that the customer can select and locate with minimal help. This merchandise usually includes a high percentage of auto accessory items, impulse-buy items, and items related to DIY and shade tree repair customers.

FRONT OF STORE OPERATIONS

For many years, the interior of a jobber store had been used only to stock merchandise, not to sell it. Upon entering a traditional jobber wholesale store, a customer had to weave in and out of a maze of cartons filled with cans of oil, cartons of windshield wiper towels, and sacks filled with floor sweep. This has changed drastically in the last decade. The jobber-retailer is now using the floor space in front of the counter to attract walk-in customers and stimulate retail trade.

Retail customers are used to seeing certain elements in a retail facility. They look for ample, well-lighted parking space; an attractive storefront with informative identification; a bright inviting atmosphere inside the store; a full line of appropriate merchandise displayed where it can be examined; handy product application information; a layout that guides them through the full range of retail merchandise; and a self-service area with wall shelving and gondolas.

Proper lighting for the retail area of the store is so important that management should consider employing a lighting specialist. The use of effective exterior lighting is also an investment that returns high dividends. Retail experience shows that people are attracted to brightly lighted parking areas and feel such lots are more secure.

Jobber owners and managers should also be aware of the advantages of traffic control in retail areas of their store. This is simply showing the customer where to go. Effective traffic control allows maximum impulse buying to take place. It can be accomplished by placing the store's parts counter or checkout counter where the customer must walk through merchandise-filled aisles or gondolas to get to it. As Figure 2-2 shows, the typical jobber store lends itself to this design.

In stores with two entrances to the retail area, traffic control is relatively easy. Mark one door "Entrance" and the other "Exit" and place the checkout counter near the exit door. In stores with only one entrance, place the checkout counter and information counter well in the back of the store. Another variation is to place a divider railing inside the front door and put the checkout counter on the other side of the railing. In this system, customers will have to move through the store to get to the checkout counter (Figure 2-3).

It is also strongly recommended that retail prices are marked on every item displayed in the retail area of the jobber store. Retail customers expect it. Pricing every item will speed up the customer's buying decision as well as save valuable time for counter personnel.

Chapter 9 gives detailed information on merchandising techniques for the retail section of the jobber store. The important concept here is that every effort should be made to stimulate traffic and circulation through the front of the counter area. Customers should feel comfortable browsing in the

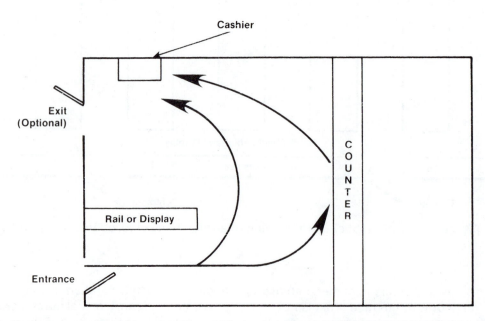

FIGURE 2-3 Store layout should be set up to maximize customer traffic through retail merchandise display areas.

FIGURE 2-4 Inventory is stored on open shelves, bins, and rack systems. All storage locations must be clearly marked and labeled.

store. Make it easy for the customer to be sold on the store's products.

BACK OF STORE OPERATIONS

The area behind the counter is reserved for hard parts storage and inventory. These parts are normally stocked on or in open shelves, bins, and racks (Figure 2-4). Parts are normally grouped according to product line such as engine, suspension, brake, and exhaust. They are also organized on shelves by individual part number for fast, efficient location and restocking.

Labor is one of the largest expenditures in operating a jobber store. The stockroom and bin layout plays an important role in keeping labor costs down by saving steps and achieving maximum efficiency when locating and pulling parts from inventory.

One of the best methods of improving stockroom efficiency is to arrange fast-, medium-, and slow-moving part lines according to their proximity to the parts counter. For example, ignition, generating, and electrical system components and brake system parts might be a store's fastest moving lines. These should be located close to the counter. If the store's slowest moving lines are gaskets, seals, and lighting system components, these should be stocked toward the rear of the stockroom.

Stockroom layout should also consider good housekeeping factors—ample aisle space (minimum of 36″) for easy handling of parts, arrangement of bins for best appearance and ease of cleaning.

Consider the arrangement of engine and other parts used in the store's machine shop. Accessibility

and ease of handling and delivering them to the machine shop will create more efficient operations.

Also consider the flexibility of the storage equipment such as bin size, shelf size, and clearance between shelves, racks, and so on. They should be flexible enough to provide for future additional storage space for increased inventory and possibly changes in parts package size.

HANDLING INVENTORY

An average jobber store has a parts inventory of between 15,000 to 20,000 items. Due to the rapid growth in parts inventories, most jobbers never find the time to re-evaluate their parts storage bins and stockroom layouts to determine if they are keeping pace with this increased volume of business.

PLANOGRAPHING

A planograph is a detailed stockroom blueprint for parts storage in bins. It describes, in pictorial or blueprint form, the physical layout of a jobber's stockroom area showing the following:

1. Number of bins
2. Arrangement of bins
3. Parts per bin (number and capacity)
4. Expandable bin space
5. Aisle space
6. Bulk item locations and dimensions
7. Overall space requirements or usage

Developing a planograph requires

1. Store sales records (or inventory guides) and judgments if part number records are not available
2. Part or box dimensions and weight
3. Bin combinations or designs available
4. Floor space limitations in existing facilities

Inventory requirements by part are calculated using the stock ceiling number for that part. As Chapter 7 will explain, the stock ceiling is the maximum number of any part the store wants to have in inventory at one time.

Sufficient storage space is allowed to accommodate the stock ceiling of each item. Approximately 25 percent of each bin's capacity should be left open to accommodate future expansion for either new parts or increased inventory levels of existing stocked parts.

Where space limitations exist, an adequate supply of high-volume, fast-moving parts should receive stocking priority. In such cases, supplemental storage should be planned and arranged for wherever possible. Planograph modernization helps correct these haphazard storage conditions:

1. Overcrowded bins and racks with no open growth or rearrangement room
2. Excessive handwritten labels
3. Parts on floor
4. Parts out of sequence
5. Mixed stock
6. Parts sticking out of bins
7. Parts on bin tops
8. Parts with multiple locations
9. Mixing bin labels
10. Unorganized special order and warranty parts
11. Lost stock and parts in wrong locations
12. Inventory location errors

Reorganization will reduce or eliminate these costs:

1. Lost sales from unable to locate and out-of-stock conditions
2. Parts damage, repair, and scrap
3. Unidentified parts
4. Labor costs—picking parts, restocking, taking inventory
5. Obsolescence and parts returnability
6. Low inventory turnover
7. Machine shop service delays
8. Pilferage and inventory shrinkage
9. Special order costs
10. Excessive parts returns to WDs and manufacturers
11. Packing and shipping errors

It will also result in improved operating performance and profit in the following ways:

1. Increased parts sales
2. Increased turnover and return on investment
3. Improved labor efficiency and better counter coverage
4. Improved parts purchasing patterns—lower special and emergency order costs
5. Higher percent of sales from stocked inventory
6. Larger stock order discounts
7. Return reserve increase—improved cash rebate and/or restocking charges
8. Increased customer satisfaction and repeat business
9. Improved gross and net profits

Table 2-1 contains approximate stockroom equipment guidelines based on available square footage of stock space.

SHIPPING AND RECEIVING

The shipping and receiving area is ideally in close proximity to the stockroom. It should be a wide-open area with direct access to outside loading docks and bays. The shipping and receiving area should be proportionately sized to the overall stockroom area. For example, a store with 3,500 to 4,000 square feet of stockroom storage requires roughly 250 square feet of shipping and receiving area to comfortably handle and sort regular shipments and deliveries.

TABLE 2-1: STOCKROOM EQUIPMENT GUIDES

Parts Stocking Area (Sq Ft)	Regular Standard Bins	Bulk Bins	Sheet Metal and Special Rack
650	25	20	2
920	38	32	5
1,170	50	42	6
1,380	57	48	7
1,530	62	52	7
1,900	70	58	8
2,290	95	79	11
2,810	108	90	13
3,000	116	96	14
3,310	129	107	16
3,710	150	124	18
4,060	167	139	21
4,500	185	154	23
4,910	201	167	25
5,290	219	182	27
5,690	236	197	29
5,990	258	215	32
6,310	275	229	34
6,755	299	249	37
7,105	316	263	39
7,555	335	279	41
8,055	358	298	44
8,620	382	318	47
9,070	402	335	50
9,520	423	352	52
9,970	442	368	55
10,480	465	387	58
10,950	486	405	60
11,400	507	422	63
11,850	527	439	65
12,350	548	457	68
12,900	573	478	71
13,310	590	492	73

The following is a general outline for checking in incoming shipments from WDs or manufacturers. These responsibilities can be carried out by shipping and receiving clerks, stockroom clerks, or counter personnel.

1. Check the freight bill or waybill for correct information.
   - Correctly addressed
   - Transportation charges correct
2. Compare number and type of cartons listed on freight bill against the number and type of cartons received.
3. Inspect all cartons for damage.
4. Record any shortages, overcharges, or damages on the freight bill.
   - Use a pen or marker, never a pencil.
   - Mark shortages clearly and accurately.
   - Record shortages in a different color ink.
5. Obtain delivery person's signature on freight bill to verify noted damages or shortages.

 SALES TALK _____

If the delivery person does not agree on damage or shortage, do not accept the order. Do not accept initials; have the complete name written. The delivery person's signature is mandatory if a valid claim is recorded.

6. Sign and date the freight bill or waybill. When you sign this document you have relieved the shipping company or carrier of responsibility for loss or damage to the merchanidise shipped. After the delivery person has left, a claim can still be filed for hidden damage or shortages found after cartons have been opened.
7. Locate the packing lists for each carton in the shipment. These packing lists name the items shipped in each carton. They are often found in plastic covers stapled to the carton, but can be packed inside the carton.
8. Open crates and cartons and save them until all items are accounted for.
9. Remove individual items from the crates and cartons. Check each item against the packing list for that carton or crate. Clearly mark the packing list when the item is accounted for (Figure 2–5).
10. Note any damaged items or shortages on the packing list.

 SALES TALK _____

Any shortages or overcharges found at this point are between the jobber store and the shipper, either a WD or manufacturer. Any problems found in the shipment at this point are the shipper's fault and should be filed immediately on the company claim form.

11. Record back orders according to company policy.
12. Inspect parts for hidden damage.
13. Notify the person responsible for making claims to the WD or manufacturer for any detected damage.
14. Remove parts and sort by stocking product line and part number grouping. Place ignition parts together, oil filters together, and so on.
15. Report special order items to the parts manager so that timely delivery and/or pickup of these rush parts can be handled.

Accurate, prompt receiving and checking-in stock is extremely important because it affects inventory control for the entire store operation. Once stock is carefully unpacked, checked in, and organized, it can be moved to the proper stockroom locations. Tools used to move merchandise include hand trucks, monorails, dollies, and caster carts. The following are safety and operational tips to follow when using these facilities:

FIGURE 2–5 Incoming shipments from WDs and manufacturers must be carefully unpacked, inspected, and checked into inventory. Always compare the shipment to the items listed on the packing slip to locate any discrepancies.

Hand Trucks

To safely use hand trucks, always do the following:

1. Items that must be moved the farthest should be placed on the hand truck first.
2. Balance the truck and load.
3. Move the truck carefully.
4. Maintain a good grip and balance.
5. Place the truck back in the storage area after use.

Monorails

When using monorail transport systems, be sure to:

1. Balance the loads.
2. Pass the hoist hook through as many chain bands as possible.
3. Securely anchor chains on all items.
4. Clear a transport path.
5. Move the monorail to the storage area.
6. Put the monorail back in its proper place after use.

Dollies and Caster Carts

To safely use this equipment, be sure to:

1. Place the heaviest loads on the bottom, lightest on top.
2. Push or pull the loads carefully.
3. Stand dollies or caster carts upright against a wall after use.

CAUTION: Never leave dollies or carts on their wheels after using.

Wagons and Basket Carts

When using wagons and basket carts, always

1. Load the heaviest items on the bottom.
2. Load the lighter items on top.
3. Move the load carefully.
4. Move the odd-shaped or bulky items with care.

CAUTION: When moving any merchandise, watch for oily and greasy spots, and use care when going around corners.

Forklifts

Forklifts are an easy method of handling large items, especially palletized materials. The following are basic forklift operating and safety considerations:

1. All starts and stops should be gradual to prevent the load from shifting. Turns should be made at a safe speed, smoothly and gradually.
2. Never carry loads in an elevated position. When in travel, the forks should be held at a level approximately 4″ above the floor.
3. Never allow anyone to ride on the load or any part of the lift.
4. Do not allow any part of the load to obstruct vision while driving. Operate the truck in reverse if forward operation obstructs the view.
5. Be sure the path is clear when raising or lowering a load or moving with a load.
6. Watch for overhead obstructions.
7. Always be sure the forklift is mechanically sound through a program of regular maintenance.
8. Never leave the forklift unattended. Always shut off the power, neutralize the controls, set the brake, remove the ignition key, and lower the forks to the floor.
9. Keep fork truck routes open and clear. Post mirrors to relieve blind spots around corners, and the like.
10. Watch for oily, wet, or otherwise slick surfaces.

LIFTING MATERIALS

Shipping, receiving, and stocking materials requires physical exertion. Knowing the proper way to lift heavy materials is important. You should always lift and work within your ability and seek help from others when you are not sure you can handle the size or weight of the material or object. Auto parts, even small, compact components, can be surprisingly heavy or unbalanced. Always size up the lifting task before beginning. When lifting any objects, do the following:

1. Place your feet close to the load. Place your feet properly for balance.
2. Keep your back and elbows as straight as possible. Bend your knees until your hands reach the best place for getting a strong grip on the load (Figure 2-6).

3. If the part or component is stored in a cardboard box, be certain the box is in good condition. Old, damp, or poorly sealed boxes will tear or otherwise fail. A heavy object could tear through the side or bottom of the container, causing injury or damage.

4. Grasp the object or container firmly. Do not attempt to change your grip as you move the load.

5. Keep the object close to your body and lift by straightening your legs. Use your leg muscles, not back muscles.

6. When changing direction of travel, do not twist your body; turn your whole body, including your feet.

7. When placing the object on an elevation, such as a shelf or counter, do not bend forward. Place the edge of the load on the surface and slide it forward, being careful not to pinch your fingers.

8. When setting down a load, bend your knees and keep your back straight. Do not bend forward—this strains the back muscles (Figure 2–7).

9. Use blocks to protect your fingers when picking up or lowering heavy objects to the floor (Figure 2–8).

STOCKING INVENTORY

Once the stock has been moved to the appropriate stockroom location or storage bin, follow these stocking tips:

1. Clean the area before stocking.

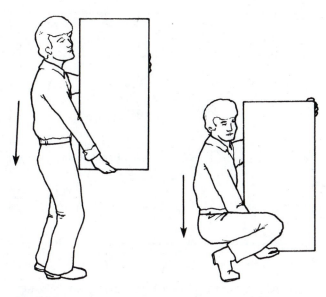

FIGURE 2-7 Put the load down as carefully as you lifted it.

FIGURE 2-8 Use blocks to protect your fingers when picking up or lowering heavy objects to the floor.

2. Check each item to be sure it is tagged, numbered, or labeled properly.

 SALES TALK ——————

If there is a change in design or a new model that contains a new number, it might be necessary to relabel the numbers on items already in the bins.

3. Place items in the bin with their labels in an upright position.

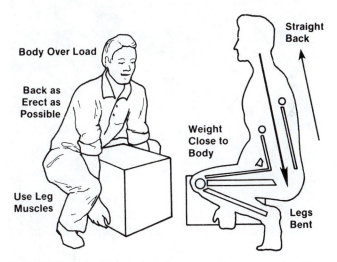

FIGURE 2-6 Use your leg muscles—never your back—when lifting any size load.

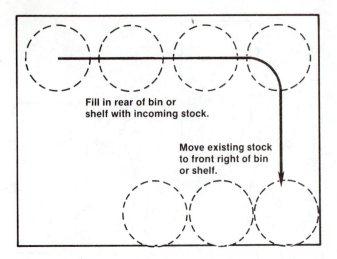

Fill in rear of bin or shelf with incoming stock.

Move existing stock to front right of bin or shelf.

Pick from front right.

FIGURE 2-9 When stocking bins or shelves, rotate the stock so older items are moved to the front shelf locations.

4. Rotate the stock; place new merchandise to the left and rear of the bin and place old merchandise to the right and front of the bin (Figure 2–9).
5. Relocate overstock and mark it as such.
6. Gaskets and rubber parts should be placed in cool, dry places where heat cannot ruin them.
7. Flammable materials should be placed in cool, dry, airy places to prevent fires.
8. Odd-shaped items should be placed in special racks and bins to prevent accidents.
9. Heavy items should be placed on lower shelves for ease in lifting and safety.
10. Fast-moving items should be placed on shelves or bins that are easy to reach and readily accessible.

MACHINE SHOP AREAS

Jobber in-store machine shops are normally considered part of the back of store operations (Figure 2-10). Depending on the type of services offered by the machine shop, equipment could include a valve seat, guide and facing machines, brake lathes, head resurfacers, rod and piston heaters and reconditioners, crankshaft grinders, polishers and balancers, power hones, welding equipment, and parts cleaning equipment.

The ability to offer machine shop services is one of the main factors that distinguishes traditional jobbers from their mass merchandise and discount house competitors. Increases in shop work sales have outdistanced increases in hard parts sales in recent years.

The jobber machine shop is the competitive edge against the mass merchandiser, the discounter, and the automotive retailers who also push hard engine rebuild parts.

Mass merchandisers have not entered the machine shop market for two reasons. First, precision equipment is a capital intensive business. It takes a large initial investment with a relatively slow return—slow enough to discourage stores used to fast returns. Second, machine shop work is a skill intensive business, not conducive to low wage employment standards or high labor turnovers. In other words, machine shop services are basically an investment in people—good people who care enough to maintain and use modern precision equipment and demonstrate their skill in a superior product.

Today's cars, because of their inaccessible engine compartments, require quality machine work for comeback-free engine repairs. Selling a complete line of reliable machine shop services means increased customer satisfaction and more profit per job.

What services do customers demand from machine shops? In farm belt areas, the demand for rebuilds on industrial and commercial engines might outstrip other market segments. In metropolitan areas, machine work for custom engine builders and fleet vehicles might be in higher demand.

FIGURE 2-10 Resurfacing rotors is a common machine shop job. Here a counterman checks for sufficient rotor thickness.

Similarly, demands for flywheel grinding, brake drum and rotor resurfacing, bearing press work, welding, lathe work, driveline repair, and numerous other services might surpass engine repair entirely. The need for engine balancing might be slight compared to the grinding of heavy truck brake drums, or vice-versa, depending upon the market in the jobber's area.

Identifying needs may be difficult. Often the need is present but unidentified because nobody asks for it. For example, crack detection might be a service seldom requested. But adding crack detection to the machine shop's list of related engine, cylinder block, or head services could uncover a profitable unfilled demand.

Likewise, related needs can be defined by offering complete services connected with the engine block. For example, cylinder boring might be good as far as it goes but also consider align boring the main bearings and decking the block-to-cylinder head surfaces in these new, easily warped, thin-wall engine blocks.

Similarly, connecting rod reconditioning goes hand in hand with crankshaft regrinding, crankshaft grinding fits in perfectly with block work, and block work fits with head work. This link-in-the-chain approach not only helps identify what the service needs are, but also helps promote the counterperson's most valuable cash flow multiplier—the related sale.

Counter personnel and outside salespeople should sell machine shop services whenever the need is indicated by sales of related parts. The purchase of brake pads or shoes, especially those that carry a lifetime warranty, requires a mandatory refinish on rotors or drums to validate the guarantee. The sale of an engine piston set will suggest a rebore job. After all, why install standard pistons on the original worn cylinder bore? Or, if the sale consists of oversized pistons, who might be performing the engine block bore job? And if the block is to be rebored, who will hot tank the engine and install camshaft bearings and block plugs? And how will the piston pins be pressed when the new pistons are installed on the connecting rods?

In any case, when special equipment is called for in the installation or fitting of a new part, jobber machine shop services should be called into play. From the pressing of the garden-variety axle shaft bearing to a complete rebuild and assembly of an engine, the purchase of parts always suggests the equivalent means of installation.

The key to selling machine shop services is selling at the point of sale, either at the counter or the installer's garage when outside sales calls are made.

Regular visits to the mechanic customer's shop reveal the type of work that flows through the shop on a routine basis. Some shops specialize in engine rebuild, other shops go for the quick service work in front suspension, brakes, and engine tune-up.

While the field is far more fertile in the engine repair shop, the light types of machine work—flywheel grinds, drum and rotor turns, press work, even driveline rebuild—will be found in all garages or service shops that do not find it economical to equip for these particular types of services.

When discussing machine shop work with customers, stress the quality and value of the work performed in the shop. Be prepared to talk about the fine points of reconditioning work; why, for example, your shop's method of valve guide reconditioning might be superior to your competitor's. Can you install bronze valve guides? Do you cut or grind valve seats, and if you do the guides and seats, does your method maintain seat and guide concentricity and preserve seat height better than that of competing shops? And what are the standards of quality for grinding crankshafts or reboring engines?

Every type of engine reconditioning method has its virtues and drawbacks. But be prepared to compare and contrast your method versus other methods. What are the similarities, differences, and, above all, the superiorities of your methods? Selling machine shop services is no different than selling brake shoes or spark plugs—the counterperson must know the product well enough to compare with the competition.

HANDLING HAZARDOUS WASTES

One of the major areas of concern in running a machine shop operation is the proper handling and disposal of hazardous chemicals and wastes.

The Environmental Protection Agency (EPA), Occupational Safety and Health Administration (OSHA), and other state and local agencies have strict guidelines for handling these materials. OSHA's Hazard Communication Standard (HCS) applies to all companies that use or store any kind of hazardous chemicals that workers might come in contact with including solvents, caustic cleaning compounds, abrasives, cutting oils, and metals. Commonly called the "Right-to-Know Law," it certainly applies to most machine shop operations.

Compliance with the law requires a system for labeling hazardous chemicals and maintaining Material Safety Data Sheets. These sheets must be made available to employees, informing them of the dangers inherent in chemicals found in the workplace. The employer must also establish a training program and written policy covering the various aspects of the HCS. Fines for violations can be very costly, so store owners and shop foreman should be fully aware of applicable laws and regulations in their area.

To cut down on the volume of hazardous wastes and solvents generated, machine shops are switching to biodegradable solvents or have begun to neutralize their caustics. It might also be wise to confer with EPA or other environmental officials before purchasing new cleaning equipment. Companies that handle waste disposal are also available.

COUNTER AREA OPERATIONS

The immediate counter area is a high-activity location in any jobber store. The counter is where customers are serviced and sales are made. Depending on store size, the counter is normally staffed with two to six countermen and women. An ideal store layout provides at least 6 feet of counter length for each counterperson. This is sufficient to accommodate catalog racks, telephones, and point-of-sale (POS) computer terminals/cash registers.

CATALOG WORK

Parts manufacturer's catalogs are normally stored in racks located on or near the counter (Figure 2-11). As explained in Chapter 5, there are several methods of organizing catalogs within a rack. The key point is that all counter personnel must be familiar with the selected system.

The catalogs list the part numbers and other vital information concerning the manufacturer's line. Catalog formats usually include vehicle manufacturer, year, and model type. However, industry catalogs are far from standardized. Each one requires a certain amount of special knowledge and familiarity (Figure 2-12).

Based on information supplied by the customer, the counterperson looks up the part number for the requested item. In general, the procedure for filling a part order using catalogs includes the following:

- Locate the part listing on the page, then locate the part number and carefully write it down.
- Suggest related sales items to the customer.
- Locate and record the part numbers for any related sales items.
- Pull the order from the inventory and assemble it at the counter.
- Write up the necessary invoice as per company policy.

FIGURE 2-11 Manufacturer's catalogs are usually stored in racks located on or near the counter.

Detailed instructions for completing a part order are given in a later chapter.

PAPERWORK

Price sheets, invoices or sales tickets, accounts receivable and lost sales reports, and other forms necessary to transact and document sales are kept near the counter area for use by counter personnel.

Price Sheets

Catalogs do not list parts prices. These are listed on separate price sheets supplied and updated by

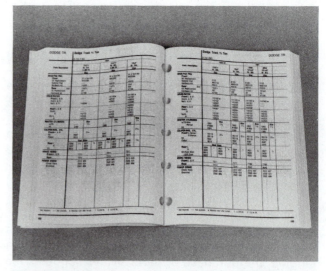

FIGURE 2-12 Most catalog formats list part number application by vehicle make, year, and model. However, formats differ from catalog to catalog.

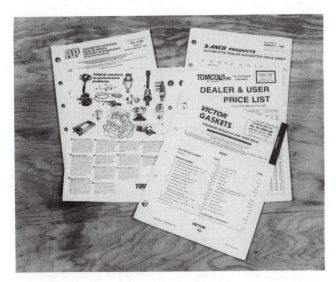

FIGURE 2-13 Separate price sheets are provided by suppliers. These sheets list prices and any number of discounts for various customer classifications.

the manufacturer or WD. Price sheets (Figure 2-13) normally list retail and professional discounted rates (offered to the store's professional trade or volume customers). Counter personnel must know how to read price sheets and use them accurately. They are also responsible for keeping the price sheet file up to date and organized.

Invoices (Sales Tickets)

Counter personnel are responsible for filling out or generating a sales ticket for each purchase (Figure 2-14). Accurate, easy-to-read sales tickets are vital for billing and accounting purposes. They are also a valuable inventory control tool.

Accounts Receivable Report

This report lists accounts that owe the store money. The routing of the accounts receivable report is usually at the discretion of the owner or parts manager. Most managers, however, provide their counterpeople with information of past due accounts and inform them of customers who must pay with cash. It is poor business practice to continue to write charge tickets for a customer whose credit has been suspended or canceled.

Lost Sales Reports

Counter personnel are also responsible for keeping an accurate record of lost sales. These lost sales reports are used to adjust and update inventory to better meet the local market needs.

In addition to these forms, counter personnel might also be responsible for filing claims for damages or shortages with WDs or manufacturers, preparing credit memorandums and price quotations, pricing warranty parts claims with manufacturers, handling core exchanges, and writing up work tickets for machine shop jobs.

More details concerning these duties are covered in other chapters. And as you will see in Chapter 6, computerized point-of-sale systems can accurately tabulate and print out invoices, lost sales reports, and many other types of documents associated with running a jobber operation. These systems have eliminated much of the tedium and error from the counterperson's daily routine, freeing time to better service customers and increase sales efficiency. The advent of computerized cataloging systems is also dramatically reducing the time and effort needed to locate part numbers and related sales items.

TELEPHONE ORDERS

The counter is also the communications center of the jobber store. Phone orders and inquiries make up a substantial segment of jobber trade, and they are fielded by counter personnel (Figure 2-15). Countermen and women must be skilled telephone communicators.

In all telephone conversations the following guidelines should be observed:

FIGURE 2-14 Accurately writing out invoices or sales tickets is an important job of counter personnel. Fortunately, more and more of this work is being done automatically through the use of advanced computer POS equipment.

FIGURE 2-15 Telephone work constitutes a large percentage of the counterperson's daily activity.

- State the name of the store and identify yourself. Always have something to write on.
- Speak clearly and distinctly.
- If the customer plans to pick up the item at the jobber store, clearly mark the package with a note giving his or her name, the part number, and price.
- Keep clear, concise notes of every phone transaction. Store the notes in a safe place so they can be referred to if a problem arises.
- If you must leave the phone to check on stock availability or other business, put the caller on hold or place the receiver gently on the counter. Tell the customer how long you will be away from the phone or offer to take the customer's number and promptly call back.
- Recite part numbers in two-digit groups to avoid confusion.
- Give letters as an example to avoid mistakes, such as "F" as in Frank, "C" as in Charlie, etc.
- When in doubt concerning a customer's discount classification, quote only the list price.
- Be sure to get proper purchase authorization such as a purchase order number or the name of the person placing the order.

More details on properly handling telephone orders are given in a later chapter.

HANDLING AND DISPATCHING DELIVERIES

If the jobber store offers delivery service, the tasks involved in running it efficiently might be handled or supervised by a separate dispatcher. Or, perhaps a member of the counter staff assists in organizing the delivery runs. To pull a delivery order from stock, follow these steps:

1. Examine the order to estimate the size and nature of the parts to be delivered.
2. Pull the order from stock and count onto an order cart.
3. Count the items into the container as they are pulled from stock. If possible, have someone else double-check the quantity.
4. Check the items pulled against the purchase order form or sales invoice for correct part number.
5. Place a checkmark by each item on the order form as that item is placed onto an order cart.
6. Move the order to the shipping area.
7. Pack separate orders into appropriate containers. Fragile items, such as glass, gauges, etc., should be packed in separate containers even if they are part of the same order. Do not pack heavy items with items subject to breakage. Many heavy or bulky items might only require tagging.
8. Check the order against the invoice as you pack each part. Be sure the quantities and part numbers match. Mark all packages with the customer's name and address. If a single order contains several packages or boxes, mark them, 1 of 3, 2 of 3, and so on. Mark packages "fragile" or "this side up," if needed.

To plan the delivery route, follow this procedure:

1. Collect all invoices for deliveries and check a map for the location of all businesses.
2. Plan deliveries by the shortest possible route and arrange invoices according to the route to be followed.
3. Arrange packages in reverse order and load the delivery truck. (The last package to be delivered is loaded first, the first to be delivered is loaded last, etc.)
4. Review the route with the driver. The dispatcher or manager keeps a copy of the route for time tracking in case a customer phones and needs to know estimated delivery time, and so on.

When enroute to stops, observe all traffic regulations. The delivery truck is a traveling sign for the store. When possible, park the delivery vehicle in low traffic areas. When making a delivery, a driver should always

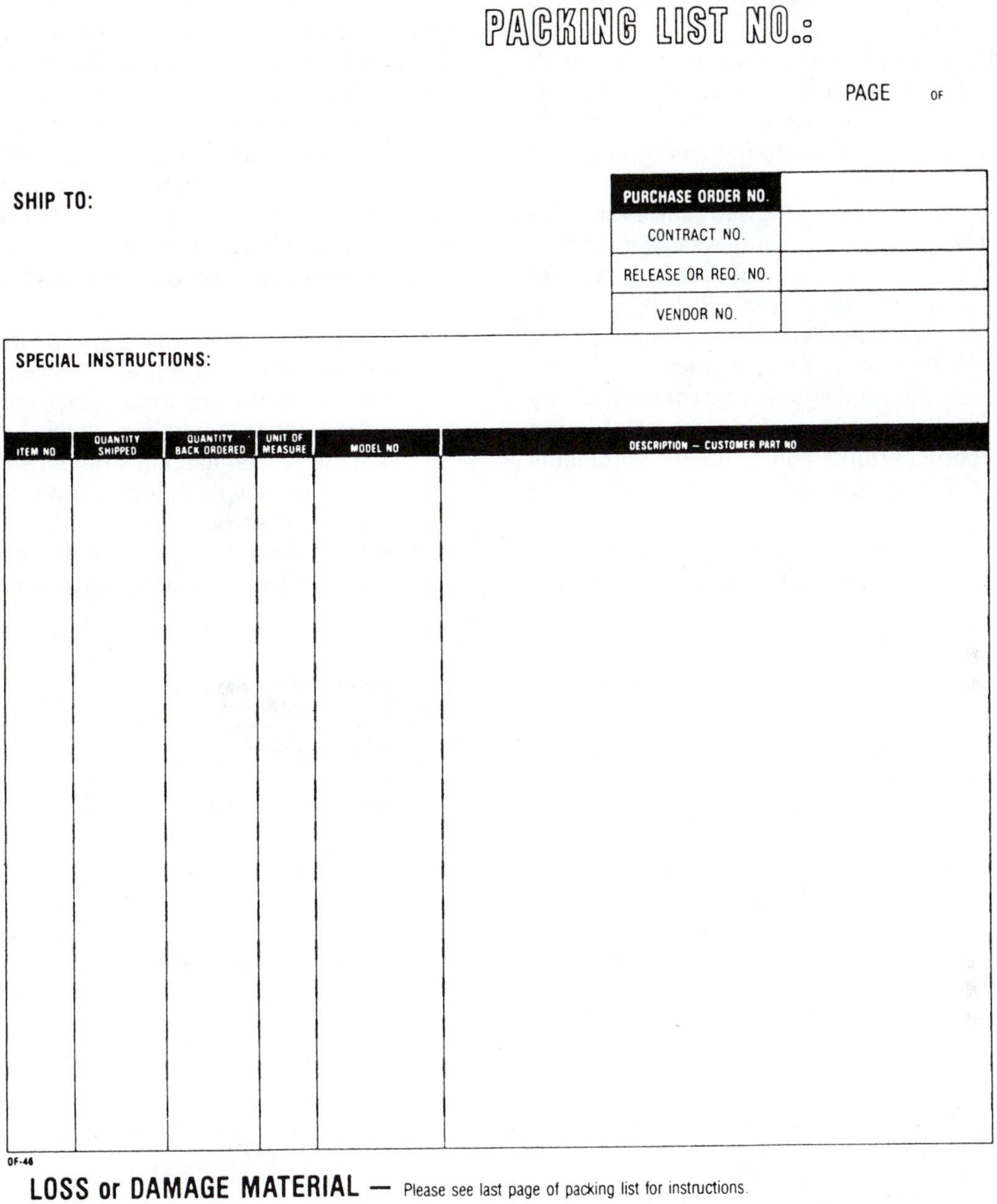

PACKING LIST NO.:

PAGE   OF

SHIP TO:

PURCHASE ORDER NO.	
CONTRACT NO.	
RELEASE OR REQ. NO.	
VENDOR NO.	

SPECIAL INSTRUCTIONS:

ITEM NO	QUANTITY SHIPPED	QUANTITY BACK ORDERED	UNIT OF MEASURE	MODEL NO	DESCRIPTION — CUSTOMER PART NO

OF-44

LOSS or DAMAGE MATERIAL — Please see last page of packing list for instructions.

FIGURE 2-16 A packing list is attached to the first carton of a multiple-carton order.

1. Greet all customers; identify self and jobber store.
2. Place delivery according to the customer's instructions.
3. Have the customer check the merchandise against the invoice.
4. Inquire about other needs the customer might have.
5. Pick up and properly tag any cores returned for credit.
6. Make notes of any problems or other tasks that should be handled when the driver returns to the store.

SHIPPING GOODS

Jobbers may also use common carriers, such as trucking, air freight, or even rail freight carriers to ship large orders to out-of-town customers, or back to WDs and suppliers.

As with deliveries, any order for shipment must be carefully picked, packaged, and documented. A packing list or slip must be included with each shipment. This list contains a detailed description of the items included in the shipment (Figure 2-16). A separate packing list can be placed in each carton of a multiple carton shipment, but often a single lengthy list is packed into or secured to the first carton of a multicarton shipment.

The packing list contains all of the information included on the sales invoice except the prices, tax, and other payments due. The sales invoice or billing is usually mailed to the customer under separate cover and is not included with the shipment.

In many cases, the packing slip (and even shipping labels) is generated when the sales invoice is typed or written up. Special multilayer forms are used, and a copy of the appropriate form is distributed to the corresponding department. Each other is picked using the packing slip as a guide.

Orders for shipment should be packed in sturdy cartons or boxes using packing material when needed. Secure cartons tightly. Label shipments clearly with the correct addresses and handling warnings (Figure 2-17).

For shipments made with common carriers, a bill of lading must be drawn up by the shipper (jobber). This is the legal agreement drawn up between the shipper and the carrier. When filling out a bill of lading, be certain to specify the number and type of packing (boxes, crates, pallets, etc.) and briefly describe the items contained in them. A standard bill of lading form is accepted by most carriers, and most carriers supply forms to their regular customers (see Figure 8-16, page 239). The carrier uses the bill of lading as a basis for preparing other shipping documents and records such as freight bills or waybills.

When the carrier arrives to pick up the shipment, make certain the carrier representative signs the bill of lading. An authorized store representative will also sign the bill of lading at pickup.

Once the carrier assumes responsibility for the goods, he usually will assign a pro number to the shipment. This is a tracking number that can be used to trace the shipment through the carrier's system in the event that problems arise. The shipping depart-

---

Date

PACKING LIST NO.

PURCH. ORDER NO.

SHIP
TO:

RELEASE OR REQ. NO.

VENDOR NO.

WEIGHT

SPECIAL INSTRUCTIONS:

ITEM NO.	QTY. SHIPPED	MODEL NO.	DESCRIPTION
2	40	D5007	DIVIDER
3	40	D5013	DIVIDER
4	40	D5005	DIVIDER

FIGURE 2-17 Individual carton packing labels are sometimes attached to each carton in an order. They list all items in that particular carton.

ment should always record each shipment and its pro number into the store's shipping log. Some carriers have systems where the shipper can check on the status of a shipment by calling a toll-free number and giving the pro number.

Specialized carrier services offer overnight and standard delivery service for emergency orders and small to medium shipments. These companies have their own forms and procedures for pickup and delivery.

Shipping personnel must become familiar with the procedures involved in working with the store's carriers. Insurance, weight and package size rates, and billing procedures can change from carrier to carrier and shippers must be aware of these details.

DEALING WITH WAREHOUSE DISTRIBUTORS

In addition to understanding in-store operations, counter personnel and parts managers must deal with warehouse distributors, or WDs, and manufacturer representatives on a daily basis. Understanding the operation of other links in the aftermarket supply chain is essential.

Understanding warehouse distributors is complicated by the fact that there is such a diversity among them. Some WDs are essentially jobbers with several lines that they buy directly from the manufacturer. These redistributing jobbers resell these lines to smaller jobbers. In fact, many large distributors began as jobbers and, as they grew, kept adding to the number of lines they bought directly until that was the only way they bought anything.

Some WDs are relatively small companies serving the local aftermarket; others are massive distribution organizations that dominate an entire state or section of the country. Then there are those WDs who exist primarily to support their own chain of retail parts stores. Others are the hubs for fleets of wagon peddlers, mobilized sales forces who call directly on professional accounts, bypassing the jobber. And there are those who are in the distribution business only and rely almost entirely on trade from independent jobbers.

A few exist as co-operative buying units for networks of independent jobbers. Still others are distribution centers for large program groups, catering almost exclusively to the needs of their member stores.

Even some engine rebuilders get involved, using their buying power to compete for the hard parts business. Some distributors offer product coverage that encompasses the entire aftermarket, with broad inventories in many different lines. Others specialize in just electrical or brake parts; some are still primarily interested in parts for domestic vehicles, and some specialize in import parts.

Although most WDs offer fairly thorough coverage in the lines they sell, carrying everything from "A" to "W" classifications, there are a few phone or mail-order giants that sell only the cream of the crop in selected fast-moving lines.

About the only thing all WDs have in common is that they buy directly from the manufacturers. Likewise, the services WDs provide range from little more than order-taking and shipping to extensive help with advertising, field sales support, cleanups and changeovers, management assistance, financial services—even help looking up part numbers.

The one major difference between warehouse distributors and jobbers is profit margin. While most parts stores enjoy average profits ranging from 30 to 45 percent, WDs work on profits that are in the low to mid 20s. This shorter margin, coupled with the need to provide some expensive services, makes it necessary for WDs to employ some operating procedures that are sharply different from those of a jobber.

The typical WD needs a sophisticated purchasing staff, good warehousing and material handling facilities, very capable data processing resources, an efficient order entry department, and effective salespeople. These are major operating expenses, and having to provide them on a short margin makes it imperative to secure a good market share and eliminate unnecessary expenses.

This is why some of the policies of some WDs might seem cold and rigid compared to the accommodations jobbers provide customers. If a jobber returns an opened set of gaskets to the WD, and the manufacturer imposes a 50 percent penalty, the WD has little choice but to pass the charge to the jobber. There is not enough room in the WD's margin to absorb that expense.

Likewise, the WD has to be vigilant with cores and returns. He or she cannot afford to buy cores that were not sold, give credit for damaged cores, or accept more returns than can be passed back to the manufacturer. The short margin and high operating costs also make it necessary to enforce stricter credit policies.

For the WD, slow collections mean less cash with which to buy parts and, therefore, a poorer fill rate. Since a low fill rate is going to cost the WD some of the market share—both through lost sales and dissatisfied customers—he or she must either keep collections moving or borrow against receivables (very unsatisfactory, given the already low profit on those sales).

Finally, the short margins and high expenses make market share one of a WD's greatest concerns. High sales volume is not merely desirable, it is absolutely necessary, and if a distributor cannot secure enough jobber support, he or she will sell wherever support can be found. This problem can only become increasingly acute as more and more retailing giants move into the automotive aftermarket, buying directly from the manufacturers, bypassing distribution channels, and carving for themselves larger and larger pieces of the aftermarket pie.

Until large numbers of independent jobbers become adapt at taking back their market share, WDs are going to feel compelled to do just about whatever it takes to keep up their volume, even if it means selling to people who have traditionally been jobber customers. For the jobber the solution is less likely to be found by hurling accusations or initiating litigation than by working together with distributors to understand the changes taking place in the market and developing joint strategies to deal with tough new competitors.

Two other areas of considerable misunderstanding between WDs and jobbers are order fill and return policies.

Fill Rates

The WD's fill rate is affected by several factors. The skill of WD purchasing agents has a tremendous influence on fill rate. No matter how much money WDs have at their disposal, they cannot maintain a balanced inventory if they are not paying attention to sales history, anticipating seasonal demands, and keeping up their ordering frequency.

Another essential force is adequate movement of each line. If a line is not selling enough, a deep inventory cannot be justified. Then, because most manufacturers have substantial minimum order requirements, the slow-moving lines will be ordered with less frequency and gaps in them will stay in place longer.

Multilocation warehouses have the advantage of being able to step up ordering frequency by pooling the requirements of several branches. This, however, does the jobber little good if there are distribution problems among WD locations. A big shipment sitting on the dock at the WD's central warehouse is of little use to a jobber store that relies on a branch outlet.

When a distributor's fill rate is down, there is a natural tendency to suspect a problem with purchasing or finances. But, problems in less obvious areas can also reduce order fill. The best purchasing does not do a lot of good if the merchandise is not put on the shelves, if it is stocked in the wrong locations, if the order pickers are incompetent, or if the goods received are not posted to the computer records.

When problems exist with a distributor's fill rate, be sure to let his or her salesperson know about it. Since few sales people get a commission on items not shipped, they are the people most motivated to look for answers to fill rate problems.

Remember to support a distributor that does a particularly nice job on order fill. Your purchases keep his or her lines moving, build sales history, and help ordering frequency. These, in turn, continue to better fills.

Returns

The more one moves up the distribution chain, the tighter the return policies become. The consumer can return just about anything he or she buys, the jobber gets to return quite a bit less, and the WD is limited by the manufacturer to a fixed percentage of previous purchases.

As a result, WDs might limit the amount of returns accepted from jobbers and also place limits on the extent to which they will clean up existing jobber stock. This makes it important for jobbers to purchase carefully. Wrong parts and parts carelessly added to jobber inventory count against the store's return allowance and limit the parts manager's ability to keep stock lean.

Still, everything is negotiable, and even when a WD explains that the manufacturer will accept only a 2 percent return, smart parts managers should be ready to point out that of the parts the store wants to return to the WD, the only parts the WD is likely to send back to the manufacturer are the few with the poorest popularity classifications. Even if you are asking him or her to take back considerably more than 2 percent, most of the parts will go right back on WD inventory shelves and be sold to other jobbers.

Placing Orders

As a counterperson, you will regularly face the problem of dealing with customers who know nothing about the part they want. They give you no part number, no engine number, and perhaps they do not even know what kind of car it is from. However, the counterperson should not fall into the same lazy mode when he or she becomes the customer ordering from the WD. Do not make the supplier do all the work.

Catalog research is the job of the counterperson or parts manager, not the WD. The distributor's lean profit margins and high operating costs do not really leave room for catalog work. For the distributor, profits are gone if he or she has to do the application research. Besides, good WDs spend time and effort to keep jobbers supplied with catalogs and price sheets.

Of course, if the WD has done a poor job of distributing current literature, counterpersons have little choice but to ask for help. But another reason jobber personnel should do their own catalog work is that if they want something done right, they should to it themselves. The more people involved in a transaction, the greater the odds of it going awry. And if a counterperson asks someone else to do what he does best, he is just inviting trouble.

Constantly asking the distributor for prices is nearly as unprofessional as expecting him or her to do application research. If the WD is supplying price sheets and counter personnel in jobber stores are filing them properly, a counterperson should rarely have to ask for a price.

But, if a price must be checked, tell the WD the date on the sheet and ask if it is still current. If it is, the counterperson is done. If not, he or she is way ahead of just asking for a single price—the person knows an entire new sheet is needed.

The best time to ask about a price is before the warehouse goes to the bins to verify stock. When the WD order clerk first keys the number into the computer and displays it to see if it should be in stock, jobber cost is right there on the screen. Later, while going to the shelves, someone else is likely to clear the terminal and use if for another task. So, if the counterperson waits until he or she gets back to check a price, extra work has been created for the WD.

Showing a little consideration for the WD goes a long way toward getting the most out of his or her resources. In the eyes of the warehouse staff, keeping counter staff requests for research and prices to a minimum can only enhance his or her reputation as a professional, and that reputation goes a long way in determining the kind of service a store receives. The way a counterperson places orders should show an understanding of the WD's operation.

Consider these steps when placing a phone order with a WD or manufacturer:

1. Think about the sequence in which the WD data entry operator has to enter the various information—account number, purchase order number, shipping instructions, back order instructions—and try to give the information in that sequence. Find out, too, whether they prefer the part number or the quantity first. This saves keystrokes and cuts down on the chances of errors creeping into an order.

2. As in dealing with phone customers, be aware of easily misunderstood characters, like "F" and "S," "sixty," and "fifty." Take the trouble to enunciate these clearly and be sure they were heard correctly.

3. If there are several items in an order, organize it before calling. Grouping parts together within their lines, rather than skipping back and forth, and putting them in numerical sequence will usually make it easier for someone to fill the order.

4. If you need some items physically checked, give the warehouse person all the numbers at once, so he or she does not have to make repeat trips to the bins.

The same professional approach should be used for written or computer link-up orders. With many computer systems, the hard copy purchase orders generated might have columns for both order quantity and shipping quantity. It is important to clearly differentiate between the two. For example, a hose order might read: order quantity 50 feet; shipping quantity 1 roll. If these are unclear or become confused, 50 rolls might arrive in the next WD shipment.

To prevent this problem, highlight the desired quantities on the items that are likely to be confused by circling them with a bright felt-tip pen.

WORKING WITH MANUFACTURER'S REPRESENTATIVES AND AGENTS

Counter personnel and part managers must work with manufacturer's representatives and agents to keep informed on the ever-changing aftermarket industry. These people can be a tremendous asset to the jobber and counter personnel.

To begin, realize that there is a difference between a manufacturer's representative and a manufacturer's agent. Even though the titles sound the same, their functions are different, and the differences are important.

Both reps and agents are supposed to feed information to the WD and the jobber from the manufacturer, and to hold dealer clinics, handle warranties, check stocks, do field work, or whatever else is required by the customer. They each have the same basic function, with one important difference. The manufacturer's representative, or factory rep as he or she is commonly called, works for one manufacturer only. Most manufacturers with a large product line, of whose product is responsible for a large portion of a WD's or jobber's business, use manufacturers' reps. Each line represents a substantial amount of business for an average jobber, and they are so large and diverse that a full-time effort is necessary to handle the business, so the manufacturers hire and train their own people and send them out into the field.

A manufacturer's agent, on the other hand, usually represents lines that are neither as deep nor as broad as the types mentioned previously. That is not to say that they are not as important to the jobber. It simply means that it is more economical for the factory to hire an agent to handle their line than to try to set up their own sales force. Many manufacturers have done this with great success because they were committed to a good product and hire good agents to work their line for them. Again, the type and size of line affects the decision greatly. In summary, a manufacturer's rep has only one product line to sell; a manufacturer's agent might represent 50 or 60 lines.

Consider how the manufacturer's rep and agent can help jobbers. They can bring a wealth of information and keep a counterperson informed about new products. They can check catalogs, supplements, and price sheets to be sure they are current, and check stock for slow movers and missing (perhaps new) good numbers. Competent reps can help tailor a store's clientele and get those parts moving off the shelves and out the door.

Besides keeping a counterperson informed, reps can help sell. They can point out features and explain little manufacturing details that can be used to make sales representations come to life. They can assist by making dealer calls and providing a depth of knowledge that neither a counterperson nor an outside sales staff can be expected to possess. Spend enough time with the really good ones and some of their selling techniques will probably rub off.

They can also help sell inside the store, too. Good reps have watched how other stores have merchandised their products and carefully noted what works and what does not. Their suggestions about merchandising can make the difference between a mediocre display and a presentation that really moves the product.

Follow a set of guidelines when dealing with any manufacturer's representative or agent. This will help maximize the benefits good reps can offer a store. It will also help distinguish good reps from poor reps, a factor that should be considered when changing product lines within the store.

Know your store's representatives on a one-to-one basis. Think of your relationship with a representative as a partnership, two people working together for a common goal—customer satisfaction. When a rep finds out a counterperson is interested in him or her, he or she will become more interested in the counterperson.

Review the rep's catalogs with him or her. It does not matter if the catalog has not changed in ten years, he or she still might show something the counterperson does not know about or reveal an easier way to look up a part.

Have the rep explain the price sheet, including what each of the classification letters or symbols means and if the sheet being looked at has them. A counterperson should know how superseding numbers are written and if any numbers appear out of sequence on the price sheet, such as specialty items.

Each time the rep calls on the store, have him or her check the current catalogs, price sheets, and supplements in the catalog rack. Have the rep also check for unneeded or unnecessary catalogs in the rack. Catalog space is at a premium in today's parts store, and only the catalogs used regularly should be on top of the counter. Keep the reference works and other similar manuals where they belong—easy to get, but out of the way.

In the area of product knowledge the factory person can really help the jobber staff. Factory reps deal with the objections, complaints, features, and benefits of their product every day.

When discussing the line, have the rep start with the features he or she believes make it the best (or one of the best) on the market.

After hearing the key features of the product, candidly tell the rep what complaints have been made about the line, or what you personally feel is wrong with it. Again, keep the discussion friendly. There could be a logical explanation for a problem you have had with the line which will help you out. Anything the rep can tell to make it easier to sell the product is to the advantage of the counterperson.

Once the counterperson understands a particular rep's line, ask about various displays of the product he or she has seen. Most reps and agents see more parts stores, good and bad, in a month than a counterperson will see in a lifetime. Many times they will see something that can benefit a store, but might be hesitant to mention it because they do not want to meddle in another person's business. However, when asked for advice or tips, most reps are eager to assist in any way they can.

While the rep is in the store, have him or her check warranties. Many times the rep can show what actually was wrong with the part. Gaining that kind of information helps a counterman do a better job, because many times warranty problems can be prevented before they happen, just by giving the customer a few precautions.

Finally, be sure to get the rep's business card. No matter how much a person thinks he or she knows about a line, there will be a time when he or she will not know the answer. Many times a call to the rep can solve the problem. For a counterperson the ability to find information is just as important as knowing the information first hand. Many counterpeople have saved their customers downtime or a costly repair just by being able to contact the right information source.

The best way for a counterperson to let the reps know he or she is interested in working with them is to contact supplying WDs. A lot of WDs will go to great lengths to encourage reps to call on their jobbers because it benefits them also. The reps are just as eager to work with jobbers as jobbers are to work with them. The end result is increased communication between the rep and the counterperson and parts manager, increased sales, and a feeling that all are working toward a common goal.

REVIEW QUESTIONS

1. What is the main difference between small and large jobber stores?
   a. Sales volume
   b. Services offered
   c. Tasks completed
   d. No differences exist.

2. Which of the following is not a main component area in the layout of a typical jobber store?
   a. back of store operations
   b. outside store operations
   c. counter area operations
   d. front of store operations

3. What is a detailed stockroom blueprint for parts storage in bins?
   a. WD
   b. back of store operations
   c. stock ceiling
   d. planograph

4. When checking in incoming shipments, Counterperson A first signs and dates the shipping receipts. Counterperson B first obtains the delivery person's initials on the shipping receipt. Who is right?
   a. Counterperson A
   b. Counterperson B

   c. Both A and B
   d. Neither A nor B

5. After the stock has been moved to the stockroom, what task must be completed first?
   a. Make sure each item is tagged, numbered, or labeled properly.
   b. Relocate overstock.
   c. Place items in bin with their labels upright.
   d. Rotate stock.

6. Jobber in-store machine shops are normally considered part of the _____ operations.
   a. counter area
   b. front of store
   c. back of store
   d. outside store

7. What does the "Right-to-Know Law" mandate?
   a. Guidelines for disposing of hazardous wastes
   b. Employers must inform those people living next to a business of any hazardous materials in use.
   c. Employers must inform employees of hazardous materials they may come in contact with in the workplace.
   d. List of hazardous materials which are not longer legal to use

8. If a customer has an old part and wants a new one to replace it, what is the counterperson's first step?
   a. Locate the part in a catalog.
   b. Locate the price on a price sheet.
   c. Suggest related sales.
   d. Locate casting or past number on old part.

9. What is another term for a sales ticket?
   a. accounts receivable report
   b. invoice
   c. warranty
   d. price sheet

10. After answering the telephone, Counterperson A first asks the customer what he or she needs. Counterperson B first gives the name of the business and his or her own name. Who is right?
    a. Counterperson A

b. Counterperson B
c. Both A and B
d. Neither A nor B

11. Which of the following should be done by a delivery person?
    a. Have the customer check merchandise against invoice.
    b. Place delivery according to each customer's instructions.
    c. Pick up and properly tag any cores returned for credit.
    d. All of the above

12. Which of the following statements about WDs is incorrect?
    a. Some WDs are large distribution organizations that dominate an entire area.
    b. WDs function on profits around 20 percent.
    c. WDs buy directly from jobbers.
    d. WDs can limit the amount of returns accepted from jobbers.

13. The one major difference between warehouse distributors and jobbers is _____ .
    a. profit margin
    b. fill rate
    c. core exchanges
    d. return policies

14. What is the difference between a manufacturer's representative and a manufacturer's agent?
    a. A rep works for one manufacturer only.
    b. An agent represents a particular line.
    c. A rep handles all lines for one manufacturer.
    d. All of the above

15. Which of the following services can be provided by a rep?
    a. Review catalogs with counter personnel.
    b. Check for unnecessary or obsolete catalogs in the rack.
    c. Suggest ways to market or merchandise products.
    d. All of the above

CHAPTER THREE

PRODUCT KNOWLEDGE: ENGINES AND THEIR SYSTEMS

Objectives

After reading this chapter, you should be able to:
- Explain the importance of product knowledge in a counterperson's job.
- List various methods for obtaining product knowledge, including working with mechanics and attending clinics and conferences.
- Explain the four-stroke cycle in an automotive gasoline engine.
- Describe the three characteristics of gasoline engine design.
- Explain the design and function of various engine parts.
- Explain the function of the main engine systems.
- Describe the components of the fuel system.
- Describe the components of the ignition system.
- Describe the components of the lubrication system.
- Describe various oil types and characteristics.
- Describe the components of the cooling system.
- Describe the components of the exhaust system.
- Describe the components of the emission control system.
- Describe the design and function of electronic engine controls (EEC).

A good, knowledgeable counterperson is the heart of an auto parts store or dealership parts department. Customers look to the counterperson to supply them with the right parts to do the job. In order to do this, the person behind the counter needs a specialized kind of knowledge about automobiles and other vehicles as well as about their systems and their parts.

It is important to remember that professional counterpeople do more than just move parts. They must answer questions and solve problems. The professional installer wants the counterperson to sell the right part to do the job, and, in addition he or she might like a quick reminder about related parts to do the best job possible for the customer.

The do-it-yourself customer relies on counter personnel (Figure 3-1) even more. They are the problem-solvers. DIY customers might need the counterperson to explain what part goes where, how

FIGURE 3-1 The do-it-yourself customer relies on the product knowledge of counter personnel.

it goes there, and why. Without counterpeople, parts are just inventory.

There are hundreds—even thousands—of parts in the average parts sales organization. Therefore, the single biggest problem is knowing enough about each automotive category to sell the right parts. Comprehensive product knowledge might seem threatening, but it is a must. It has a single purpose: to help the counterperson become a better, more professional salesperson. This means the counterperson must collect facts about the store's merchandise that will help him or her do an efficient job of selling.

The greatest single use made of product information is that of showing prospects how and what they will gain by buying. A minor use of product information, yet one that is worthwhile, is that of supplying facts that have conversational or curiosity value. Such facts might be an indirect aid in selling.

What should the counterperson know about the products he or she sells? The knowledge should include major improvements identification (brand name, trademark, packaging), competitive position, availability, talking points, related products, uses and applications, user benefits or satisfactions, performance potential, limitations, operation and service, cost of operating and maintaining, cost of replacement, sizes, models, designs, specifications, prices, discounts, terms, and profit. In addition, such services as credit, shipping and delivery, installation, instruction for use, maintenance, and support should be included.

All of this is a lot to learn, but that does not mean it is difficult. This chapter and Chapter 4 give an overview of how the various automotive systems work. In the remaining chapters of this book, the other phases of counterperson activities are fully covered. But as in any other phase of learning, familiarity is acquired over time through experience and practice. Product familiarity is gained by working hand-in-hand with the mechanic customer in a problem-solving format. Research the products by reading tech bulletins, sales brochures, and being totally familiar with the parts (see Chapter 5). Be sure to study the back pages and important footnotes. Tech clinics, marketing conferences, and outside sales training schools are offered by various product manufacturers or distributors for one reason—to help the counterperson become familiar with the use of their product and verbally illustrate the product's features and benefits in an effective manner. The counterperson can also attend a speech class at a local college or seek help from a successful colleague. It is important to learn to communicate effectively.

When it comes to being familiar with products, having direct nuts and bolts mechanical experience

FIGURE 3–2 ASE certification patch

helps. Many counterpersons have worked in shops and garages. If you have worked there also, capitalize on this area of expertise by taking the appropriate ASE certification tests administered by the National Institute for Automotive Service Excellence (Figure 3–2). To appeal to the mechanic customers' sense of experience and logic, the counterperson must be professionally credible.

THE BASIC ENGINE

In a passenger car or truck, the engine provides the rotating power to drive the wheels through the transmission—usually by a clutch or torque converter—and driving axle. All automobile engines, both gasoline and diesel, are classified as "internal combustion engines" because the combustion or burning that creates heat energy takes place inside the engine. These systems require an air/fuel mixture that arrives in the combustion chamber with exact timing and an engine constructed to withstand the temperatures and pressures created by thousands of fuel droplets burning.

The combustion chamber is the space between the top of the piston and cylinder head. It is an

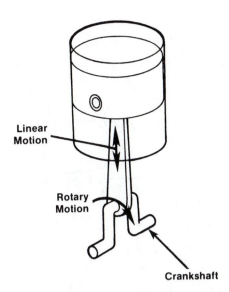

FIGURE 3-3 Reciprocating motion must be converted to rotary motion by the crankshaft

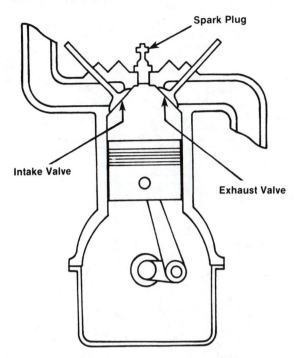

FIGURE 3-4 Valves and spark plug control combustion

enclosed area in which the gasoline and air mixture is burnt. The piston is a hollow metal tube with one end closed that moves up and down in the cylinder. This reciprocating motion is produced by the burning of fuel in the cylinder.

The reciprocating motion must be converted to rotary motion before it can drive the wheels of a vehicle. As shown in Figure 3–3, this conversion is achieved by linking the piston to a crankshaft with a connecting rod. The upper end of the connecting rod moves with the piston as it moves up and down in the cylinder. The lower end of the connecting rod is attached to the crankshaft and moves in a circle. The end of the crankshaft is connected to the transmission to continue the power flow through the drivetrain and to the wheels.

For the explosion or combustion action in the cylinder to take place completely and efficiently, precisely measured amounts of air and fuel must be combined in the right proportions. The carburetor (or in most cases, a fuel injection system) makes sure that the engine gets exactly as much fuel and air as it needs for the many different conditions under which the vehicle must operate: starting, idling, power accelerating, or cruising.

There are usually two valves at the top of the cylinder. The air/fuel mixture enters the combustion chamber through an intake valve and leaves (after having been burned) through an exhaust valve. The valves are accurately machined plugs that fit into machined openings. A valve is said to be seated as closed when it rests in its opening. When the valve is pushed off its seat, it opens.

A rotating camshaft, connected to the crankshaft, opens and closes the intake and exhaust valves (Figure 3–4). Cams are raised sections of the shaft, or collars, with high spots called *lobes*. As the camshaft rotates, the lobes rotate and push away a spring-loaded valve tappet. The tappet transfers the motion to a pushrod and perhaps a rocker arm to open the valve by lifting it off its seat. Once the lobe on the cam rotates out of the way, the valve, forced by a spring, moves down and reseats.

In summary, the essentials for the complete combustion process include the following:

- Admit a proper mixture of air and fuel into the cylinder.
- Compress (squeeze) the mixture so it will burn better and deliver more power.
- Ignite and burn the mixture.
- Remove the burned gases from the cylinder so that the process can be completed and repeated.

With the proper timing of the action of the valves and spark plug to the movement of the piston, the combustion cycle takes place in four strokes of the piston. The basis of automotive gasoline engine operation is the four-stroke cycle. A stroke is the full travel of the piston up or down. There are four strokes in this cycle: the intake stroke, the compression stroke, the power stroke, and the exhaust stroke.

- *Intake Stroke.* As the piston moves away from top dead center (TDC), the intake valve

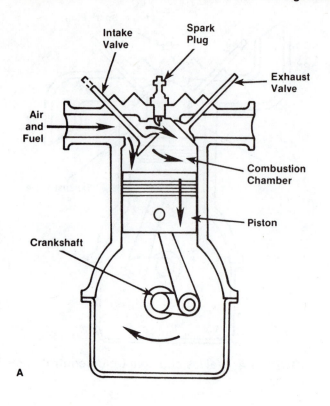

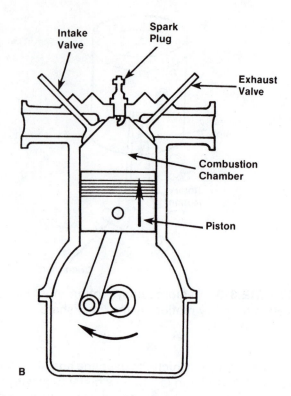

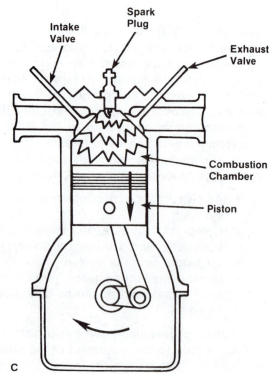

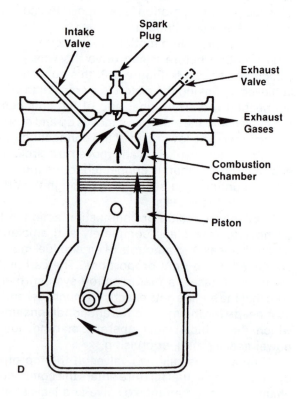

FIGURE 3-5 Strokes in a four-stroke cycle

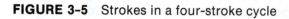

opens (Figure 3–5A). The downward movement of the piston increases the volume of the cylinder above it. This, in turn, reduces the pressure in the cylinder below atmos-

pheric pressure. The reduced pressure causes atmospheric pressure to push a mixture of air and fuel through the open intake valve. As the piston reaches the bottom of its stroke, the reduction in pressure stops and the intake of air/fuel mixture nearly ceases. But due to the weight and movement of the air/fuel mixture, it will continue to enter the cylinder until the intake valve closes. The delayed closing of the intake valve increases the volumetric efficiency of the cylinder by packing as much air and fuel into it as possible.

- *Compression Stroke.* The compression stroke begins as the piston starts to move from bottom dead center (BDC). The intake valve closes, trapping the air/fuel mixture in the cylinder (Figure 3–5B). Upward movement of the piston compresses the air/fuel mixture. At TDC, the piston and cylinder walls form a combustion chamber in which the fuel will be burned. The volume of the cylinder with the piston at BDC compared to the volume of the cylinder with the piston at TDC determines the compression ratio of the engine.

- *Power Stroke.* The power stroke begins as the compressed fuel mixture is ignited in the combustion chamber (Figure 3–5C). An electrical spark across the electrode of a spark plug ignites the air/fuel mixture. The burning fuel rapidly expands, creating a very high pressure against the top of the piston. This drives the piston down toward BDC. The downward movement of the piston is transmitted through the connecting rod to the crankshaft. Up-and-down movement of the piston on all four strokes is converted to rotary motion of the crankshaft.

- *Exhaust Stroke.* The exhaust valve opens just before the piston reaches BDC on the power stroke (Figure 3–5D). Pressure within the cylinder when the valve opens causes the exhaust gas to rush past the valve and into the exhaust system. Movement of the piston from BDC pushes most of the remaining exhaust gas from the cylinder. As the piston nears TDC, the exhaust valve begins to close as the intake valve starts to open. The exhaust stroke completes the four-stroke cycle. The opening of the intake valve begins the cycle again. This cycle occurs in each cylinder and is repeated over and over, as long as the engine is running.

The four strokes of the cycle require two full revolutions of the crankshaft. Also, the piston is being acted on by combustion pressure during only about half of one stroke, or about one-quarter of one revolution. This makes it easier to understand the function of the flywheel. Even if the engine has multiple cylinders, a certain amount of power it produces has to be stored momentarily in the flywheel. From there, it is used to keep the piston in motion during about seven-eighths of the total cycle and to compress the fuel mixture just before combustion.

CHARACTERISTICS OF ENGINE DESIGN

There are many variations in engine design. Generally, modern automotive engines have four, six, or eight cylinders arranged either in-line or with two banks of in-line cylinders in a "V" or "Y."

In-line means that the cylinders are all in one straight row around the crankshaft. Most four- and six-cylinder engines are built with the cylinders in line. Figure 3–6 shows a typical arrangement of four- or six-cylinder engines. In the past there have been eight-cylinder in-line engines on automobiles, but the modern eight-cylinder engine is a V-8 (Figure 3–7), which has two in-line banks of four cylinders, each arranged in a "V" pattern with a common crankshaft. A V-6 engine on some units has the same

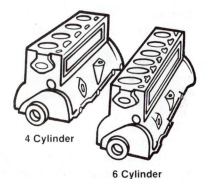

FIGURE 3–6 In-line arrangement of four- and six-cylinder engines

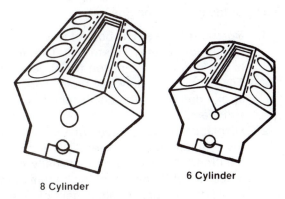

FIGURE 3–7 V-8 and V-6 engines

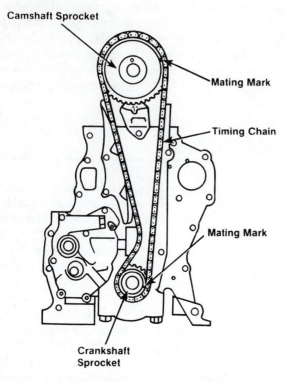

FIGURE 3-8 Overhead cam engine design

basic arrangement, but has only three cylinders in each bank. The V-4 has two banks of two cylinders each.

Opposite air-cooled engines have been used in compact vehicles in the past. In such an arrangement, the two rows of cylinders are located opposite the crankshaft.

Most modern engines have the valves mounted above the cylinders in the cylinder head. This arrangement is called an overhead valve (OHV) design. A variation of this arrangement has the camshaft, which operates the valves, also mounted above the cylinders and is called an overhead cam (OHC) design (Figure 3-8).

There are three design characteristics that are the basis of the gasoline engine.

1. *Bore and Stroke.* The bore of a cylinder is simply its diameter (Figure 3-9). The stroke is the length of the piston travel between TDC and BDC. Between them, bore and stroke determine the displacement of the cylinders.
2. *Displacement.* Displacement is the volume the cylinder holds between TDC and BDC positions of the piston. It is usually measured in cubic inches, cubic centimeters, or liters. The total displacement of an engine (including all cylinders) is a rough

indicator of its power output. Displacement can be increased by opening the bore to a larger diameter or by increasing the length of the stroke.

3. *Compression Ratio.* Compression ratio is the amount the air/fuel mixture is squeezed or compressed, based on the cylinder and combustion chamber volume at BDC and TDC. In the example shown in Figure 3-10, the volume before compression is eight times the volume after compression, so the compression ratio is 8 to 1. Increasing compression ratio usually increases the power output of the engine. However, high-compression engines require premium gasoline, which might be higher in lead content. Modern compression ratios are in the range of 8.5 to 1 or less to take advantage of the lower lead content of regular and lead-free gasoline.

Gasoline engines are made of many different parts and components (Figure 3-11). Those that the counterperson is concerned with include the following:

- Intake manifold
- Cylinder block
- Cylinder head
- Intake and exhaust valves
- Valve seat
- Valve guides
- Camshaft
- Lifters
- Crankshaft
- Vibration damper
- Flywheel

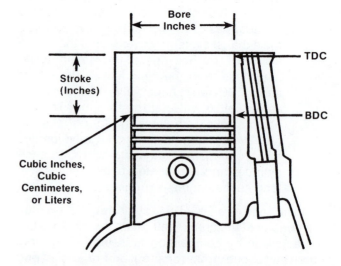

FIGURE 3-9 Bore and stroke of a cylinder

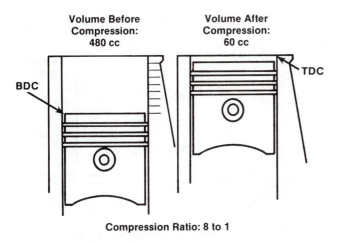

FIGURE 3-10 Compression ratio is a measure of volume to volume

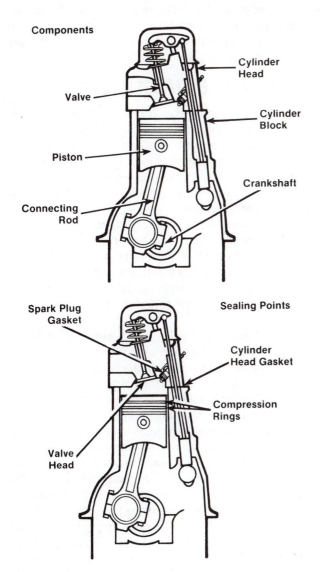

FIGURE 3-11 Compression system seals the combustion chambers

- Piston and piston rings
- Connecting rod
- Gaskets

INTAKE MANIFOLD

The intake manifold (Figure 3-12) is the mounting point for the carburetor or fuel injection components. It distributes fuel as evenly as possible to each cylinder, helps to prevent condensation, and assists in vaporization of the air/fuel mixture. Smooth and efficient engine performance depends on mixtures entering each cylinder that are uniform in strength, quality, and degree of vaporization. This is partly the job of the intake manifold. The ideal air/fuel mixture is completely vaporized when it goes into the combustion chamber. Complete vaporization requires high temperature and high temperature increases volume and decreases the volumetric efficiency of the engine. Therefore, the best alternative is to introduce an air/fuel mixture into the manifold that is vaporized above the point where fuel particles will be deposited on the manifold and below the point where excess heat results in power losses. Intake manifolds are carefully designed to meet these requirements. The walls of the manifold must be

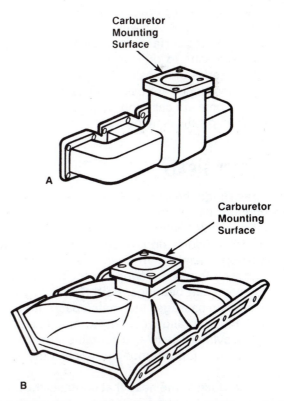

FIGURE 3-12 (A) Typical four-cylinder and six-cylinder in-line intake manifold; (B) typical V-6 and V-8 intake manifold

smooth and offer no obstruction to the flow of the air/fuel mixture. Design must also prevent collecting of fuel at the bends in the manifold.

The intake manifold should be as short and straight as possible to reduce the chances of fuel condensation between the carburetor and cylinders. To assist in vaporization of the fuel, some intake manifolds are heated by exhaust or engine coolant. This helps to vaporize the fuel, a particularly important action in cold conditions when fuel vaporization is slow.

CYLINDER BLOCK

The cylinder block is normally one piece, cast and machined so that all the parts contained in it fit properly. The word *cast*, with regard to the engine block, refers to how it is made. To cast is to form molten metal into a particular shape by pouring or pressing it into a mold. This molded piece must then undergo a number of machining operations to make sure all the working surfaces are smooth and true. The top of the block must be perfectly smooth as the cylinder head will later be attached at this point. The base or bottom of the block must also be machined as the oil pan attaches here. All block sealing areas are also machined. The cylinder bores must be smooth and a proper diameter to accept the pistons.

The main bearing area of the block must be line bored to a diameter which will accept the crankshaft. Camshaft bearing surfaces must also be bored. The word *bore* means to drill or machine a hole; line boring is a series of holes in a straight line.

Cylinder blocks are cast from several different materials: some are cast iron, some cast aluminum. Today, even plastic is being tested as a material for block construction.

CYLINDER HEAD

The cylinder head is normally made of cast iron or aluminum. It contains, on overhead valve engines, the valves, valve seats, valve guides, valve springs, rocker arm supports, and a recessed area called the combustion chamber. In overhead camshaft engines can be found the supports for the camshaft and the camshaft bearings. Both types of cylinder heads contain passages that match passages in the cylinder block to allow coolant to circulate in the head. The cylinder head also contains tapped holes in the combustion chamber to accept the spark plugs.

The surface of the head that contacts the block must be perfectly smooth because this area must contain the force of the burning fuel mixture. To aid in the sealing, a gasket is placed between the head and block. This gasket is called the head gasket. The

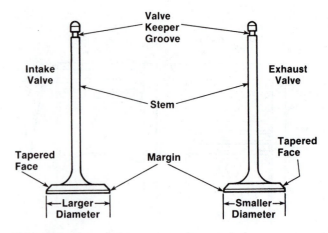

FIGURE 3-13 Intake and exhaust valves

gasket is made of special material that can withstand high temperatures, high pressures, and the expansion of the metals around it. The head also serves as the mounting point for the intake and exhaust manifolds and contains the intake and exhaust ports.

INTAKE AND EXHAUST VALVES

Every cylinder of a four-stroke cycle engine contains at least one intake valve to permit the air/fuel mixture to enter the cylinder and one exhaust valve to allow the burned exhaust gases to escape (Figure 3-13). The intake and exhaust valves, along with the cylinder head gasket, must also seal the combustion chamber.

The type of valve used in automotive engines is called a *poppet*. The word *poppet* is derived from the popping action of the valve as it opens and closes. A poppet type valve has a round head with a tapered face, a stem that is used to guide the valve, and a slot that is machined at the top of the stem for the valve spring retainers.

The head of the valve is the large diameter end and is used to seal the intake or exhaust port. This seal is made by the valve face contacting the valve seat. The valve face is the tapered area machined on the head of the valve. The angle of this taper is determined by design and manufacture of the engine. The taper will vary from one engine family to another and between intake and exhaust valves in the same engine. The area between the valve face and the head of the valve is called the *margin*. The margin allows for some machining of the valve face, which is sometimes necessary to restore its finish, and allows the valve an extra capacity to hold heat.

The intake and exhaust valve heads are different diameters, the intake being the larger of the two. The size or diameter of the valves is determined by the engine design and its uses. As mentioned, the stem guides the valve during its up-and-down movement and serves to connect the valve to its spring through its valve spring retainers and keepers. The stem rides in a guide that is either machined into the head or pressed into the head as a separate replaceable part.

The valve seat is the area of the head contacted by the face of the valve. The seat may be machined into the head or it may be pressed in like the valve guide.

The valves found in today's modern engines are made from special high-strength steel that is highly heat resistant. Heat resistance is very critical in exhaust valves as they must withstand working temperatures of between 1500 and 4000 degrees Fahrenheit. Heat resistance is much less a problem for intake valves as they receive extra cooling from the fuel mixture during the intake stroke.

There are two ways for the exhaust valve to rid itself of its heat. First, when the valve face is in contact with its seat, the heat from the valve will be transferred to the cylinder head, which is liquid cooled. The second is through the valve stem to the valve guide and again to the cylinder head. To aid in this second method of heat transfer, some exhaust valve stems are hollow. This hollow section is filled with sodium. Sodium is a silver-white alkaline metallic chemical element that transfers heat much better than steel.

 **SALES TALK**

Warn customers about sodium-filled valves. That is, never cut them open for any reason. Sodium will burn violently when it contacts the air.

VALVE SEAT

The valve seat is the area contacted by the valve face when the valve is in the *closed* position. Both intake and exhaust valves have seats. The valve seat area must be hard enough to withstand the constant closing of the valve and supply good heat transfer. Due to corrosive products found in the exhaust gas the seats must be highly resistant to corrosion. When the cylinder head material meets these requirements the seats are machined directly into it. When it does not, the seats are then made of material that will meet the requirements and the seats are pressed into the head.

VALVE GUIDES

The valve stem moves up and down through the valve guides (Figure 3-14). The guide must be ma-

chined properly so that the valve will contact its seat accurately and must also have the proper clearance to ensure the valve stem is lubricated and to provide for heat transfer.

For the engine to function properly, the intake and exhaust valves must open and close at exactly the precise moment. The proper timing of these events is dictated by the position of the pistons in their cylinders, and a precise connection must be made between the crankshaft and the camshaft so that the valves respond at exactly the right time. This is accomplished through timing gears, a chain and sprocket system, or a timing belt. These methods provide a positive, nonslip connection between the crankshaft and camshaft that synchronizes the valve opening and closing with the piston action. Two types of gears are used in engines: helical (angle cut) and spur (straight cut).

The timing mechanism in a four-stroke engine activates the camshaft at a predetermined speed of two revolutions of the crankshaft to one revolution of the camshaft. To obtain the 2 to 1 ratio, the crank gear, or sprocket, has half the number of teeth as the cam gear or sprocket.

A timing gear system consists of a helical camshaft gear and a mating crankshaft gear (Figure 3-15). There are timing marks on the gears to help synchronize their operation.

The chain and sprocket system (Figure 3-16) consists of a spur type camshaft sprocket and a crankshaft sprocket, both connected by a timing chain (either a silent inverted tooth type or a roller type). As shown in Figure 3-8, this arrangement is often used in overhead cam engines.

CAMSHAFT

As mentioned earlier in this chapter, the camshaft—usually referred to as a cam—will be found in

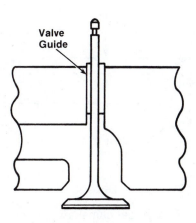

FIGURE 3-14 Valve stem moves up and down through the valve guides

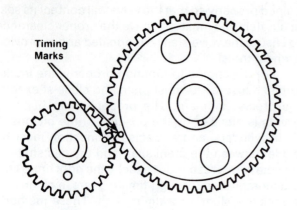

FIGURE 3-15 Timing gears

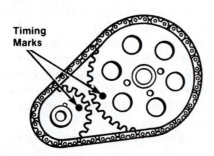

FIGURE 3-16 Timing chain and sprocket system

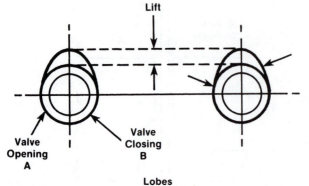

Lobes

Long Duration   **Typical Camshaft**   Short Duration

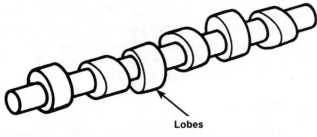

FIGURE 3-17 Typical camshaft and lobes

the block above the crankshaft on overhead valve engines. In overhead cam engines the cam is mounted in special cam towers fitted to or machined as part of the cylinder head. The camshaft resembles a series of eggs fastened together with the small end of each egg pointing in a different direction.

Most camshafts used in today's engines are made of cast iron, nodular iron, or cast steel and are hardened for long life. The egg shape is called the cam lobe. There are the same number of lobes as there are valves in an engine; for example, a four-cylinder engine normally has eight valves, the camshaft for this engine will have eight cam lobes.

Designing the camshaft for a given engine is a very exacting science. The positioning of the lobe on the shaft determines when the valve will open with relation to piston travel (Figure 3–17). The design of the lobe determines how high the valve will open and how long it will remain open. How high the camshaft will open the valve is called the valve *lift*. How long the camshaft keeps the valve open is called the *duration*. Valve lift and duration will vary from one engine family to another. It will also vary within the same family depending on the engine use. For example, a certain engine family might be used in automobiles, trucks, and commercially. Each would have a camshaft with a different design (Figure 3–18). Most automobile engines are equipped with a camshaft that will deliver a moderate lift and short duration. This design will allow smooth idling and good low-speed torque. The same engine design for higher speeds, such as a racing engine, would use a cam with high lift and long duration. This configuration does not allow for smooth idling or good low speed operation.

The proper nomenclature for a camshaft and lobe (Figure 3–19) is

- *Diameter.* The actual measurement of the camshaft base.
- *Heel.* The portion of the cam lobe contacted by the lifter when the valve is closed.
- *Flank.* There are two flanks, one for opening the valve, one for closing. The flank is the ramp between the cam base and its nose.
- *Nose.* The top or highest point of the cam lobe.
- *Lift.* The distance the lobe will move the lifter; measured from the base to the top of the nose.
- *Duration.* The distance from the start or the lift flank, across the nose to the heel of the closing flank; measured in degrees of crankshaft rotation.

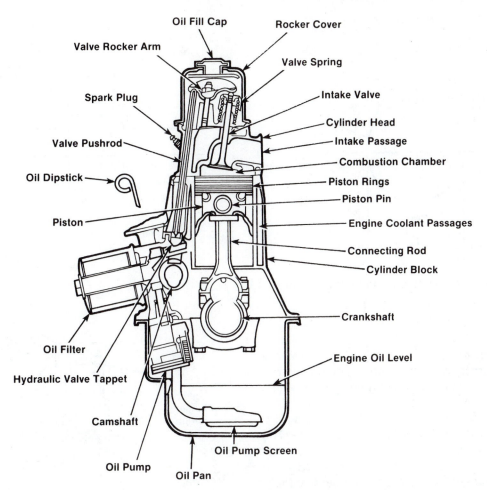

A

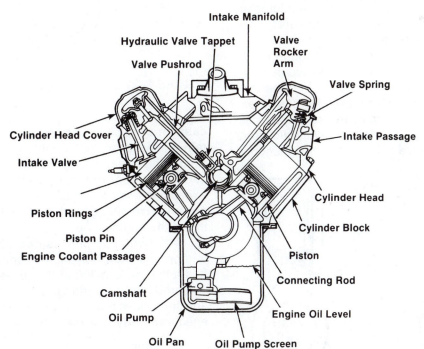

B

FIGURE 3-18   (A) Typical in-line six-cylinder engine (front view); (B) typical V-8 engine (front view)

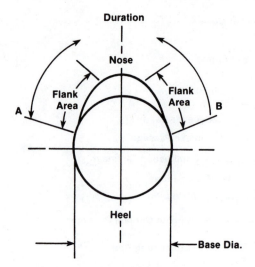

FIGURE 3-19   Camshaft and lobe nomenclature

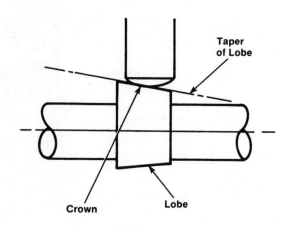

LIFTERS

Valve lifters or cam followers are found in a bore above each cam lobe. As there are the same number of lobes for each valve in the engine, there are also the same number of lifters.

There are two basic types of lifter designs (Figure 3-20). The first is the solid or mechanical. The solid lifter is simply an iron cylindrical piece machined at one end to contact the cam lobe and at the other end to accept the pushrod. There are no moving parts found in the solid lifter. Engines equipped with solid lifters require periodic adjustments to correct for wear in the valve train and to eliminate valve noise. This type of lifter is usually not found in today's automobiles. It is, however, still used in commercial and racing engines. Some high-performance engines use a third type, called a *roller lifter*.

In most of today's engines, the solid lifter has been replaced by what is called the hydraulic lifter. The hydraulic lifter does not require adjustment be-

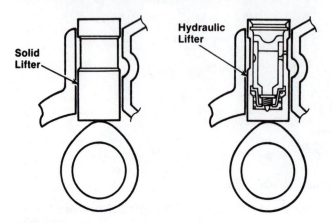

FIGURE 3-20   Two lifter designs

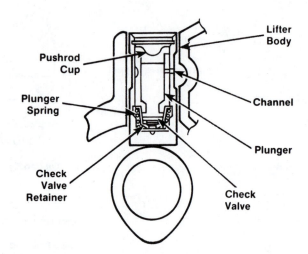

FIGURE 3-21   Valve closed

cause it uses oil under pressure to correct for any minor clearance that might develop in the valve train. This clearance is called *lash*. Lash can be caused by wear to the valve train or by the simple expansion and contraction of the valve train components that might occur due to the heating and cooling of the engine.

The hydraulic lifter is one of the most highly engineered, precision engine components. In effect, it is a hydraulic cylinder that uses engine oil pressure and internal spring pressure to provide a quiet and snug valve train. The unit is contained in a hardened iron body. It is made of the following:

- Plunger
- Oil metering disc valve
- Pushrod seat
- Snap ring

- Check valve disc
- Check valve spring
- Plunger return spring

The base of the body is not flat; it is ground spherically with a slight crown in the center. This design aids in the lifter to lobe contact and causes the lifter to rotate as it circles up and down in its bore. By rotating, the lifter wear is more evenly spread over the face of the lifter and cam lobe.

When the lifter is on the heel of the cam, the spring under the plunger pushes the plunger up to eliminate any valve train lash. By pushing up the plunger, the check valve in the lifter opens and engine oil under pressure fills the hydraulic cylinder. As the cam lobe begins to lift the lifter, the check valve closes (Figure 3–21), trapping the oil under the plunger. The oil will not compress. The cam lobe pushes the lifter up and, by means of the pushrod and rocker arm, the valve is open (Figure 3–22). As the camshaft continues to rotate, the valve closes and the process starts again.

When the camshaft is in place and the lifters are installed, they are offset from the cam lobes. This offset, combined with the taper of the cam lobe and spherically ground lifter body, causes the rotation of the lifter necessary for long life of the camshaft and lifter. In use the lifter will wear to its bore and the face of the cam lobe. The pushrod will wear to the lifter as well as the rocker arm. For this reason, if the engine is dismantled, all parts of the valve train must be kept in order and reinstalled in their same location.

There are several terms used regarding camshaft and valve parts that counter personnel should know.

- *Valve Train.* These are the parts necessary to transfer camshaft movement to valve movement. The parts involved are the lifter, pushrod, rocker arm, and valve spring.
- *Valve Timing.* The valve timing—when and how much the valves open—is determined by the design of the camshaft.
- *Camshaft Timing.* Camshaft timing is the relationship between the camshaft and the crankshaft. The camshaft is driven by the crankshaft directly, by means of gears, or indirectly, by means of a timing chain or timing belt. This relationship is critical and must be done with great care because it will affect total engine operation.
- *Super or Hot Camshaft.* This is a camshaft that uses specially designed hydraulic lifters.

 SALES TALK ——————

Sell new camshafts only with new lifters. Customers sometimes want to replace hydraulic lifters with solid components. That is fine. Racing-type hydraulic and roller lifters constitute another choice because they feature a drilled bleed hole to prevent lifter pump-up, which leads to valve-float. Give the customer what he or she wants, but be certain he or she does not use old lifters with a new cam. It is acceptable to replace a single lifter or two without replacing an entire camshaft, but most good mechanics like to use new only with new. The counterperson should sell "hot" cams only when reasonably sure they will improve performance by a meaningful amount. A super cam in a choked-down engine that restricts breathing is not a good buy in itself.

CRANKSHAFT

The crankshaft is used to convert the up-and-down movement of the pistons to rotary movement, which can be used to supply power. Crankshafts are generally made of cast iron, forged cast steel, nodular iron, and then machined. At the centerline of the crankshaft are the main bearing journals (Figure 3–23). These journals must be machined to a very close tolerance because the weight and movement of the crankshaft will be supported at these points. The number of main bearings is determined by the design of the engine. V-block engines generally have fewer main bearings than an in-line engine with the same number of cylinders because the V-block engine will use a shorter crankshaft. Offset from the crankshaft centerline are the connecting rod bearing journals. The degree of offset and the number of

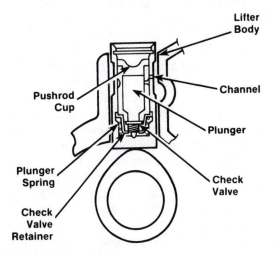

FIGURE 3–22 Valve open

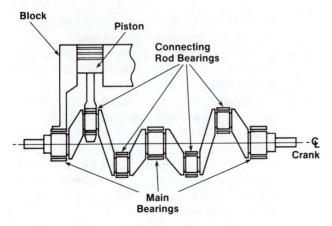

FIGURE 3-23 Main bearings at the centerline of the crankshaft

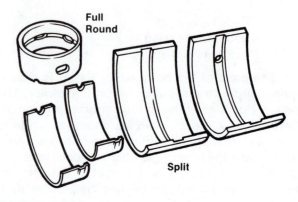

FIGURE 3-25 Insert or sleeve bearings

journals are determined by the engine design. An engine having six cylinders in-line will have six connecting rod journals; a V-8 engine will have only four because each journal will be connected to two connecting rods, one from each side of the V. The connecting rod journal is also called the *crank pin*. This area is machined to a very close tolerance just like the main bearing journal. The machining of the main and rod bearing journals must be done to have a very smooth surface at the bearing area. The bearings must fit tightly enough to eliminate noise but must also have a clearance between them and their bearings for an oil film of 0.0015 and 0.002 to form. This is usually attained by the proper use of torque wrench specifications and checked by the use of Plastigage that is available at most jobber stores.

The crankshaft does not turn directly on the bearings. It turns on a film of oil trapped between the

bearing surface and the journal surface (Figure 3-24). This oil is supplied by the engine's oil pump. Should the crankshaft journals become out of round, tapered, or scored, the oil film will not form properly and the journal will contact the bearing surface. This causes early bearing and/or crankshaft failure. The main and rod bearings are generally made of lead coated copper or tin aluminum. Both of these are softer material than that used to form the crankshaft. By using the soft material, any wear at the crankshaft journals will appear first on the bearings. Early diagnosis of bearing failure most often will spare the crankshaft and only the bearings will have to be replaced (Figure 3-25).

As mentioned in this chapter, the connecting rod journals are offset from the centerline of the crankshaft. This puts weight and piston pressure off center on the crankshaft. To balance this for smooth engine operation, counterweights must be added to the crankshaft. These weights can be cast as part of the crankshaft. They will be found opposite the connecting rod journals.

The journals of the crankshaft must be smooth and highly polished. Bearings surround each journal and are fed oil under pressure. In order for the oil to reach these bearings oil passages must be drilled into the crankshaft (Figure 3-26). Each main bearing will receive oil under pressure from the pump. Each main bearing journal will have a hole drilled into it with a connecting hole or holes leading to one or more rod bearing journals. In this way all bearing journals receive oil under pressure to protect both the bearing and the journal. The crankshaft configuration determines the engine block design, or the positioning of the connecting rod journals around the centerline ($\mathcal{C}$) of the crankshaft (Figure 3-27).

The crankshaft has two distinct ends (Figure 3-28). One is called the flywheel end and, as its name implies, this is where the flywheel is connected to it. The front end or belt drive end of the crankshaft contains a threaded snout or is drilled and tapped. This is for attaching a vibration damper.

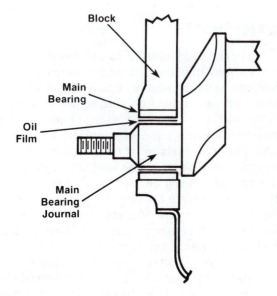

FIGURE 3-24 Crankshaft turns on oil film

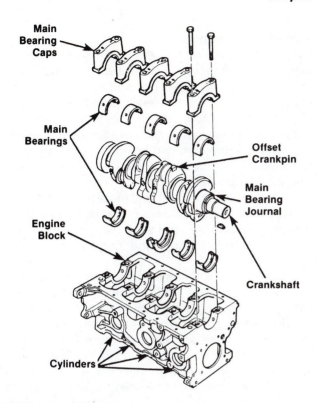

FIGURE 3-26 Crankshaft assembly showing main engine bearings

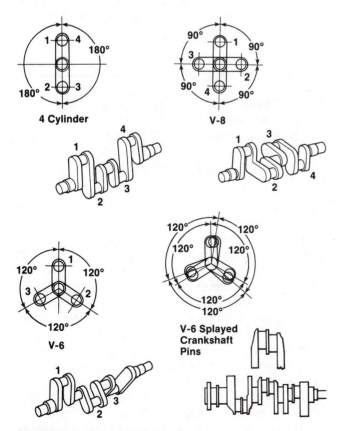

FIGURE 3-27 Crankshaft configurations for four-cylinder, V-6, and V-8 engines

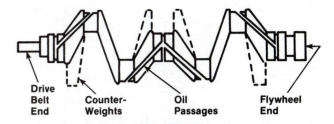

FIGURE 3-28 Flywheel and drive belt ends

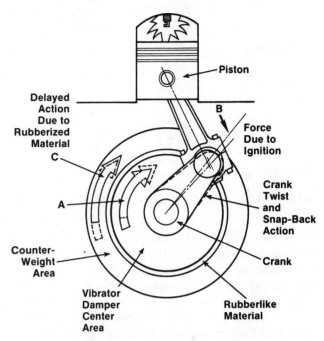

FIGURE 3-29 Operation of vibration damper

CAUTION: Never lay down a crankshaft to store it without proper support in the middle. Remember that all engine parts must be carefully stored.

Vibration Damper

The purpose of the vibration damper is to dampen crankshaft vibration. Vibration is a back-and-forth motion. The crankshaft of a running engine will have a back-and-forth or twisting motion each time a cylinder fires. The force applied to the crankshaft can be more than two tons. This causes the crank to momentarily twist and snap back. The vibration damper (Figure 3-29) consists of a center section, which is attached to the crankshaft. Surrounding the center section is a strip of rubber-like material. Attached to the material is a grooved counterweight. As the crankshaft twists, the center section "A" applies a force to the material. The material

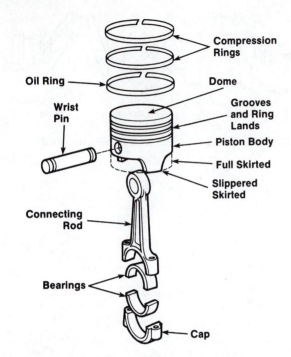

FIGURE 3–30 Piston and its components

must then apply this force to the counterweight "C." The weight is snapped in the direction of the crankshaft rotation to counter balance the crankshaft connecting rod journal snapping back against the force due to ignition "B." The back-and-forth movement of the crankshaft is counter balanced by the back-and-forth movement of the vibration damper, and the engine runs smoothly.

Flywheel

The flywheel also adds to the smooth running of the engine by applying a constant moving force to carry the crankshaft from one firing stroke to the next. Because of its large diameter, the flywheel makes a convenient point for the starter to connect to the engine. The large diameter supplies good gear reduction for the starter, making it easy for the starter to turn the engine against its compression. The surface of a flywheel may be used as part of the clutch. On an engine that drives an automatic transmission, a flexplate is used and the automatic transmission converter provides the weight required to attain flywheel functions.

PISTON AND PISTON RINGS

The piston forms the lower portion of the combustion chamber. The force of the expanding air/fuel mixture at the time of ignition is exerted against the head or dome of the piston. This force then pushes the piston down in the cylinder. The force

applied to the piston can be as high as 2-1/2 tons in a gasoline engine. For this reason the piston must be very strong. In the past, pistons were made primarily of cast iron or a mixture of iron and steel. This type of piston is strong but is also heavy. Due to advances in aluminum technology, the pistons in most modern automobile engines are made of aluminum or aluminum alloy. Aluminum can be made strong enough to withstand combustion pressure and has the advantage of light weight. The aluminum piston's disadvantage is that it will expand more than the iron or iron and steel pistons. To counteract this, aluminum pistons usually have steel struts or inserts cast into them to help hold heat and contain the piston expansion.

The top of the piston is called the *head* or *dome*. Just below the dome on the side of the piston is a series of grooves. The grooves, sometimes called the ring lands, are used to contain the piston rings. Below the grooves, as shown in Figure 3–30, there is a bore, or hole, which is used for the piston pin, sometimes called the wrist pin. The pin is used to connect the piston to the connecting rod. This hole is not centered in the piston; it is offset to one side. The piston pin offset is toward the major thrust side of the piston, the side that will contact the cylinder wall during the power stroke. By offsetting the pin, piston slap caused by the piston changing direction in the cylinder is eliminated.

 **SALES TALK** _____

The term piston slap or bang is used to describe the noise made by the piston when it contacts the cylinder wall. This noise is usually heard only in older, high-mileage engines that have worn pistons or cylinder walls.

To ensure that the piston is installed correctly so the offset is on the proper side, the top of the piston will have a mark. The most common mark is a notch, machined into the top edge of the piston. This mark must always face the timing chain end of the engine when the piston is installed. The base of the piston, the area below the piston pin, is called the *piston skirt*. The area from just below the bottom ring groove to the tip of the skirt is the piston thrust surface. There are two basic types of piston skirts: the slipper type and the full skirt. The full skirt is used primarily in truck and commercial engines; the slipper type is used for automobile engines. The slipper type skirt allows the piston enough thrust surface for normal operation and has the advantage of allowing the piston to be lighter. This design also

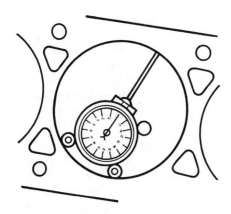

FIGURE 3-31 Measuring cylinder bore with a dial gauge

FIGURE 3-32 Measuring with a micrometer

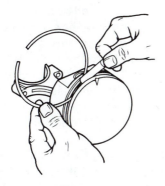

FIGURE 3-33 Measuring piston ring clearance

cuts down on piston expansion because there is less material to hold heat. When an engine is designed, the piston expansion determines how much piston clearance will be needed in the cylinder bore. Too little clearance will cause the piston to bind at operating temperature; too much will cause piston slap. The normal piston clearance for an automobile engine, using slipper skirted pistons, is about 0.001 to 0.002 inch. This clearance is measured cold between the piston skirt and the cylinder wall. Because of this necessary expansion clearance, the piston does not

seal the combustion area. This is the job of the piston rings. The piston clearance also supplies a space for piston lubrication.

The fitting or matching of the piston to its cylinder bore is a very important step in engine servicing. If the clearance between the piston and cylinder is too large, the piston will "slap;" if the clearance is too small, the piston will bind in the bore and cause early piston failure.

Bore pistons are usually available in two or three standard sizes. Always refer to the specifications of the engine being serviced for the standard piston dimensions. To select the proper piston for each bore, the bore must first be measured using a dial gauge or a telescope gauge (Figure 3-31) and micrometer (Figure 3-32). The bore must be checked for clearance (Figure 3-33), out of round, and diameter. If taper and out of round are within specifications on the engine being serviced, select a standard piston that will allow the proper piston-to-cylinder wall clearance. Always refer to the service manual for the engine being serviced.

 SALES TALK _____

A seemingly simple sale often turns out to be an exercise in applied partsmanship because there are so many variables to consider. For example, suppose a do-it-yourselfer wants a set of replacement rings for a 1979 Chevy 350 V-8. Does he or she want standard or oversized rings and, if oversized, what size of oversize? If he or she is a typical DIY mechanic who is trying to save some money by doing the work, the customer might not know what size rings he or she needs or the other parts or machine work that might be necessary to do the job right.

Furthermore, the customer might not be aware of the common pitfalls that can foul up a "simple" ring job. He or she will likely need help in figuring out which parts to reuse, which parts have to be replaced (and why), what oversize to use, what kind of machine work might be necessary, what kind of tools will be needed to complete the job and, finally, some suggestions on installation and assembly.

The first problem is determining whether or not the engine has standard-size pistons. The counterperson must depend on the customer's ability to determine correct dimensions. Even so, counter personnel should know something about measuring pistons and comparing specs to figure out if a piston is standard or oversize.

Piston Rings

Piston rings are used to fill the expansion gap between the piston and cylinder wall. It is the piston rings that seal the combustion chamber at the piston. Piston rings, which are available in several coatings, must serve three functions:

1. Seal the combustion chamber at the piston
2. Remove oil from the cylinder walls. This is necessary to keep oil from reaching the upper cylinder where it will be burned.
3. Carry heat from the piston to the cylinder walls to help cool the piston

There are two basic ring families: compression rings and oil control rings. In most modern automobile engines the pistons will be fitted with two compression rings and one oil control ring (Figure 3–34). The compression rings are found in the two upper grooves closest to the piston head. The oil ring is fitted to the groove just above the wrist pin.

Compression Rings. The compression rings form the seal between the piston and cylinder walls (Figure 3–35). They are designed to use combustion

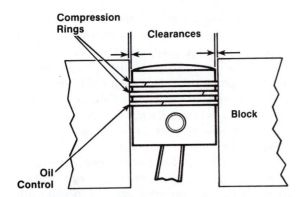

FIGURE 3–34 Piston is fitted with one oil control ring and two compression rings

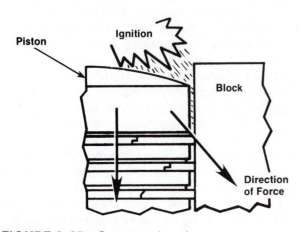

FIGURE 3–35 Compression ring

pressure to force the ring against the cylinder wall and against the bottom edge of the ring groove. The top ring is the primary seal with the second ring used to seal any small amount of pressure which may reach it. The compression rings are also used to remove excess oil from the cylinder walls and aid in cooling the piston carrying its heat to the cylinder walls.

During the power stroke, the pressure caused by the expanding air/fuel mixture is applied between the inside of the ring and the piston groove. This forces the ring into full contact with the cylinder walls. The same force is applied to the top of the ring forcing it against the bottom of the ring groove. The combination of combustion pressure and the compression ring join together to form a tight ring seal.

Compression rings are generally made of cast iron. They come in a number of variations in cross section design. Most compression rings have a coating on their face, which will aid in the wear in process. Wear in is the time necessary for the rings to conform to the cylinder wall. Typical soft coatings are graphite, phosphate, iron oxide, and molybdenum. Some compression rings have a hard coating, such as chromium.

Oil Control Rings. Oil is constantly being applied to the cylinder walls. The oil is used for lubrication as well as to clean the cylinder wall of carbon and dirt particles. This oil bath will also aid somewhat in cooling the piston. Controlling this oil is the primary function of the oil ring. The two most common types of oil rings are the segmented oil ring and the cast iron oil ring. Both types of rings are slotted so that excess oil from the cylinder wall can pass through the ring. The oil ring groove of the piston is also slotted. After the oil passes through the ring it can then pass through the slots in the piston and return to the oil sump through the open section of the piston.

Segmented oil rings are made up of three pieces: an upper and lower scraper and an expander. The scrapers and expanders are made of spring steel. Segmented oil rings are a free fit design. This means that the expander is larger than the diameter of the cylinder. They must be carefully compressed when installed. The expander is used to form a tight seal of the scrapers at the cylinder walls.

Connecting Rod

The connecting rod is used to transmit the pressure applied on the piston to the crankshaft (Figure 3–36). The rod must be very strong and at the same time be kept as light as possible. Connecting rods are generally forged from high-strength steel. The center section is made in the form of an "I" for maximum strength with minimum weight. The small end or piston pin end is made to accept the piston pin. The pin is used to connect the piston and connecting rod. The rod must be free at the piston to move back and forth as the crankshaft rotates. The piston pin can be a pressed fit in the piston and free fit in the rod. When this is the case, the small end of the rod will be fitted with a bushing. The pin can also be a free fit in the piston and pressed fit in the rod; in this case no bushings are used. The pin simply moves in the piston using the piston hole as a bearing surface. A third mounting allows the pin to move freely in both the piston and the rod. In this case the rod's small end contains a bushing and the piston is fitted with a retainer at both ends of the piston pin. The retainers are used to keep the pin within the piston.

The larger end of the rod is used to attach the connecting rod to the crankshaft. This end is made in two pieces. The upper half is part of the rod, the lower half is called the *rod cap* and is bolted to the rod. The connecting rod and its cap are manufactured as a unit and must always be kept together. The large crankshaft end of the rod is fitted with bearing half shells made of the same material as the main bearings (Figure 3–37). As mentioned earlier in the chapter, the crankshaft has oil passages that lead to the crank pins for lubricating the rod bearings. Connecting rods have a hole drilled through the big end to the bearing area. The bearing insert might have a hole, which will align with this drilling. This hole is used to supply oil for lubricating and cooling the piston skirt. When the rod is properly

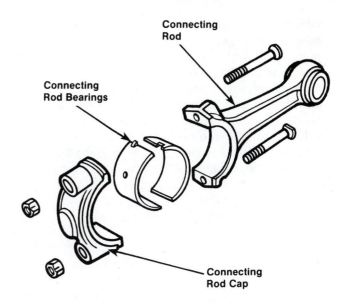

FIGURE 3-37 Connecting rod assembly showing bearings

installed, the oil hole should be pointing to the major thrust area of the cylinder wall.

There are other terms that the counterperson should keep in mind:

Crankcase The lower portion of the cylinder block that encloses the crankshaft.

Sump The oil pan that is bolted to the engine block at the crankcase.

Crank Throw The distance from the crankshaft main bearing centerline to the connect rod journal centerline (Figure 3–38). The stroke of any engine is twice the crank throw.

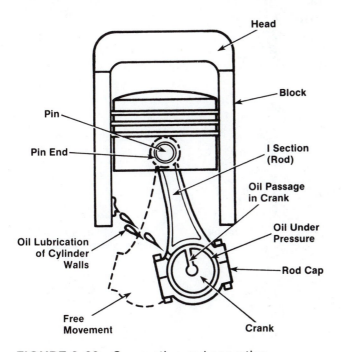

FIGURE 3-36 Connecting rod operation

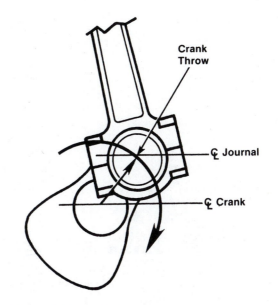

FIGURE 3-38 Crank throw

Valve Train The components of an engine necessary to convert camshaft movement to valve movement.

Flat-Head Engine Any engine with both intake and exhaust valves located in the cylinder block.

Overhead Valve Engine Any engine with the camshaft located in the block and the intake and exhaust valves located in the cylinder head.

Overhead Cam Engine Any engine with the intake valves, exhaust valves, and camshaft located in the cylinder head.

ENGINE GASKETS AND OIL SEALS

Figure 3–39 shows the various gaskets and oil seals used on a typical engine. To the uneducated eye, a gasket is just a piece of die-cut metal or other material. Its sole purpose is to be sandwiched between two pieces of metal to seal the mating surfaces. What product knowledge does the counterperson need to help the customer to select the proper gasket for his or her vehicle? Plenty, if the job selection is to be a good one.

The most technical gasket to work with is the head gasket (Figure 3–40). There are basically two types of head gaskets available: one-piece embossed steel gaskets and multilayer composition gaskets. The one-piece stamped steel gaskets have raised ribs, or embossing, to increase the clamping pressure around cylinders, water jackets, and oil passageways. This type of gasket is the least expensive to manufacture and is used as original equip-

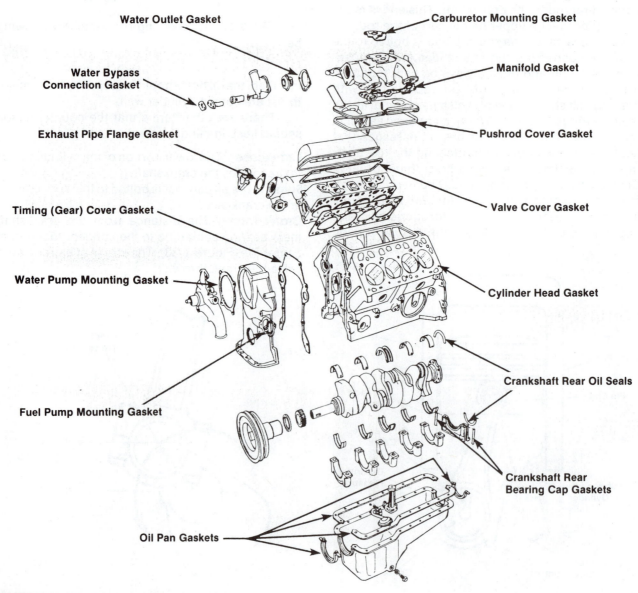

Water Outlet Gasket

Carburetor Mounting Gasket

Water Bypass Connection Gasket

Manifold Gasket

Exhaust Pipe Flange Gasket

Pushrod Cover Gasket

Timing (Gear) Cover Gasket

Valve Cover Gasket

Water Pump Mounting Gasket

Cylinder Head Gasket

Fuel Pump Mounting Gasket

Crankshaft Rear Oil Seals

Crankshaft Rear Bearing Cap Gaskets

Oil Pan Gaskets

FIGURE 3–39 Engine gaskets and oil seals

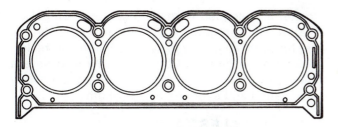

FIGURE 3-40 Cylinder head gasket

ment on most engines today. The composition gasket uses a sandwich type of construction with either a solid or perforated steel core between two layers of steel. Those with the steel core and soft surface layers can also have raised rubber beading around certain key areas to improve sealing. The added thickness increases the clamping torque on both sides of the gasket for a better seal. Those with steel surface layers might have an additional coating to help improve sealing. Another approach is to use a nonstick top surface layer to improve sealing and make disassembly easier.

The steel embossed head gasket is basically an OE gasket because it works best with relatively smooth flat surfaces. A newly machined engine block deck and cylinder head on an assembly line might have a surface finish as smooth as 30 to 35 microinches. But as time and miles take their toll, warpage and corrosion can destroy the once smooth mating surface between head and block. Also, if the block is cast iron and the head is aluminum, the difference in thermal expansion between the two produces a scrubbing action that puts a great deal of shear stress on the gasket. Because of these factors, the gasket might not last the life of the engine.

When a gasket is replaced, the block and head surfaces must be thoroughly cleaned, then checked for flatness using a straightedge and feeler gauge. If either surface is warped more than 0.003 inch side to side or within a 6-inch span, or more than 0.003 inch end-to-end on a V-6, 0.004 inch on a four-cylinder or 0.006 inch on an in-line six, the head and/or block deck should be resurfaced. The recommended surface finish for best sealing is 60 to 120 microinches.

Now the question becomes one of picking either an embossed steel or composition replacement gasket. If both the head and block surfaces are smooth and flat, free from scratches and in like-new condition, or if one or both surfaces have been refinished to recommended specs, an embossed steel gasket should work fine. But a bare steel gasket will only seal surface imperfections and scratches to a depth of about 0.0003 inch. Applying gasket sealer

to the surface (which is highly recommended with steel-faced gaskets) extends the gasket's ability to seal surface irregularities to a depth of about 0.0013 inch. Yet, even that might not be enough to cope with the warpage that is frequently encountered on many of today's thin-wall blocks and aluminum heads.

When a block or head is in less than ideal condition or the surfaces have not been refinished, a composition head gasket is usually the best choice because it offers much greater conformability. It is certainly not a cure-all and it will not seal properly if surface warpage is greater than the limits described earlier or if there are severe scratches, but a composition gasket can tolerate much more warpage and imperfection than an embossed steel gasket. This type of gasket is usually preferred by mechanics doing valve jobs or other types of internal engine work short of a complete overhaul.

 SALES TALK —————————

One pitfall to advise a customer against is never use a sealer on a coated gasket. This can damage the surface layer and cause the gasket to fail. Sealer should only be used on bare metal gaskets or when specifically recommended by the gasket manufacturer. On composition gaskets that use a soft antistick surface coating, an added plus is that the surface coating allows a certain amount of shear action between the head and block, which is important on bimetal engines. Since aluminum expands nearly twice as much as cast iron, having a gasket that can tolerate shear forces helps eliminate early gasket failure.

Another difference between composition and embossed steel gaskets is that of installed thickness. A composition gasket is thicker because of the multiple layers of material in it so its installed thickness might be around 0.035 to 0.040 inch compared to 0.020 to 0.025 for an embossed steel gasket. Increasing the thickness of the head gasket lowers the compression ratio, but it does not amount to much on a gasoline engine. A change of 0.005 inch decreases the compression by only a tenth of a point. On the other hand, using a thicker gasket prevents the compression ratio from being increased if the head or block heis resurfaced, and detonation can be a real problem on many late-model engines if the compression is increased too much.

Another question that do-it-yourselfers often ask is about retorquing head gaskets. At one time retorquing was a problem. But because of the improved materials used in today's composition gaskets, there is less settling and loss of torque over time. As a result, retorquing is usually not necessary. But on some engines, regardless of the type of gasket used, it might be required because of inherent design problems in the engine itself. Engines with aluminum heads often fall into this category. The best advice is to always follow the gasket manufacturer's recommendations for retorquing. They know best how their products perform in any given application, so if they say it is not needed, advise your customer of that fact. But if they advise retorquing, make sure the bolts are retorqued at the suggested mileage interval.

There are so-called "soft" gaskets for things such as valve covers, oil pans, and timing covers; the counterperson often gets an argument about what type of gasket material is best. Cork has long been used for such applications because it has a closed cell structure, a natural sponginess that enables it to tolerate a lot of variation between mating surfaces, and it is cheap compared to most other gasket materials.

If properly installed—which means mounting the gasket between clean, dry, flat surfaces and not overtightening the bolts—a cork gasket works reasonably well. But it does have a tendency to wick oil. Over time, oil will soak through the cork and create a wet spot on the outside of the engine. Although technically not a leak, it produces the same external symptoms as a leak, allowing dirt and grime to build up on a wet spot. Tightening the bolts to eliminate the problem usually only succeeds in deforming the stamped steel pan or cover flange, which reduces compression on the gasket between the bolt holes, encouraging real leaks to form, or it crushes the gasket under the bolts and squeezes it out of place.

Cork also has a limited shelf life. It tends to lose moisture over time, which deforms the gasket and makes it brittle. A cork gasket can also dry out and shrink on an engine if it is not run for a long period of time.

One alternative to the cork gasket is a rubber (synthetic or natural) gasket. Rubber does not wick, and it has a long shelf life. However, it tends to become very slippery when wet and squishes out of place rather easily. Rubber gaskets are also very floppy, which can make installation difficult at times.

The solution to this dilemma is to combine the best of both materials, and that is what most gasket manufacturers have done. A combination of cork and rubber (or rubber-like synthetics, such as ure-

thane) are blended to make a composition gasket that will not shrink, wick, dry out, or become brittle, yet holds its shape for easy installation and resists deterioration. Some of these gaskets also have a steel core to make the gasket more rigid for easy installation and to improve its strength.

 SALES TALK _____

A sealer should be used with soft gaskets to help make a leak-free seal. The gasket sealer can be a good add-on sale. Head gaskets should always be stored flat.

Many new engines are assembled at the factory without precut gaskets. Room temperature vulcanizing (RTV) silicone sealers and anaerobics (adhesives that cure in the absence of oxygen) have become popular with carmakers because they offer considerable cost savings over conventional gaskets. Yet the products have received bad reports from mechanics because they have not held up as well as they should in some applications.

When a leaky valve cover, oil pan, water pump, exhaust manifold, or intake manifold on a "gasketless" engine turns up, conventional replacement gaskets are available as problem-solvers for such applications. In most cases, the installation is simple and straightforward. But on some engines, longer bolts might be needed to compensate for the added thickness of the conventional gasket. If these parts are not provided with the gasket, they should be included with the sale to save the installer the trouble of having to hunt down the necessary parts later.

Oil Seals

Oil seals are used for the rear of the crankshaft and timing gear cover to prevent oil from escaping around a moving shaft. They are pressed into a stationary bore in the block or cover. These seals generally consist of a metal casing to which a synthetic rubber sealing element is bonded. As shown

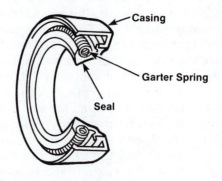

FIGURE 3-41 Oil seal

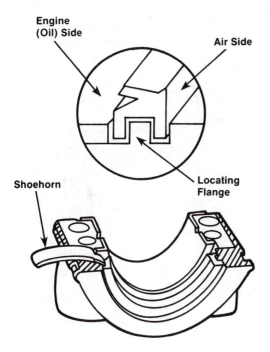

FIGURE 3-42 Split oil seal

in Figure 3-41, most oil seals utilize a garter spring to provide the proper pressure between the seal and rotating shaft.

The most popular type of rear main bearing seal is the so-called split seal (Figure 3-42). The seal is manufactured in two halves so that it can be installed on the crankshaft without removing the engine from the chassis. The seal fits into a channel machined into the block and the main bearing cap. The split seal cannot utilize a garter spring.

ENGINE SYSTEMS

Besides the major power-generating system in the engine, there are several other systems essential to engine operation.

- *Air/Fuel System.* This system ensures that the engine gets the right amount of both air and fuel needed for efficient operation. These two systems join in the carburetor, which blends the fuel and air and supplies the resulting mixture to the cylinder. Some automobiles have a fuel injector system that replaces the carburetor.
- *Ignition System.* This system supplies a precisely timed spark to ignite the compressed air/fuel mixture in the cylinder at the end of the compression stroke.
- *Lubrication System.* The system supplies oil to the various moving parts in the engine.

The oil lubricates all parts that slide in or on other parts, such as the piston, bearings, crankshaft, and valve stems. The oil enables the parts to move easily so that little power is lost and wear is kept to a minimum.

- *Cooling System.* This system is also extremely important. Coolant circulates in jackets around the cylinder and in the cylinder head. This removes part of the heat produced by the combustion of the air/fuel mixture and prevents the engine from being damaged by overheating.
- *Exhaust System.* This system efficiently removes the burned gases and limits noise produced by the engine.
- *Emission Control System.* Several control devices, which are designed to reduce emission levels of combusted fuel, have been added to the engine. Engine design changes, such as reshaped combustion chambers and altered tune-up specs, have also been implemented to help control the auto's smog-producing byproducts. These devices and adjustments have reduced emissions considerably but have changed automotive engine servicing to a great extent.

In this book, the major concern is to provide the necessary product knowledge to train the counterperson to help customers with these engine systems. To understand the systems, a person must have an overview of their functions and components.

FUEL SYSTEM

The fuel system supplies to the engine cylinders a combustible mixture of gasoline and air. To do this, it must store the fuel and deliver it to the fuel metering and atomization system, where it is mixed with air to provide the combustible mixture that is delivered in a manner that meets the varying load requirements of the engine.

The system uses several components to accomplish this task: a fuel tank to store the gasoline in liquid form, fuel lines to carry the liquid from the tank to the other parts of the system, a pump to move the gasoline from the tank, a filter to remove dirt or other harmful particles that might be in the fuel, a carburetor or electronic fuel injector to mix the liquid gasoline with air for delivery to the cylinders after the air has passed through an air filter, and an intake manifold through which the air/fuel mixture from the carburetor or fuel injector is directed to each of the engine cylinders.

The components of a fuel system, shown in Figure 3-43, are as follows:

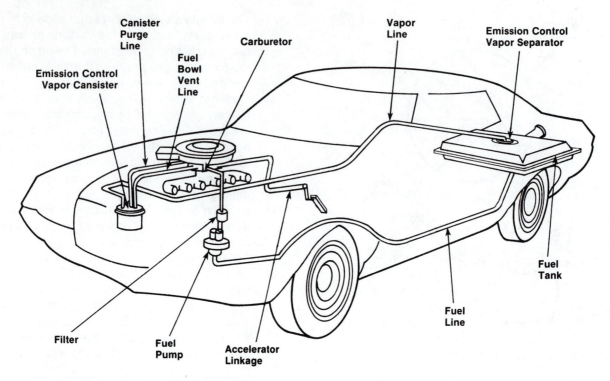

FIGURE 3-43 Components of a fuel system

- Fuel tank
- Fuel lines
- Fuel pump
- Fuel filter
- Air filter
- Fuel metering and atomization system (carburetor or electronic fuel injector)

Fuel Tank

The fuel tank is usually located at the rear of the vehicle, aft of the rear axle. It is mounted by straps or brackets to keep it in place. It is constructed of either pressed sheet metal with welded seams and special coating to prevent corrosion or fiberglass-reinforced plastic material. The latter tanks conform better to unusual underbody contours.

The fuel tank has several openings (Figure 3-44), the largest being the fill pipe through which the gasoline is pumped into the tank from an external source. Adjacent to it is a vent line to permit air vapors in the tank to escape while it is being filled. The gasoline exits the tank through an outlet pipe that usually enters from the top of the tank. The outlet pipe extends down into the tank to about 1/2 inch from the bottom. This prevents sediment and debris at the bottom of the tank from being pulled into the outlet line. Some outlet lines also have a screen to prevent particles from being pulled in.

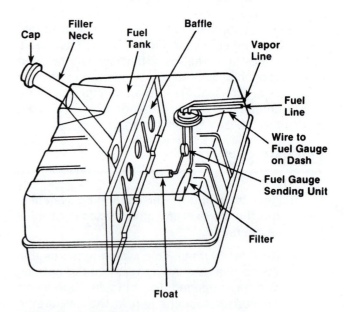

FIGURE 3-44 Parts of a fuel tank

Prior to 1971, self-venting filler caps were used to vent the tank. Since the introduction of emission controls, several different internal venting devices have been used to extract vapors from the tank (Figure 3-45). In the majority of these devices, the vapors are carried by a tube to a charcoal canister (Figure 3-46). Here they are absorbed by the char-

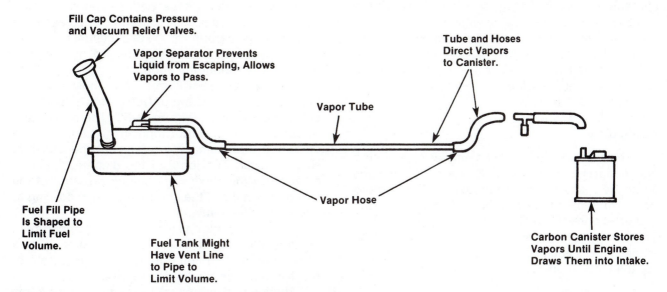

Fill Cap Contains Pressure and Vacuum Relief Valves.

Vapor Separator Prevents Liquid from Escaping, Allows Vapors to Pass.

Vapor Tube

Tube and Hoses Direct Vapors to Canister.

Vapor Hose

Fuel Fill Pipe Is Shaped to Limit Fuel Volume.

Fuel Tank Might Have Vent Line to Pipe to Limit Volume.

Carbon Canister Stores Vapors Until Engine Draws Them into Intake.

FIGURE 3-45 Internal venting devices used to extract vapors

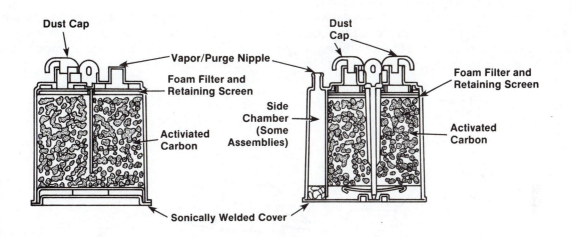

Dust Cap

Vapor/Purge Nipple

Foam Filter and Retaining Screen

Activiated Carbon

Dust Cap

Foam Filter and Retaining Screen

Side Chamber (Some Assemblies)

Activated Carbon

Sonically Welded Cover

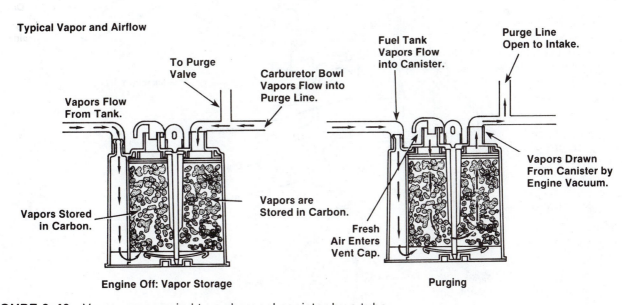

Typical Vapor and Airflow

To Purge Valve

Carburetor Bowl Vapors Flow into Purge Line.

Vapors Flow From Tank.

Vapors Stored in Carbon.

Vapors are Stored in Carbon.

Engine Off: Vapor Storage

Fuel Tank Vapors Flow into Canister.

Purge Line Open to Intake.

Vapors Drawn From Canister by Engine Vacuum.

Fresh Air Enters Vent Cap.

Purging

FIGURE 3-46 Vapors are carried to a charcoal canister by a tube.

coal and later drawn with the air/fuel mixture into the combustion chambers of the engine.

Fuel tanks usually have some type of device to indicate the amount of fuel that is in the tank. The indicating fuel gauge is located on the vehicle's dashboard, while the fuel information sensing unit can be found in the tank.

Fuel Lines

The fuel lines are small-diameter steel tubing and neoprene hoses. Usually the steel line runs along the chassis from the tank to the fuel pump and has flexible neoprene attachments to the fuel tank pickup at one end and to the fuel pump at the other. This arrangement absorbs the vibrations when the engine moves and the frame does not, thus preventing the metal tubing from cracking or breaking.

Fuel Pump

The fuel metering and atomization system is located higher in the vehicle than the fuel tank. A mechanical or electrical pump is employed to draw the fuel from the tank and deliver it to the carburetor or fuel injectors. It must have sufficient capacity to supply the engine with fuel under all operating conditions. It should also maintain sufficient pressure in the line to the carburetor to keep the fuel from boiling and causing vapor lock due to high engine temperatures under the hood.

Mechanical Fuel Pump. The most common type of fuel pump is the universal mechanical diaphragm-type pump (Figure 3–47). The rocker arm rides against an eccentric lobe on the camshaft. The amount of fuel pumped increases as the camshaft rotates faster. The force of the diaphragm spring establishes the maximum working pressure of the fuel pump. It limits the amount of fuel according to engine requirements.

Mechanical pumps in the engine compartment are subject to heat. A low-pressure area is also created in the fuel line during fuel intake. Both of these conditions can lead to vapor lock.

Electric Fuel Pumps. Electric fuel pumps come in four basic types.

1. *Diaphragm.* This type works the same way as a mechanical pump, with the exception that an electromagnet moves the diaphragm.

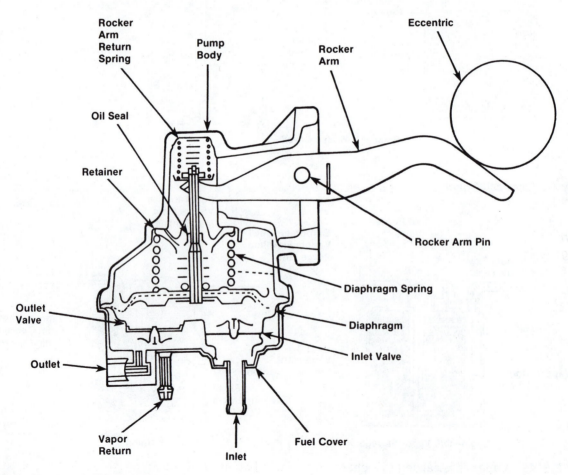

FIGURE 3-47 Mechanical fuel pump

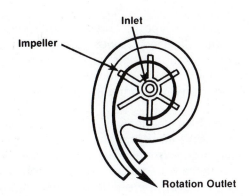

FIGURE 3-48 Electric fuel pump

2. *Plunger.* An electromagnetic switch controls a plunger or piston that moves up and down in this type.
3. *Impeller.* This pump has no input and outlet valves. Instead it has a revolving armature that works much the same way as a fan blows air (Figure 3–48). The impeller creates suction that draws fuel into the pump and pushes it out to the carburetor.
4. *Bellows.* This pump is similar to the diaphragm type. However, a metal accordian-pleated bellows is used instead of a diaphragm.

The distinct advantage of the electric fuel pump is its capability of mounting the electric fuel pump anywhere between the fuel tank and the carburetor. When installed close to the fuel tank or in the fuel tank, it eliminates vapor lock because with fuel under pressure, it is difficult for vapor to form even if the fuel lines get hot.

 SALES TALK

Carburetor and fuel injected engines require different fuel pump pressures. They range from approximately 4 to 6 psi to 30 to 60 psi and are not interchangeable.

Fuel Filters

Fuel filters are in several locations in the fuel system between the tank and the fuel metering and atomization system intake. Moisture and contaminants can collect in the fuel tank. Consequently, particles of rust from the tank can clog the fuel lines and small passages in the carburetor or fuel injectors. There are three basic types of fuel filters.

1. *Strainer.* These fuel filters consist of small fibers molded into a suitable shape for installation in a see-through glass bowl or

metal canister. This type of filter is usually mounted in the fuel pump (Figure 3-49).
2. *Screen.* These fuel filters are usually mounted in the carburetor fuel inlet fitting. The fuel line must be disengaged to replace the filter. Screen filters are usually made of porous bronze or woven brass wire. The filter openings are small enough to stop contaminants from entering the carburetor.
3. *Paper.* Although these filters are sometimes found in other parts of the system, they are usually mounted in the fuel line between the pump and carburetor. They remove smaller sized particles and larger amounts of them than it is possible to remove with the strainer or screen types.

Some fuel filters screw into the fuel inlet fitting of the carburetor and connect to the fuel line with a hose and clamps (Figure 3-50). Others are located inside the carburetor itself. Often these are small pleated paper filters with a built-on gasket to provide

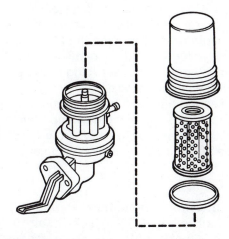

FIGURE 3-49 Strainer type fuel filter

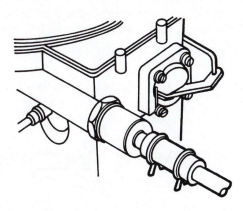

FIGURE 3-50 Some fuel filters are connected to the fuel line by a hose and clamps

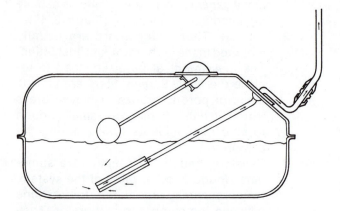

FIGURE 3-51 Some fuel filters are in the gas tank

a positive seal. Sometimes, however, they are sintered bronze, which would be considered a strainer type filter. Some of the sintered bronze filters contain a built-in check valve to limit fuel leakage if the vehicle overturns.

Some cars and light trucks use two gasoline filters. The first filter, made of fine woven fabric, is found in the gasoline tank (Figure 3-51). Besides preventing large pieces of dirt or other contaminants from damaging the fuel pump, this filter also prevents most water from going to the carburetor. The tank filter will not require servicing or replacement under normal circumstances.

The second filter can be found in a number of different locations in the engine compartment. This filter must be serviced regularly. Figure 3-52 illustrates one type, called an in-line fuel filter.

Air Filter

An internal combustion engine typically "breathes in" about 10,000 gallons of air for every one gallon of fuel it consumes. To avoid damage to the engine, this air brought into the system must be very clean. The most common contaminants that an air filter must remove are dirt, exhaust soot, vegetable matter, and insects. Naturally, geographic loca-

tion will affect the type of contaminants encountered. An autmobile used in a wet, humid climate will have different filtration problems than one used in a dry, dusty climate. An off-road vehicle will encounter more dirt than an automobile used only on the highway.

The oil bath air cleaner was used for many years to clear the intake air of the engine. When the airstream was forced to make an abrupt 180-degree turn, dirt particles were trapped on the surface of a pool of oil. However, the efficiency of such an air cleaner (that is, the percentage of contaminants it removed from the air) was approximately only 95 percent. At low engine speed, efficiency dropped by as much as 25 percent. Thus, the invention of the dry air filter was an important step toward prolonging engine life.

There are both light- and heavy-duty dry types of air filters. The light-duty air filters (Figure 3-53) are used most often on passenger car and pickup truck engines. Because of the limitations of space under the hood, these filters are usually small. Nevertheless, these filters allow the free flow of air for even the largest gasoline engines while providing a minimum efficiency of 98 percent.

Heavy-duty air filters (Figure 3-54) have a minimum efficiency rating of 99.9 percent. However, they are much larger and bulkier than light-duty types. Therefore, they are seldom found in passenger vehicles.

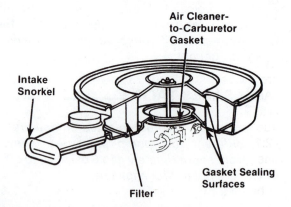

FIGURE 3-53 Light-duty air filter

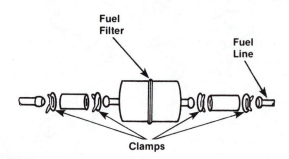

FIGURE 3-52 In-line fuel filter

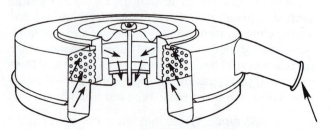

FIGURE 3-54 Heavy-duty air filter

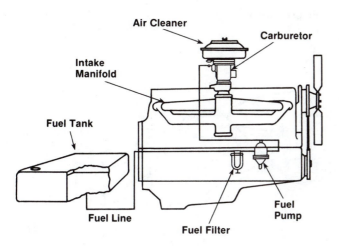

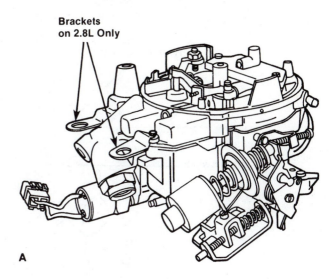

FIGURE 3-55 Fuel and air system

A

Fuel Metering and Atomization System

Technicially, it is inaccurate to say that cars run on gasoline. A vehicle actually runs on a precise mixture of gasoline and air. The carburetor or fuel injector can be considered the "chemist" of the vehicle. It mixes the proper amount of fuel and air and releases it to the engine according to the needs of the engine in various situations (Figure 3-55).

The carburetor and fuel injectors subdivide or "atomize" the fuel and mix these fine particles of fuel with air. This air/fuel mixture is expressed as a number of parts of air to 1 part of gasoline. The chemically correct mixture (at sea level) is 14.7 parts of air to 1 part of gasoline, measured by weight. It may be expressed by the ratio 14.7:1. A chemically correct fuel mixture is called *stoichiometric*. Because air at higher elevtions contains less oxygen, more parts of air are required for complete combustion.

Air/fuel ratios having high concentrations of air, such as 16:1, 18:1, or 20:1, are called *lean mixtures*. Air/fuel ratios having low concentrations of air, such as 12:1, 10:1, or 8:1, are called *rich mixtures*. The engine will not run if the air/fuel ratio goes beyond these rich and lean limits.

The engine requires a mixture as rich as 11.5:1 during warm-up and on heavy acceleration. But a mixture as lean as 18:1 can be used after warm-up and during low-load cruising. An air/fuel ratio of 14.7:1 (at sea level) provides the most power for the amount of fuel consumed.

Carburetor. The purpose of a carburetor on a gasoline engine is to meter, atomize, and distribute the fuel throughout the air flowing into the engine. These functions are designed into the carburetor and are carried out by the carburetor automatically over a wide range of engine operating conditions such as varying engine speeds, load, and operating

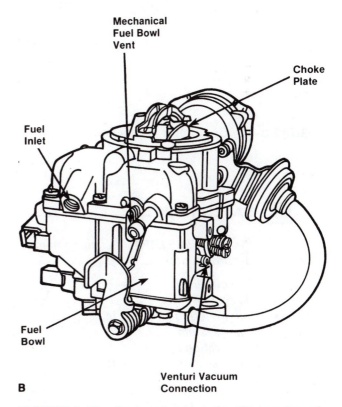

FIGURE 3-56 Carburetor barrels: (A) two barrel; (B) single barrel

temperature. The carburetor also regulates the amount of air/fuel mixture that flows to the engine. It is this mixture flow regulation that gives the driver control of the engine speed. Typical two-barrel and single-barrel carburetors are shown in Figure 3-56.

Although the carburetor performs a comparatively simple job, it does so under such varied conditions that it must have several systems to alter its functions so it can adjust to various situations.

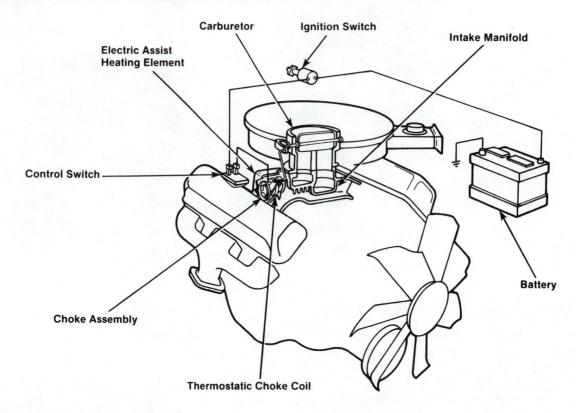

FIGURE 3-57 Carburetor installation

 SALES TALK

Use the acronym FIMPAC to help remember the carburetor circuits.

- *Float circuit*   *= F*
- *Idle system*   *= I*
- *Main metering circuit = M*
- *Power system*   *= P*
- *Acceleration circuit*   *= A*
- *Choke system*   *= C*

A typical carburetor installation is shown in Figure 3-57. Because of the cost of a new replacement carburetor, especially the electronic type, remanufactured and carburetor kits are successful sales items in most auto parts stores and dealerships.

Electronic feedback control is a sophisticated approach to regulating the air/fuel ratio to maintain optimum fuel economy, performance, and emissions. Fuel metering in an electronic carburetor is controlled by a Mixture Control (M/C) solenoid or smaller stepper motor that receives its commands from the engine control computer. An oxygen sensor in the exhaust manifold tells the computer if the engine is running lean or rich, so corrections can be made as circumstances dictate.

Most electronic carburetors also have some kind of throttle position sensor that signals the computer how far open the throttle is so fuel mixture changes can be made if needed. The throttle position sensor is nothing more than a variable resistor, but its adjustment is critical for accurate operation. Some carburetors also have a wide open throttle switch and/or an idle switch to tell the computer when either of these two operating conditions exist. Again, accurate adjustment of the switch or switches is needed for correct functioning of the system. Except for these unique components, an electronic carburetor is no different than any other carb. It can fall prey to the same types of failures as the ordinary ones: vacuum leaks, weak or inoperative accelerator pumps, worn throttle shafts, bent or damaged linkages, warpage, cracking, bad needle and seat, heavy float, and so on.

Fuel Injection. Fuel injection systems use pressure to force fuel through a small opening into the air. Unlike a carburetor, however, the pressure forcing the fuel into the air is not atmospheric pressure. Instead, the pressure is produced by a pump in the injection system itself. The benefits of fuel injection

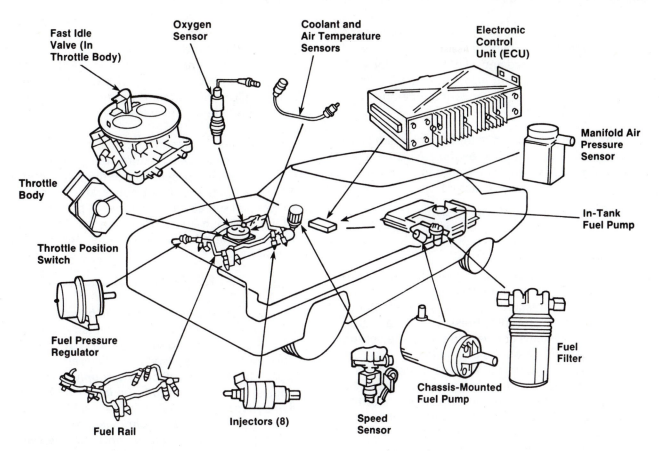

Fast Idle Valve (In Throttle Body)

Oxygen Sensor

Coolant and Air Temperature Sensors

Electronic Control Unit (ECU)

Manifold Air Pressure Sensor

Throttle Body

Throttle Position Switch

In-Tank Fuel Pump

Fuel Pressure Regulator

Fuel Rail

Injectors (8)

Speed Sensor

Chassis-Mounted Fuel Pump

Fuel Filter

FIGURE 3-58 EFI system

include a more precise control of the air/ fuel ratio, improved emission control, fuel efficiency, and driveability. On the other hand, fuel injection systems are more expensive, usually considered harder to service, and require clean fuel.

In late-model cars, Electronic Fuel Injection (EFI) is the most widely used fuel system (Figure 3–58). There are currently two major designs of electronic injection systems used in passenger vehicles.

1. Throttle Body Injection (TBI) uses one or two injectors centrally mounted in a throttle body unit located where the carburetor would normally be. TBI (Figure 3–59A) provides precise fuel control, but is still subject to manifold problems that are common to a carbureted system. If the TBI unit is equipped with two injectors, they are opened alternately to prevent the fuel pressure from dropping too low. TBI systems are also known as Central Fuel Injection (CFI), Single-Point Fuel Injection (SPFI), and Digital Fuel Injection (DEFI).

2. Port Fuel Injection (PFI) uses one injector for each cylinder in the engine, which is located in the intake runner near the intake

valve port (Figure 3–59B). By positioning the injector in the intake port, manifold problems associated with the TBI or carbureted systems are eliminated. Both TBI and PFI use similarly designed, electroni-

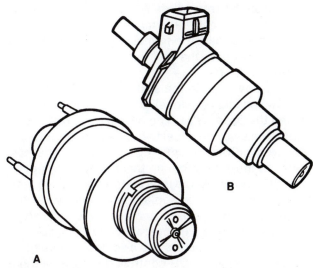

B

A

FIGURE 3-59 (A) Throttle body injection; (B) port fuel injection

cally controlled injectors. The difference between the two systems is the number of injectors used and where they are located. PFI systems are also known as Multiport (MFI), Rail, and Sequential Fuel Injection (SFI).

The electronic fuel injector is a solenoid-operated valve that is controlled by the car's computer. The amount of fuel that is delivered to the engine is controlled by the following factors:

- The size of the orifice in the injector is determined when the system is designed. This factor is fixed; it cannot be changed or modified.
- The amount of fuel pressure is also predetermined by system design. The fuel pump and fuel pressure regulator set the system's operating pressure. Pressure regulators on some systems can vary the pressure setting depending upon the engine load.
- The length of time the injector is held open is controlled by the car's computer and can be altered. The computer varies this amount of time the injector is held open in accordance with various computer inputs and specifications from the Programmable Read Only Memory (PROM).

The typical EFI system provides a pulse of fuel to get the system underway. The length of the pulse, that is, the length of time the injector is held open, is controlled by the electronic system (Figure 3–60).

An electrical fuel pump pressurizes the fuel. Fuel pressure is regulated according to manifold absolute pressure, with excess fuel returned to the tank through a return line. The fuel damper prevents pulsations in the line and maintains fuel pressure in the system. Because extremely clean fuel is required for fuel injection, the fuel filter is a very important component.

Air flowing into the system is measured by an airflow sensor. The signal from the airflow sensor is called the *base pulse*. The base pulse provides the minimum amount of fuel to be injected. A solid state Electronic Control Unit (ECU) modifies the base pulse to increase the time the injector stays open. The ECU receives and interprets signals from other sensors that monitor coolant temperature, throttle positions, engine speed, and engine load to adjust the injectors to engine operating conditions.

Electronic injectors use a solenoid to open the nozzle valve. The ECU controls the pulse width at the injector by the length of time the solenoid receives a voltage signal.

A cold start injector valve, a thermo-time switch, and an auxiliary air device are used to make provision for cold starting and warm-ups. Extra fuel is delivered by the cold start injector while the engine is cranking. During start and warm-up, the auxiliary air device provides extra air to overcome cold engine friction.

A feedback or closed loop fuel control system can meter the air/fuel mixture more closely. Because a closed loop system continually monitors itself and adjusts the inject pulse rate, the air/fuel ratio can be kept as close to ideal as possible. The ECU in an open loop operation sends preprogrammed signals for warm-up, idle, and full-load enrichment.

▮ SALES TALK ⎯⎯⎯⎯⎯⎯

As a fuel injector ages, it will become partially restricted. This can be corrected either by cleaning or replacing the injector(s). Any time the injectors are serviced all O-rings should be replaced. Most manufacturers supply these O-rings separately or in tune-up kits. To prevent damage to the O-rings when the fuel injector is reinstalled, it should be lubricated. Any time a fuel injector is serviced or replaced, the fuel injector filter must also be replaced. Inform the customer of these sales add-ons.

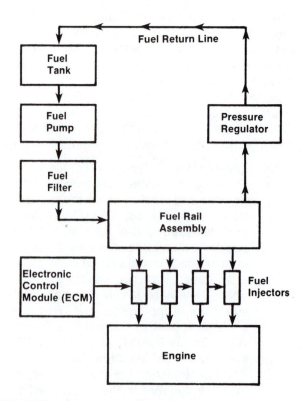

FIGURE 3–60 Fuel system supply

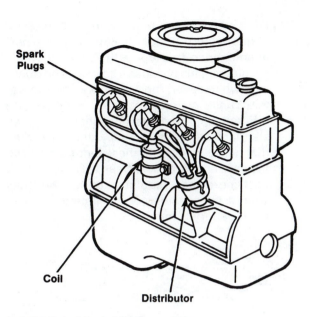

FIGURE 3-61 Ignition system

IGNITION SYSTEM

More "no-start" problems and on-the-road engine troubles can be traced to the ignition system than to any other system in the car. But if this system is kept in good operating condition with regular tuning, it is as reliable as most other components of the car.

To "fire" the compressed air/fuel mixture in the combustion chamber, an electric spark must be provided near the top of each compression stroke. This spark is furnished at the proper timing by the ignition system (Figure 3-61). Several types of ignition systems are found on automobiles. The most common are

- Standard or breaker point
- Electronic

Standard Ignition System

The standard or breaker point ignition system has been used since batteries were first installed in automobiles. It is a very simple system. Its main components (Figure 3-62) are

- *Spark Plugs.* Provide the gap for the ignition spark to take place.

 SALES TALK _____

Check the vehicle's VIN number (see Chapter 4) and be sure the correct plug has been selected for the customer. If the plug is too hot, it will have a short electrode life. If too cold, excessive deposits will form on the insulator and short-circuit the spark. If the gap is too wide, it will stress the ignition system; if it is set too close it could prevent the ignitable mixture from entering. Check the specifications and adjust accordingly if necessary. Remember most modern vehicles use resistor type spark plugs because they reduce radio and TV interference (Figure 3-63).

- *Plug Cable or Wires.* Provide a path for the high voltage energy.
- *Cap.* Distributes the spark energy toward each plug.
- *Rotor.* Rotates under the cap and directs the spark energy to the cap terminals.
- *Points.* Make and break the coil low-voltage circuit.

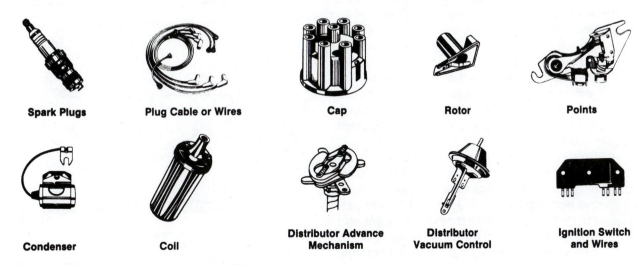

FIGURE 3-62 Components of the ignition system

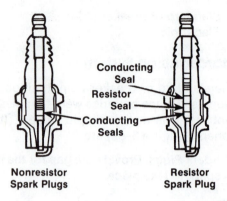

FIGURE 3–63 Nonresistor and resistor-type spark plugs

- *Condenser.* Provides a path of least resistance or an electrical shock absorber to reduce arcing at the points.
- *Coil.* Steps up the low voltage provided by the battery. Creates the high voltage needed by the plugs.
- *Ballast Resistor.* Added to the ignition circuit to compensate for variations in primary circuit current. These variations occur due to change in engine rpm and load on the charging system.
- *Distributor Advance Mechanism.* Tailors the spark event to the needs of the engine. It changes when the spark occurs so that the engine operates properly; related to engine speed.
- *Distributor Vacuum Control.* Modifies the moment at which the spark occurs according to engine load.
- *Ignition Switch and Wires.* Route battery voltage to the hot side of the coil.
- *Battery Cables.* Link between the battery and electrical system of the vehicle (Figure 3–64). Because of the relative low voltage and high amperage needed to supply these electrical systems, the battery cable is a large, stranded, and jacketed cable. Although some manufacturers use aluminum or copper-clad aluminum, most use all copper, which is an all-around superior conductor for battery cables.
- *Battery.* Provides the low voltage power necessary to produce current flow in the primary circuit of the ignition coil. Modern automotive electrical systems employ a 12-volt battery, which is available in several designs and capacities. They all have a leakproof case of either hard rubber or plastic and two external terminals, a positive (+) and a negative (–). In

addition to the top terminal, regular-maintenance battery (Figure 3–65), there is a new maintenance-free type with either top or side terminals. This type has a case with no openings whatsoever; there is no way to add water. These maintenance-free batteries use different components and chemicals than standard batteries. Actually, the only thing that has to be done to these batteries is keep the case and terminals clean. Of course, they cost more than regular batteries, but they should be considered any time a replacement battery is needed.

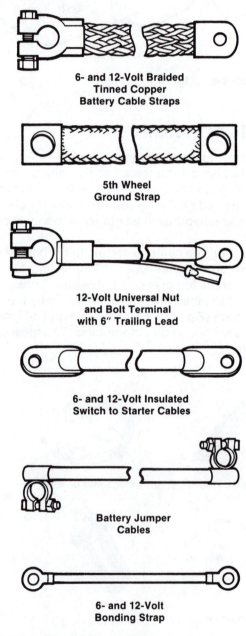

6- and 12-Volt Braided Tinned Copper Battery Cable Straps

5th Wheel Ground Strap

12-Volt Universal Nut and Bolt Terminal with 6" Trailing Lead

6- and 12-Volt Insulated Switch to Starter Cables

Battery Jumper Cables

6- and 12-Volt Bonding Strap

FIGURE 3–64 Battery cables

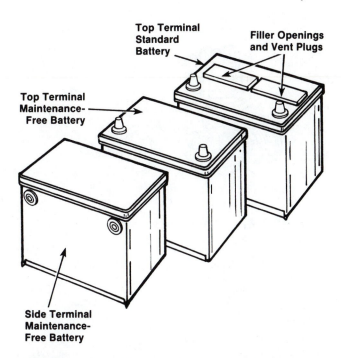

Top Terminal Standard Battery

Filler Openings and Vent Plugs

Top Terminal Maintenance-Free Battery

Side Terminal Maintenance-Free Battery

FIGURE 3-65 Three standard types of batteries

 SALES TALK

Most auto supply centers stock, in addition to a standard line(s) of batteries, such types as used in heavy-duty commercial, marine, cold weather start, lawn and garden vehicles, wheel chairs, golf carts, cable TV, floor scrubbers, and other utility applications. Most batteries fall into a class of deep cycle batteries, especially designed to go through many deep discharges, sometimes drained to practically nothing before being recharged. It is important for the counterperson to know the complete line of batteries that his or her establishment handles.

Another consideration that the counterperson must keep in mind is that batteries are perishable. When inventorying batteries, use the FIFO system: first in, first out. Use shelves that allow flow-through, which means stock them from the back and sell from the front. Batteries should never be stacked one on top of the other unless they are in cartons or have protective layers of cardboard between them.

All batteries will self-discharge in storage. How much they will discharge depends on the type of battery and the storage temperature. If it is summer or if in a hot climate, store batteries in as cool a place as possible. In the winter or in a colder climate, however, protect a discharged battery from freezing,

which can cause serious damage to the case and cover.

Wet batteries in inventory should be checked when they have been in stock more than four or five months—or two to three months in warm climates. Check the voltage with a digital voltmeter. If the battery voltage is below 12.4 volts, charge the battery at a rate of 5 to 10 amps until the specific gravity comes back up to 1.265. Then mark the date of recharge with chalk or grease pencil. Naturally, when selling a battery, make sure that the voltage is 12.4 or higher. If it is not, boost charge it. Also, do not forget the batteries on display. Boost charge them periodically as well.

WARNING: When being charged, all batteries generate hydrogen gas, which is highly flammable. If ignited by a spark or flame, the gas can explode violently causing spraying of acid, fragmentation of the battery, and possible severe personal injuries, particularly to the eyes. Always wear safety glasses and charge in a ventilated area. The charger should be shut off while being connected to the battery.

Operation of Standard Ignition System

The ignition system is divided into two circuits: the primary or low-voltage circuit and the secondary or high-voltage circuit. The primary circuit contains the battery, ignition switch, resistor, ignition coil primary windings, ignition points, and condenser. The secondary circuit contains the coil secondary windings, high-tension leads, distributor cap, distributor rotor, and spark plugs.

The system begins to operate as the ignition switch is turned on and the engine is cranked. As the ignition points close, current flows through the primary circuit. This current flow, through the primary winding of the coil (Figure 3–66), produces a strong mangetic field within the coil. As the ignition points open, current flow in the primary circuit stops, causing the magnetic field around the primary winding to collapse. The condenser aids in this collapse by furnishing a low resistance path and a current storage place during the field collapse. The magnetic lines of force around the primary winding cut through the hundreds of turns of the secondary winding and induce a high voltage in it. This voltage can be 20,000 volts or more and can jump the gap between the distributor cap and rotor. The current flows through the rotor to one of the spark plug cable towers in the cap, through the high-tension lead to the spark plug, and jumps across its gap. This

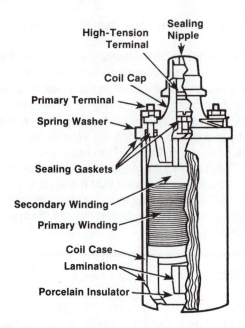

FIGURE 3-66 Parts of typical ignition coil

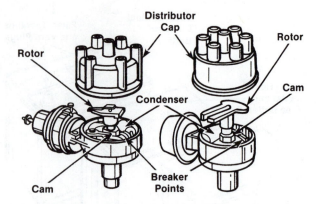

FIGURE 3-67 Distributor assembly

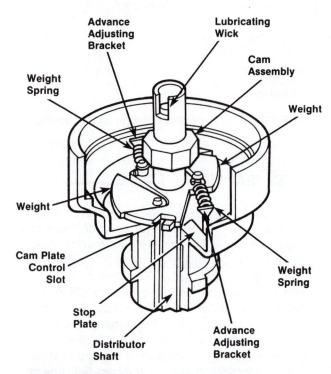

FIGURE 3-68 Centrifugal or mechanical advance system

high voltage causes an arc across the spark plug electrodes, which ignites the fuel mixture in the combustion chamber.

When the ignition points separated and the magnetic field started to collapse, a voltage was also induced in the primary winding. This voltage can be as high as 250 volts and can cause an arc across the face of the ignition points. The condenser is connected to the ignition points to divert the current flow from the arc. This brings the primary current to a quick controlled stop for efficient coil operation. The condenser also reduces the arc at the ignition points and in doing so prolongs the life of the points.

Ignition point dwell is a term used to describe the time the ignition points stay closed in a standard ignition system. The proper term for dwell is *dwell angle*. Dwell angle describes the number of degrees the distributor cam turns while the ignition points remain closed.

The distributor assembly (Figure 3-67) serves three purposes:

1. Controls the primary circuit by opening and closing the ignition points.
2. Directs the high voltage current in the secondary circuit to the proper spark plug.
3. Times the delivery of the high voltage to meet the engine requirements of speed and load. The third function is the job of the advance systems. Normally, two systems are used: the centrifugal advance and the vacuum advance. Both are integral parts of the distributor assembly.

Centrifugal Advance. The centrifugal advance is sometimes referred to as the mechanical advance system (Figure 3-68). It is made up of fly weights and springs and is housed in the distributor. The centrifugal advance is connected to the upper part of the distributor shaft and will advance the distributor cam based on distributor rpm. As the rpm increases, the weights move away from the distributor shaft and overcome spring tension; the higher the rpm the more the weights move. The distributor shaft is made as two pieces: the lower portion is the shaft, the upper is the cam assembly. The more the weights move, the more the cam assembly is ad-

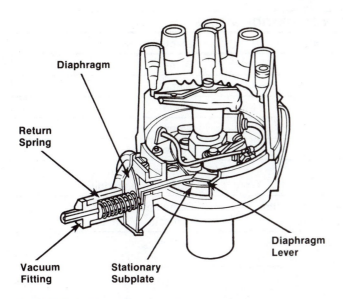

FIGURE 3-69 Vacuum advance system

vanced. How fast it will advance is based on spring tension.

Vacuum Advance System. Distributor assemblies are equipped with a vacuum advance system for better fuel economy when the engine is operated at part throttle. The vacuum advance consists of a spring-loaded diaphragm connected to the distributor plate by a line (Figure 3-69).

A hose runs from the spring side to either the intake manifold or the carburetor. Vacuum is used to overcome the spring tension and rotate the distributor plate opposite to distributor cam rotation. This causes the points to open sooner in the power stroke

to allow a longer burn time for the fuel mixture. When the vacuum hose is connected directly to the intake manifold there will be advance at idle. When the unit is operated by carburetor vacuum there is no advance at idle. Both systems work at part throttle and neither system works at full throttle. This is due to low manifold vacuum, which cannot overcome the spring tension in the unit. Tune-up specifications list a total ignition advance, which is to be checked at a specific engine rpm. This is the maximum advance reached by combining both the vacuum and centrifugal advance systems.

ELECTRONIC IGNITION SYSTEM

An electronic ignition system is one that does not contain ignition points. Instead, the primary circuit is controlled electronically by an electronic switching unit (Figure 3-70). In an electronic ignition system, the primary circuit contains

- Battery
- Ignition switch
- Primary coil windings
- Magnetic pickup assembly, often called the stator assembly
- Electronic switching unit

The battery, ignition switch, and primary coil windings operate the same as in the standard system. The magnetic pickup replaces the movable parts of the ignition points. They are the distributor cam, cam follower, and movable point arm. The electronic switching unit replaces the ignition point contacts and the condenser.

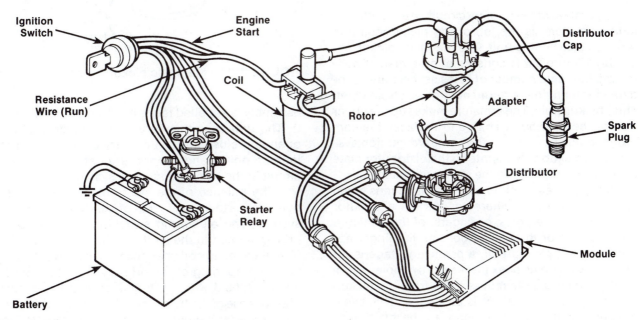

FIGURE 3-70 Electronic switching unit

TABLE 3–1: COMPARISON OF IGNITION SYSTEMS

Function	Secondary System	Electronic System
Provide spark	Spark plug	Same
Carry high energy voltage	Plug wires	Same
Distribute high energy voltage	Cap	Same
Distribute high energy voltage	Rotor	Same
Step up low voltage to high voltage	Coil	Same
Trigger the collapse of the magnetic field in the coil by circuit	Points	Trigger mechanism and control module
Control spark event	Mechanical and vacuum advance mechanisms	Same and/or computerized controls
Route electrical current from battery to coil	Ignition switch and wires	Same

The secondary circuit in an electronic ignition system contains the same components as the standard ignition system. They are

- Secondary coil winding
- Distributor cap
- Distributor rotor
- High tension leads
- Spark plugs

All of the functions performed by the standard ignition system are also performed by the electronic ignition system. Table 3–1 compares the two systems.

Operation of a Typical Electronic Ignition System

Operation of the primary circuit, in an electronic ignition system, is the job of the pickup coil and electronic switching unit. The magnetic pickup assembly (Figure 3–71) consists of an armature containing the same number of poles as there are cylinders in the engine, a permanent or electromagnet, and a pickup pole. The armature makes up the upper part of the distributor shaft and replaces the cam found in this position on the standard ignition system. The pickup coil mounts on the distributor plate as the ignition points in the standard system.

The magnetic pickup assembly functions much like a self-energized alternator. As the distributor shaft rotates, the poles, or fingers, of the armature pass in front of the pickup coil. As the finger approaches the pickup coil, a positive voltage is induced, reaching its peak just as the two poles align. As the armature pole moves away from the pickup coil, the negative current is produced. This cycle happens continuously as long as the distributor shaft is turning. The voltage produced by the pickup

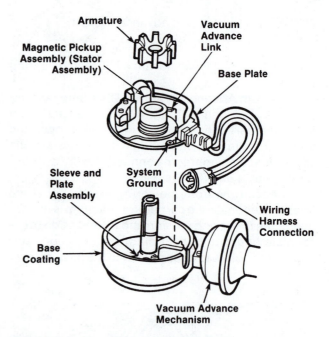

FIGURE 3–71 Magnetic pickup assembly

assembly is supplied to the electronic switching unit. The unit uses this very small voltage signal to open and close the coil primary circuit just as the ignition points and condenser control the primary circuit in the standard ignition system.

The balance of the electronic ignition system operation is identical to that of the standard ignition system. The electronic switching unit opens the primary circuit in the ignition coil just as the signal from the pickup coil changes from positive to negative. The opening of the primary circuit causes its magnetic field within the ignition coil to collapse. The collapse of the field induces a high voltage in the coil secondary winding. This high voltage is then sent to the spark plugs through the high tension leads, distributor cap, and rotor. When the high voltage reaches the spark plug, it jumps the gap between the spark plug electrodes and ignites the air/fuel mixture in the combustion chamber.

In a standard ignition system, the dwell angle is set by adjusting the ignition points. It will vary as the cam follower wears and the point gap becomes smaller. The dwell angle in an electronic ignition system is controlled by the switching unit and cannot be adjusted. The dwell angle will remain proper for the life of the switching unit.

Different names for the electronic ignition parts by different OEM with similar functions are given in Figure 3–72. On some electronic systems, a capacitor may be used to reduce static.

The counterperson or parts manager is often asked what are the advantages of electronic ignition systems. The more important ones are as follows:

- No ignition points
- Ignition timing, once set, remains correct
- High secondary voltage
- No misfire at high engine rpm due to ignition point bounce
- Longer spark plug life due to higher secondary voltage

- Easier engine starting under all conditions
- Better fuel economy
- Cleaner emissions

A few electronic systems now on the market use sensors and an elaborate coil system in place of the distributor.

LUBRICATION SYSTEMS

An engine's lubrication system (Figure 3–73) must perform several important functions:

1. Hold an adequate supply of oil
2. Remove contaminants from the oil
3. Deliver oil to all necessary areas of the engine

The main components of a typical lubrication system are:

- *Oil Pump.* The oil pump is the heart of the lubrication system. Just as the heart in your body circulates blood through your veins, the oil pump circulates oil through the engine.

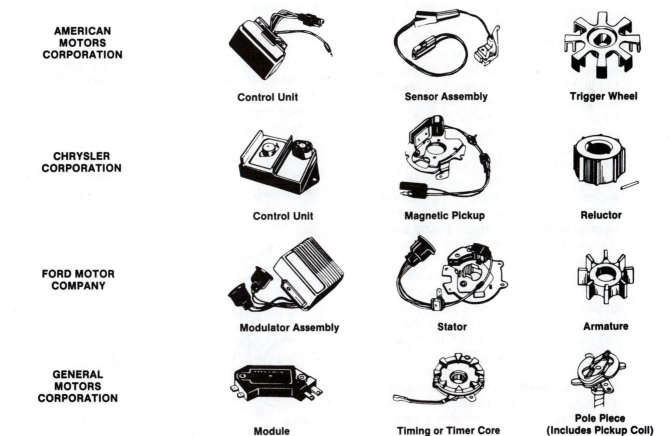

AMERICAN MOTORS CORPORATION	Control Unit	Sensor Assembly	Trigger Wheel
CHRYSLER CORPORATION	Control Unit	Magnetic Pickup	Reluctor
FORD MOTOR COMPANY	Modulator Assembly	Stator	Armature
GENERAL MOTORS CORPORATION	Module	Timing or Timer Core	Pole Piece (Includes Pickup Coil)

FIGURE 3-72 Electronic ignition parts

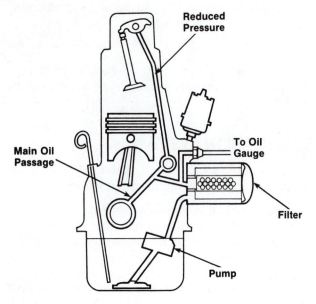

FIGURE 3-73 Lubrication system

- *Oil Pump Pickup.* The oil pump pickup is a line from the oil pump to the oil stored in the oil pan. It usually contains a filter screen, which is submerged in the oil at all times. The screen serves to keep large particles from reaching the oil pump. This screen should be cleaned any time the oil pan is removed.
- *Oil Pan or Sump.* The oil pan attaches to the crankcase or block. It serves as the reservoir for the engine's lubricating oil. It is designed to hold all the oil necessary to lubricate the engine when it is running, plus a reserve. The oil pan helps to cool the oil through its contact with the outside air.
- *Engine Oil Passages.* Oil passages are designed to carry engine oil to all necessary areas. They are drilled or machined into the cylinder block, cylinder head, crankshaft, and camshaft. In some applications these passages might be external piping added to the engine assembly to feed engine-driven accessories.
- *Oil Filter.* The engine oil filter (Figure 3-74) is designed to remove minute particles of metal and dirt from the engine oil. This is necessary so they will not circulate through the engine and cause premature wear. Filtering also increases the usable life of the oil.
- *Crankcase Ventilation.* Crankcase ventilation is necessary because of pressure that can build in the crankcase due to combustion pressure, which passes the piston rings. Piston rings do not provide a complete positive seal of the combustion area and some

combustion gases reach the oil pan. These gases can contaminate the oil and apply unwanted pressure to gaskets and seals.

- *Oil Pressure Indicator.* This system can be in the form of a gauge, which indicates the engine oil pressure at all times, or it can be a warning light that will come on whenever the engine is running with insufficient oil pressure. The warning light is the most common oil pressure indicator.
- *Oil Seals and Gaskets.* As mentioned earlier, oil seals and gaskets are used throughout the engine to prevent both external and internal oil loss. The most common materials used for sealing are synthetic rubber, soft plastics, fiber, and cork. In critical areas these materials might be bonded to metal.
- *Dipstick.* The dipstick is used to measure the level of oil in the oil pan. The end of the stick is marked to indicate when the engine oil level is correct. It has a mark to indicate the need to add oil to the system.

The lubricant system has many major components with which the counterperson should be familiar.

Oil Pumps

The oil pump is usually located in the oil pan and is driven by the distributor through the drive shaft. The two most common types of oil pumps are the gear and the rotor (Figure 3-75). When a replacement is necessary, check the service manual as to the type needed.

With most types of oil pump systems, oil pressure will be high when the oil is cold, such as cold start-up, and it will become higher as the pump turns faster, such as high engine rpm. To keep the pump from supplying oil pressure, which would be too

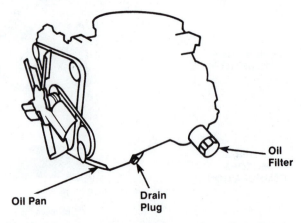

FIGURE 3-74 Location of a typical oil filter

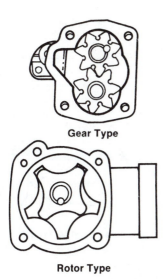

Gear Type

Rotor Type

FIGURE 3-75 Gear and rotor type oil pumps

high and damage the oil filter and seals, a pressure relief valve is incorporated into the pump housing. This relief valve is a spring type valve; when the pressure from the pump is greater than the spring pressure, it pushes the valve off its seat. Excess oil pressure passes the valve, and the oil is returned to the crankcase.

While the amount of the lubrication system pressure varies with engine design, the oil under pressure passes first through the oil filter then onto the main oil gallery (Figure 3–76). From the gallery it is directed to the main bearings, the connecting rod bearings, camshaft bearings, and hydraulic tappets. This pressure also supplies the oil sprayed on the cylinder walls through the holes in connecting rods. Reduced oil pressure is supplied to other parts of the engine. This reduction is achieved by restricting the

size of the oil passage. This is necessary to eliminate engine bearing starvation and oil consumption. The areas that receive this reduced oil pressure are the pushrod ends, shaft pivoted rocker arms, and some distributor shafts. The oil returning to the oil pan is also used for lubrication. Some areas receiving this nonpressure lubrication are the timing gears, sprockets, timing chain, solid tappets, and cam lobes.

Oil Filter

As previously stated, all oil leaving the oil pump is directed to the oil filter. The oil flows through the filter then on to lubricate the engine. This type of filtering is called full-flow. No oil that is not filtered reaches the engine. This insures that very small particles of dirt and metal carried by the oil will not reach the close-fitting engine parts. The filter element and container are made as a unit, with a seal built in at the point the filter assembly contacts the block. The filter assembly threads directly on to the main oil gallery tube, eliminating external oil leaks and the possibility of oil leakage under pressure. The oil from the pump enters the filter can on the outside of the element, passes through the element to the center of the filter and into the main gallery.

The filter unit itself (Figure 3–77) is a disposable metal container filled with a special type of treated paper or other filter substance (cotton, felt, etc.) that catches and holds the oil's impurities. Should these impurities become so great that the oil cannot pass through the filter, a bypass valve in the filter permits unfiltered oil to flow to the engine. Because of this, the oil filter must be changed at regular intervals.

There are two basic types of oil filters.

1. *Surface-Type Spin-on Filter (Figure 3-78A)*. Pleated filter paper is placed around the

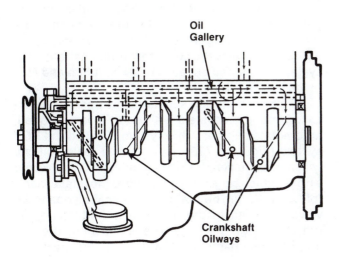

FIGURE 3-76 Oil paths in the engine

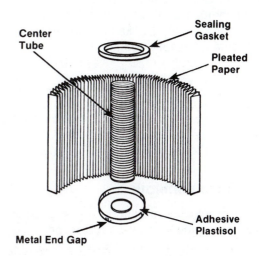

FIGURE 3-77 Oil filter construction

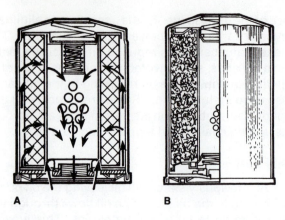

FIGURE 3–78 (A) Surface-type filter; (B) depth-type filter

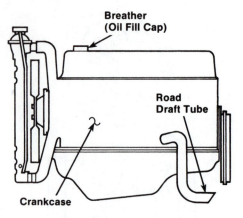

Old Road Draft Method

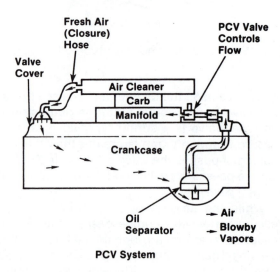

PCV System

FIGURE 3–79 (A) Road draft method of crankcase ventilation; (B) PCV system

center tube. By folding the paper into pleats, the maximum amount of surface area can be exposed. The pleated paper is composed of cellulose (and sometimes synthetic) fibers bonded together with phenolic resin. The correct porosity, thickness, and other properties are chosen for each particular application and will vary from filter to filter.

2. *Depth-Type Filter.* In this type, which is less commonly used on automobiles, the filter media is a blend of selected cotton thread and resilient supporting fibers (Figure 3–78B). The depth media has a high efficiency and capacity for dirt retention, but the filter is limited in flow capacity when compared to surface filters of the same size.

In recent years, the trend toward smaller filters has prompted the development of dual filtration systems (two separate filters in series) by some aftermarket companies. Dual filtration systems are common on heavy-duty trucks but are a relatively new development for automobiles.

 SALES TALK _____

When selling a replacement oil filter always follow the vehicle manufacturer's recommendations as to type.

Crankcase Ventilation

The crankcase ventilation system is considered part of the engine lubrication system because one of its functions is to prevent blowby gases from contaminating and breaking down the engine oil. The

system is also used to prevent pressure from building in the engine, which would damage the engine gaskets and seals.

On earlier models of gasoline engines, blowby gases were vented through the breather cap and a road draft tube (Figure 3–79A). A road draft tube is a pipe used to carry the gases to the atmosphere. This system works well but adds to air pollution and also tends to drip oil.

To meet emission standards the positive crankcase ventilation system is fitted to most, if not all, automobile engines. This system is called the PCV system (Figure 3–79B). It contains a sealed oil cap with a hose that leads to the air filter, a PCV valve and a hose, which runs from the valve to the intake manifold. As pressure builds in the crankcase it opens the valve. Vacuum in the intake manifold draws the fumes through the valve. The fumes then enter the combustion chamber with the air/fuel mixture and are burned. As the fumes are removed they must be

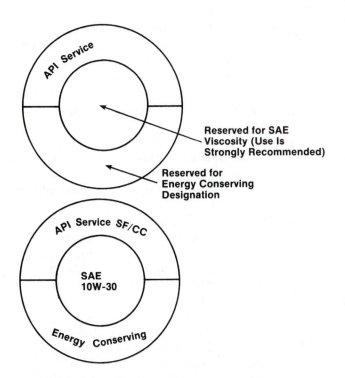

FIGURE 3-80 API designation and SAE rating

replaced by fresh air. If this is not done, a vacuum will develop in the crankcase, which would draw the oil from it. The fresh air is drawn from the air filter through the oil filler cap and the air cleaner. The PCV valve serves to regulate the flow of crankcase fumes and acts as a safety valve. Should the engine backfire, the pressure in the intake manifold will close the valve, keeping the backfire from reaching the crankcase.

The fuel induction systems on engines equipped with PCV systems are set up to compensate for the air supplied through the system. For this reason the system must be kept clean and the proper valve must be installed. Each engine type requires a different valve; the wrong valve will affect engine performance.

Oil Types

It is very important that a counterperson understands that all oil is not created equal. The top of the oil can or container will tell two important things: the viscosity of the oil and how it should be used. The SAE ratings tell the oil's viscosity and the API designation tells the oil's intended use (Figure 3-80).

Viscosity is an expression of the internal friction of a fluid and is one of the most important properties of lubricating oils. The Society of Automotive Engineers' (SAE) viscosity number system is used worldwide to designate the degree of viscosity of engine and transmission oils. The viscosity characteristic curve distinguishes the viscosity temperature behavior of a lubricating oil. In simple terms, the SAE numbers tell the temperature range that the oil will lubricate best. An SAE 10 lubricates well at low temperatures but becomes too thin and breaks down at high temperatures. SAE 30 lubricates well at mid temperatures but becomes too thick at low temperatures to supply good lubrication. Multigrade oils are engine oils that cover more than one SAE viscosity number. Their short designations are made up from the designations of the two viscosity numbers that the oil has met. For example, an SAE 10W-30 oil meets the requirements of a 10 weight oil for cold start and cold lubrication and the requirements of a 30 weight oil for mild temperature lubrication (Table 3-2).

Most automobile manufacturers recommend a multigrade oil for their engines because this insures good lubrication over a wider temperature range. The American Petroleum Institute (API) service designations are used in the United States to determine the engine service qualifications of engine oils. These qualifications are indicated by the letters found on the containers (Table 3-3).

TABLE 3-2: SAE GRADES OF MOTOR OIL		
Lowest Atmospheric Temperature Expected	**Single-Grade Oils**	**Multigrade Oils**
32° F (0° C)	20, 20W, 30	10W-30, 10W-40, 15W-40, 20W-40, 20W-50
0° F (–18° C)	10W	5W-30, 10W-30, 10W-40, 15W-40
–15° F (–26° C)	10W	10W-30, 10W-40, 5W-30
Below –15° F (–26° C)	5W*	*5W-20, 5W-30

*SAE 5W and 5W-20 grade oils are not recommended for sustained high-speed driving.

TABLE 3-3: API DESIGNATION

SC	Service typical of gasoline engines in 1964 through 1967. Oil designed for this service provides control of high and low temperature deposits, wear, dust, and corrosion in gasoline engines.
SD	Service typical of gasoline engines in 1968 through 1970. Oils designed for this service provide more protection against high and low temperature deposits, wear, rust, and corrosion in gasoline engines. SD oil can be used in engines requiring SC oil.
SE	Service typical of gasoline engines in automobiles and some trucks beginning in 1972. Oil designed for this service provides more protection against oil oxidation, high temperature engine deposits, rust, and corrosion in gasoline engines. SE oil can be used in engines requiring SC or SD oil.
SF	Service typical of gasoline engines in automobiles and some trucks beginning with 1980. SF oils provide increased oxidation stability and improved antiwear performance over oils that meet API designation SE. It also provides protection against engine deposits, rust, and corrosion. SF oils can be used in engines requiring SC, SD, or SE oils.
SG	Service typical of gasoline automobiles and light-duty trucks, plus CC classification diesel engines beginning in late 1980. SG oils provide the best protection against engine wear, oxidation, engine deposits, rust, and corrosion. It can be used in engines requiring SC, SD, SE, or SF oils.

 SALES TALK _____

When selling motor oil to a customer, a good rule for API designation is: it is all right to go higher than required, but do not go lower. API designations CA, CB, CC, and CD are oils for diesel engines.

COOLING SYSTEM

On most gasoline engines, the method of cooling is to circulate coolant through the cylinder block and cylinder head (Figure 3–81). Liquid cooling is preferable to air cooling because it is less noisy and better able to maintain a constant temperature at the cylinders. It also lets the engine operate more efficiently and makes a ready supply of hot coolant available to operate a heater for the passenger compartment.

The cooling system is made up of the following:

- *Pump.* Circulates the cooling liquid through the system. The liquid is a mixture of water and antifreeze; referred to as *coolant.*
- *Water Jackets.* Cored passages in the cylinder block and cylinder head that carry the coolant around the cylinders and combustion chambers.
- *Radiator.* Transfers heat in the coolant to the outside air as coolant flows through its tubes.
- *Fan.* A device that pulls cool outside air through the fins of the radiator to pick up heat.
- *Pressure Cap.* Maintains a pressure in the system to raise the boiling point of the coolant to a higher temperature. Also provides relief from excess pressure or vacuum.
- *Hoses.* Connect the components of the system to one another.
- *Thermostat.* Blocks off circulation in the system until a preset temperature is reached to speed engine warm-up. Also controls engine temperature at a predetermined level.
- *Temperature Indicator.* Warns the driver in case of overheating.

Operation of the Cooling System

The water pump is belt driven by the engine crankshaft. The engine cooling fan can be mounted on the water pump or operated electrically. Coolant is pumped or circulated by the water pump from the bottom or side of the radiator through internal passages in the engine block and heads. The internal passages are often called the *water jacket.* The water pump then pumps the coolant through the thermo-

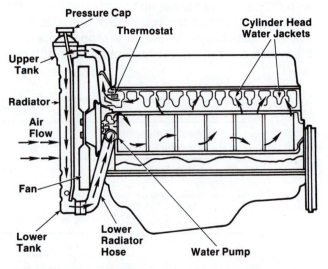

FIGURE 3–81 Cooling system circulates coolant through the engine.

stat to the top tank of the radiator. From the top tank the coolant flows slowly through a system of tiny tubes surrounded by cooling fins. These tubes and fins are called the radiator core. When the coolant reaches the opposite radiator tank the process begins again. Actually, from the point the thermostat opens, the coolant is constantly circulated through the engine and radiator. As the coolant travels through the engine block and heads, unwanted heat caused by combustion is transferred to the coolant. As the coolant flows through the radiator core this heat is transferred to the air as the air passes through the core and around the fins. The fan, usually mounted on the water pump, is used to move air through the radiator when the engine is idling or operating slowly, as in traffic. This insures sufficient air movement and good heat transfer at slow speeds. The cooling system's job is two-fold.

1. It carries away unwanted heat from the engine.
2. It must be able to maintain the engine temperature at the right point for efficient operation.

Water Pump

Engine water pumps are generally the impeller type (Figure 3–82). An impeller is something that is designed to push, drive, or move something; in this case, engine coolant. The impeller or water pump is mounted on a shaft and designed like a rotating paddle. The shaft is mounted in the water pump housing and rotates on bearings. The pump contains a seal to keep the coolant from passing through it. At the drive end, the exposed end, a pulley is mounted to accept the belt, which is driven by the crankshaft. The pump housing usually contains the mounting point for the lower radiator hose.

When the engine is started, the crankshaft turns the water pump. The pump impeller pushes the wa-

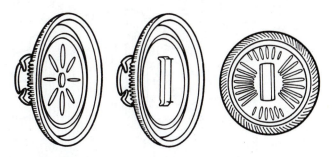

FIGURE 3-83 Fan clutches

ter from its pumping cavity into the engine block. When the engine is cold the thermostat will be closed. This stops the coolant from reaching the top of the radiator. In order for the water pump to circulate the coolant through the engine during warm-up, a bypass passage is added below the thermostat, which leads back to the water pump. This passage must be kept free to eliminate hot spots in the engine during warm-up. It also allows the hot coolant to pass through the valve, which will open the thermostat.

Attached to the water pump at its drive pulley is the cooling or engine fan. The fan is driven or turned at approximately crankshaft speed. It is used to move air through the radiator at idle or low speed and has very little effect at high speed. Most modern fans will have four or more blades to supply good air movement. On many engines a fan shroud is used. The fan shroud is attached to the radiator and surrounds the fan. This concentrates the airflow to add to the system's cooling efficiency. Fan shrouds are found on most automobiles equipped with air conditioning.

On some engines, the fan is attached to the pulley through a viscous fan drive clutch (Figure 3–83). The purpose of the clutch is to allow the fan to cut out or run at lower speed when there is enough cold air available. This reduces the noise of the fan and conserves some of the power it uses up, especially at higher road speeds.

The viscous drive is a type of fluid coupling. The amount of coupling is regulated by a bimetal thermostat, which responds to the temperature of the air stream. The thermostat moves a piston, or valve, which controls the amount of fluid available to couple the clutch. The term *viscous* refers to the thickness or viscosity of the fluid, which makes the coupling occur. Fan drive clutch kits are generally sold by many auto parts dealers.

Some late-model engines have electrically driven radiator fans. When replacement is necessary, follow the instructions in the service manual.

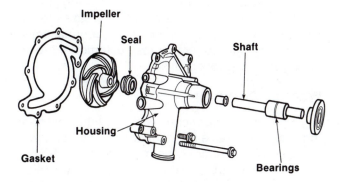

FIGURE 3-82 Impeller-type water pump

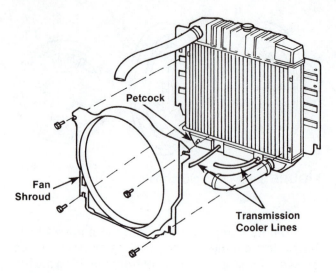

FIGURE 3-84 Components of downflow radiator system

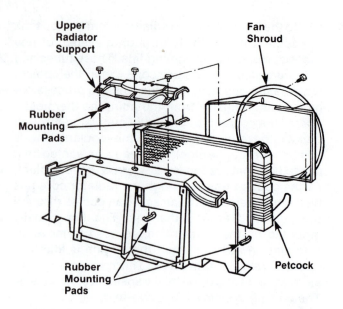

FIGURE 3-85 Components of crossflow radiator system

Radiators

There are two basic types of radiator designs, which are distinguished from one another by the direction of the coolant flow and location of the two tanks. In the downflow radiator, the coolant flows from the top tank downward to the bottom tank (Figure 3-84). In the crossflow type, the tanks are located at either side, and the coolant flows across the radiator core from tank to tank (Figure 3-85).

 SALES TALK ——————

When radiator replacement is necessary, measure only the core to determine the size (Figure 3-86). The core is the central part of the radiator (between the tanks and mounting brackets) and consists of parallel rows of tubes and fins. Measure the height from the top edge of the core to the bottom edge and from side to side to determine the width. To measure the thickness of the core, take a piece of straight wire, for example, and insert it through the core until the end becomes flush with the other side. Mark the other end of the wire with the thumb and forefinger (at the point where it is flush with the core), withdraw the wire, measure it, and obtain the core thickness.

Cool coolant flows from the radiator outlet tank to the engine, and hot coolant is returned to the inlet tank. Turbulence created by the water pump sometimes traps minute bubbles in the coolant, which can act to reduce its efficiency. This and the fact that hot

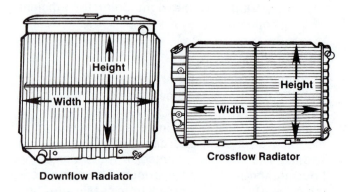

FIGURE 3-86 How radiators are measured

coolant expands necessitates an air or expansion space at the top of the radiator. Some cars have a separate expansion tank for the bubbles to dissipate and the coolant to expand into, thereby allowing the radiator to contain only coolant, not air. Coolant recovery systems do much the same thing and are available at auto supply stores to increase the efficiency of the radiator. When selling a replacement radiator, be sure to remind the customer of the advantages of a recovery system.

Pressure Radiator Cap

Pressure radiator caps are used with most cooling systems to raise the permissible operating temperature. Where there is no pressure, water boils at just over 200 degrees. Increasing the pressure raises the boiling point so that the system can operate in the 250-degree range. This has the effect of permitting a higher engine operating temperature.

The three basic pressure cap designs are:

1. *Constant Pressure Type (Figure 3-87A).* The lower seal or pressure valve is held closed by a spring until the coolant gets hot enough to build enough pressure to open it (within the preset pressure range). Pressure builds gradually as the coolant becomes hot.

2. *Pressure Vent Type (Figure 3-87B).* The lower seal or pressure valve is held closed as with the constant pressure type, but the vacuum release valve installed in the pressure valve is held open by a weight. The cap is open to atmospheric pressure through the vacuum release valve until the coolant gets hot enough to cause a surge of coolant or steam to hit it, causing it to

close to atmospheric pressure. From this point, the pressure vent cap works in the same way as the constant pressure cap.

3. *Closed System Type (Figure 3-87C).* These caps operate in the same manner as the constant pressure caps except that they are designed for a closed system (one where the heated coolant overflows from the cooling system into a reservoir and returns when the engine coolant has cooled and contracted). The closed design keeps the cooling system full at all times without losing coolant due to expansion or overflow. On closed systems, coolant is added to the reservoir instead of to the radiator, and the radiator caps are not designed to be removed. Some late-model cars use this type of radiator cap, and accessory kits are available to install them in other cars.

Safety pressure caps are also available with pressure release devices, usually a lever or a button, that allow the cooling system pressure to be released before removing the radiator cap.

💻 SALES TALK ────────

Frequently, a counterperson might be asked to check a radiator cap. There are pressure testers used to test the system for leaks. The tester (Figure 3-88) is attached to the radiator filler neck and pumped up to see if the system will hold pressure. The cap also is tested separately the same way and must be replaced if it does not hold its specified pressure (Figure 3-89).

Thermostats

As already noted, the thermostat restricts the circulation of the coolant until a preset temperature

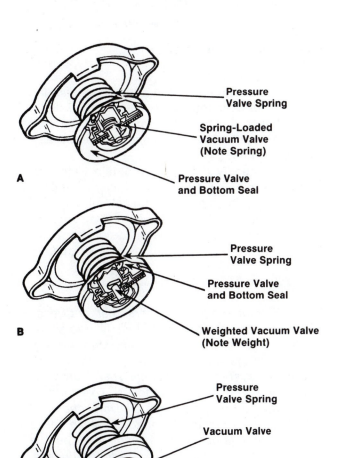

A

Pressure Valve Spring

Spring-Loaded Vacuum Valve (Note Spring)

Pressure Valve and Bottom Seal

B

Pressure Valve Spring

Pressure Valve and Bottom Seal

Weighted Vacuum Valve (Note Weight)

C

Pressure Valve Spring

Vacuum Valve

Pressure Valve and Bottom Seal

FIGURE 3-87 Radiator caps (A) constant pressure type; (B) pressure vent type; (C) closed system type

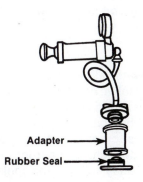

Adapter

Rubber Seal

FIGURE 3-88 Pressure tester

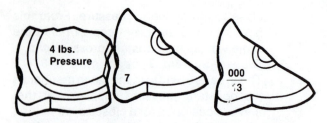

FIGURE 3-89 Pressure rating on radiator caps

is reached. This helps the engine warm quickly, which is very important to fuel economy, emission control, and keeping the oil clean of unburned blowby fuel contaminants.

The cooling system thermostat (or engine thermostat) is nothing more than a valve that is controlled by heat. It consists basically of a heat-sensitive, wax-filled pellet and a sleeve-type valve. When the pellet expands, the valve opens so that the coolant can circulate freely.

Thermostats are all set to operate at specific temperatures. The rating is the temperature where the thermostat starts to open. A 180-degree thermostat, for instance, will begin to open between 177 and 182 degrees, and will be fully open at about 200 degrees. There are two basic types of thermostats on the market today (Figure 3–90):

- Balanced sleeve
- Reverse popped

Both types function in the same manner, but always check the vehicle manufacturer's recommendation as to type and temperature range.

 SALES TALK _____

When replacing the thermostat, remind the customer that it is a good practice to replace the gasket, which seals the thermostat in place, positioned between the water outlet casting and the engine block. Generally, these gaskets are made of a composition fiber material and are die cut to match the thermostat opening and mounting bolt configuration of the water outlet.

Coolants

Although water alone was used for years as a coolant in automotive cooling systems, the fact that it has a high freezing point (32 degrees Fahrenheit), a relatively low boiling point (212 degrees Fahrenheit), evaporates, creates rust and corrosion, and leaves mineral deposits made it less than an ideal coolant alone. Today, to lower the coolant's freezing point and raise its boiling point, a chemical called ethylene glycol is added to water generally in a 50-50

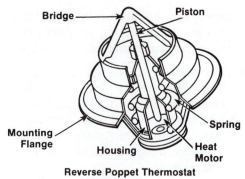

FIGURE 3-90 Thermostats

mixture to create antifreeze. Since ethylene glycol raises the boiling point of antifreeze to 227 degrees Fahrenheit and lowers its freezing point to –27 degrees Fahrenheit, in a 50-50 mixture it should also be called anti-boil.

Good quality antifreezes contain water pump lubricants to help maintain the efficiency of the water pump, rust inhibitors to keep scaly mineral deposits from building up inside the engine, and acid neutralizers to help protect the inside of the radiator, hoses, and heater core from chemical corrosion.

 SALES TALK _____

Remind your customers that antifreeze manufacturers recommend that the coolant be changed at least every two years. Many drivers, however, get careless and allow their cars' cooling systems to become rusty, dirty, or clogged with mineral deposits. To eliminate potential problems, there are several chemical solutions available designed to clean these contaminants out of the system. Most people refer to these solutions as radiator flushes. These solutions are generally poured into the system and allowed to circulate 10 to 15 minutes while the car is being run standing still. During this time, they dissolve much of the accumulated rust and grime and allow it to be drained away. A new mixture of water and antifreeze is then added to the system.

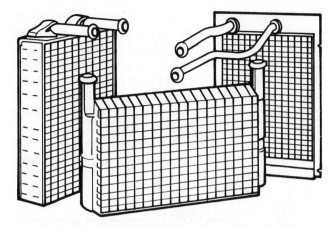

FIGURE 3-91 Heater cores and connections

Over time, as cars are driven under varying conditions, leaks can develop in the cooling system. If these leaks are not stopped, eventually enough coolant can be lost to cause an overheating problem. Although some leaks can be stopped by mechanical means (tightening a hose clamp), some leaks in the radiator or heater core cannot be repaired short of replacing the entire unit. Chemical additives called stop-leaks or sealers have been developed to stop these minor leaks from the inside. Most stop-leaks are chemical solutions that contain thousands of tiny particles that are carried by the coolant into the area of the leak. There, they tend to collect in sufficient quantities to seal the hole and stop the leak. The remaining particles are left in the coolant solution, ready to stop other leaks that might occur. Remember that these various cooling materials can mean a profit to the counter operation.

Other Cooling Sales Items

Radiator hoses, hose clamps, and fan belts are other cooling system items that are sold by auto parts stores. Remember, always select types and designs recommended by OEM. It is a good idea to recommend the replacement of belts in pairs so that new belts do not wear prematurely.

EXHAUST SYSTEM

The exhaust system is used to collect and discharge the exhaust gases caused by the combustion of the air/fuel mixture within the engine. The exhaust system found on today's automobile must also aid in the control of exhaust emissions.

The engine exhaust consists mainly of nitrogen (N_2). In addition it also contains carbon monoxide (CO), carbon dioxide (CO_2), water vapor (H_2O), oxygen (O_2) nitrous oxides (NO_x), and hydrogen (H_2). Also found are unburned hydrocarbons (HC). Three of these exhaust components, CO, NO_x, and HC, are major air pollutants. The emission of these components to the atmosphere must be controlled.

The catalytic converter (CC), mounted in the exhaust system, plays a major role in the emission control system. The converter works as a gas reactor, and its catalytic function is to speed up the heat producing chemical reaction between the exhaust gas components in order to reduce the air pollutants in the exhaust. There are two basic designs of catalytic converters.

1. *Monolithic.* This design, used on Chrysler and Ford vehicles (Figure 3-91), contains a ceramic honeycomb-shaped substrate, or carrier, on which the precious metal catalyst is deposited. The catalyst is usually platinum and/or palladium. This substrate is housed in a stainless steel casing that is insulated to produce a skin temperature of between 250 and 300 degrees Fahrenheit.

2. *Pelletized.* This design (Figure 3-92), used in GM and some AM vehicles, is wide and flat (looks like a hot water bottle) and contains a substrate made up of thousands of tin porous beads or pellets that are coated with the same precious metals as those found on the honeycomb or the monolithic design. This pellet design is housed in a stainless steel casing, which is also insulated.

There are three types of catalytic converters used on vehicles.

1. Single-bed, two-way, introduced on vehiin the mid 1970s.
2. Dual-bed, two-way

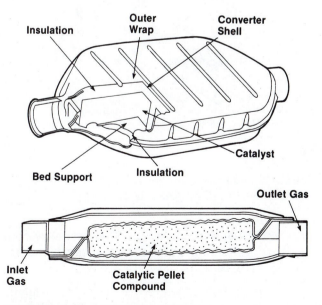

FIGURE 3-92 Catalytic converter

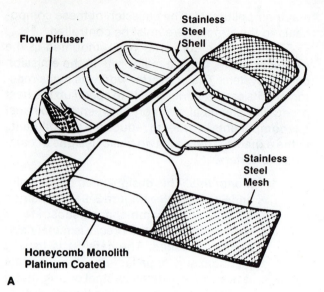

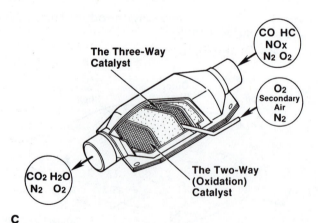

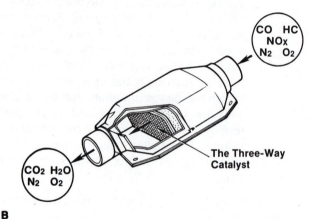

FIGURE 3-93 (A) Single-bed converter; (B) dual-bed converter; (C) three-way converter

3. Dual-bed, three-way. The three-way converters were introduced on cars in the late 1970s.

A single-bed converter has all of the catalytic materials in one chamber (Figure 3-93A). A dual-bed converter has a reducing chamber for breaking down NOx and an oxidizing chamber for oxidizing HC and CO. The exhaust passes through the reducing chamber first (Figure 2-93B). Additional oxygen is introduced into the oxidizing chamber to make it more efficient. A two-way converter oxidizes HC and CO; a three-way converter oxidizes HC and CO and reduces NOx (Figure 3-93C).

The converter itself is located either under the floor of the automobile (Figure 3-94) or in the exhaust manifold.

 SALES TALK _____

Because of constant change in EPA catalytic converter sales and installation requirements, check with the CC manufacturer or EPA for the latest data regarding replacement. Of course, when making a sale, always be sure to follow OE recommendations.

In some exhaust systems, an air injection works in conjunction with the catalytic converter. The air pump or aspirator forces air into the exhaust manifold so oxygen will combine with the hot exhaust gases to reduce HC and CO. The converter adds the extra kick needed to reburn the pollutants. The result is significantly lower HC and CO emissions.

In addition to the catalytic converter, the other important components of a typical exhaust system (Figure 3-95) that are generally sold by auto parts sales organizations include the following:

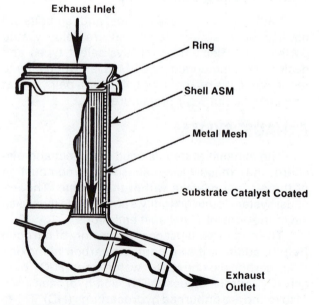

FIGURE 3-94 Floor model converter

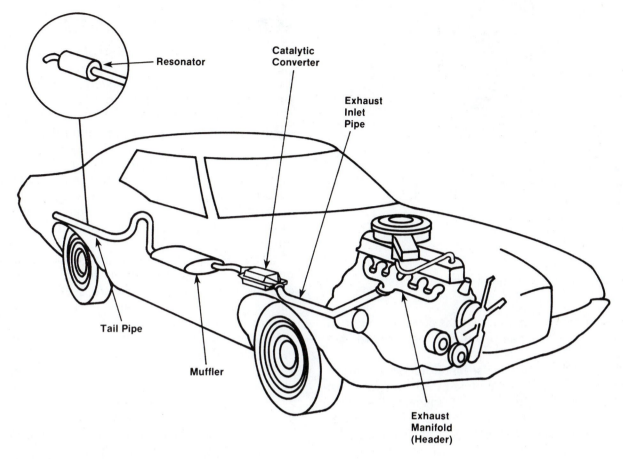

FIGURE 3-95 Exhaust system components

- Exhaust manifold and gasket
- Exhaust pipe, seal, and connector pipe
- Intermediate pipes
- Muffler
- Resonator
- Tail pipe
- Heat shields
- Clamps, gaskets, and hangers

EXHAUST SYSTEM OPERATION

The collection of exhaust gas begins in the exhaust manifold or header. This is the first part of the exhaust system and bolts directly to the cylinder head. The exhaust manifold is designed so that it will mate correctly with the exhaust ports of the cylinder head. All in-line engines are equipped with a single exhaust manifold. While V-design engines have one manifold for each cylinder head, in older automobiles, it is common for a V-design engine to be equipped with a complete dual exhaust system. Each side of the engine had its own system to collect and discharge exhaust. On late-model automobiles the two exhaust manifolds are connected by the exhaust or inlet pipe, which is manufactured in the form of a "Y." All exhaust gas is then discharged through a single muffler, tail pipe arrangement.

In all types of engines the exhaust gas flows from the manifold to the exhaust or inlet pipe. The inlet pipe then connects to a series of exhaust pipes made up of a catalytic converter, muffler, and interconnecting pipes.

The muffler helps quiet the sound of combustion. Some large engine exhaust systems also have a second muffler or resonator. This is used to further muffle the noise of the exhaust. The muffler is simply a can or container with an inlet and outlet pipe attached. Inside the container is a series of baffles designed to deflect the exhaust within the container. The baffeling or deflecting of the exhaust quiets the sound of combustion as it leaves the tail pipe.

Manifold Heat Function

There are two "helping" functions of the exhaust systems on some engines, which involve the transfer of exhaust gas heat to the induction system.

With modern emission, the exhaust manifold provides a "stove" to heat some of the air taken into the carburetor air cleaner. A hot-air pipe or hose

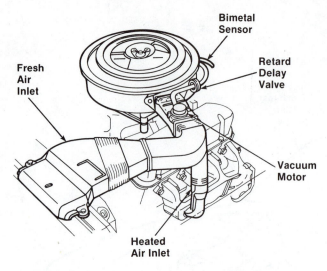

FIGURE 3-96 Vacuum-controlled system

leads up from the manifold to the air cleaner snorkel and carries heated air to mix with the normal intake air.

Some engines may use manifold heat to warm a section of the intake manifold by directing some of the exhaust gas through it. This is done through a butterfly type heat riser or exhaust control valve located between the exhaust pipe and manifold. Opening and closing of the valve is controlled by a thermostatic spring or vacuum motor (Figure 3-96). On thermostatic spring type valves, a weight is installed on the valve shaft to minimize valve flutter as the exhaust gas discharges from the cylinders. This system is used to warm the intake manifold for better fuel vaporization during engine warm-up.

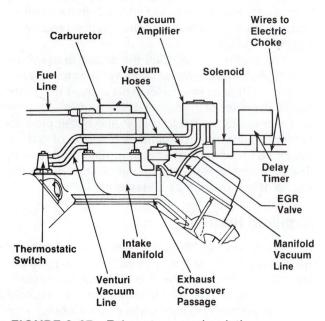

FIGURE 3-97 Exhaust gas recirculation

EMISSION CONTROL SYSTEM

Of the six major components of a vehicle's emission control system, four—positive crankcase ventilation (PCV), evaporative emission controls, heated air intake, and air injector and catalytic converter (CC)—have been covered in other systems. This leaves only exhaust gas recirculation (EGR) to be covered here.

EGR reduces NO_x by diluting the air/fuel mixture and lowering combustion temperatures. Recirculating a small amount of exhaust back into the intake manifold keeps combustion temperatures below the 2500 degree Fahrenheit NO_x formation threshold.

The heart of the system is the EGR valve, which opens to allow engine vacuum to siphon exhaust into the intake manifold (Figure 3-97). The EGR valve consists of a poppet valve and a vacuum actuated diaphragm. When ported vacuum is applied to the diaphragm, it lifts the valve off its seat. Intake vacuum then siphons exhaust into the engine. Like a PCV valve, the EGR valve is a kind of calibrated vacuum leak.

On many late-model vehicles, a positive back pressure or negative back pressure EGR valve is often used. This type of valve requires a certain level of back pressure in the exhaust before it will open when vacuum is applied.

The EGR valve uses ported vacuum as its primary vacuum source. Ported vacuum is used because EGR is not needed at idle. (NO_x levels are low at idle.) If the EGR valve sticks open at idle or if the vacuum line is mistakenly connected to a source of manifold vacuum, it will cause a rough idle.

The vacuum control plumbing to the EGR valve usually includes a temperature vacuum switch (TVS) or solenoid to block or bleed vacuum until the engine warms up. On some late-model vehicles with computerized engine controls, the computer actuates the solenoid to further modify the opening of the EGR valve.

 **SALES TALK** _____

Aftermarket sales of replacement EGR valves are growing rapidly and are now in excess of 3.5 million units a year. Add in all the EGR-related hardware and plumbing such as ported vacuum switches, back pressure transducers, vacuum amplifiers, vacuum pumps, control solenoids, and counterpersons begin to appreciate the significant sales opportunity in knowing about EGRs.

ELECTRONIC ENGINE CONTROLS (EEC)

Electronic engine controls have given the automotive manufacturer the capability to meet federal government regulations by controlling various engine systems accurately. In addition, electronic control systems have fewer moving parts than old style mechanical and vacuum controls. Therefore, the engine can maintain its calibration almost indefinitely. As an added advantage, the EEC system is very flexible. Because it uses microcomputers, it can be modified through programming changes to meet a variety of different vehicle engine combinations or calibrations. Critical quantities that describe an engine can be changed easily by changing data that is stored in the system's computer memory. In other words, the EEC system is an assembly of electronic and electromechanical components that continuously monitor engine operation to meet emission, fuel economy, and driveability requirements. Because of warranties required, most parts of the EEC system are available at a dealership parts store.

DIESEL ENGINES

Gasoline and diesel engines are similar in basic construction. The main difference is the fuel system and the means of initiating combustion. The two types are different in combustion; what is desirable in a diesel is totally undesirable in a gasoline engine.

With a gasoline engine, vaporized fuel and air are mixed before going into the cylinder for combus-

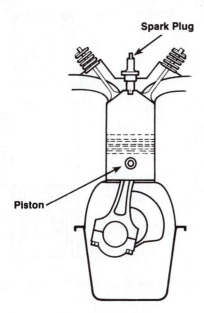

FIGURE 3-98 Spark ignition

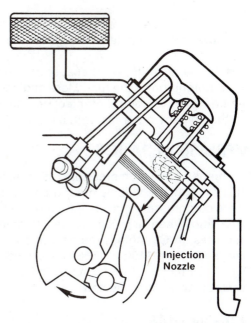

FIGURE 3-99 Compressed ignition engine

tion. Combustion is started by a spark ignition at the spark plug.

The diesel engine brings air into the cylinder, and it is compressed to increase its temperature. This heat of compression is significantly higher than in a gasoline engine because of the high compression ratios commonly used on diesels. Fuel is then injected under high pressure into the heated air of the cylinder. The vaporized fuel ignites because of the high temperature of the compressed air. Diesel combustion is started by the same conditions causing knock in a gasoline engine.

Diesel engines are called *compression ignition (CI)*; gasoline engines are called *spark ignition (SI)*. In the CI engine, knock is desirable. In contrast, in an SI engine, knock is undesirable and leads to destruction of the engine. Even though both engines are constructed essentially the same, this difference in the effect of knock is not contradictory. Knock in an SI engine occurs at the end of combustion. Knock in a CI engine occurs at the beginning of combustion. This is the key difference.

In an SI engine (Figure 3-98), combustion is completed while the piston is at the upper end of travel. This means the volume of the mixture stays about the same during most of the burning process. When the piston moves down and the volume increases, there is little additional combustion to maintain pressure.

In a CI engine (Figure 3-99), there is continuous combustion during most of the power stroke. The expansion of gases increases at the same rate the volume of the cylinder increases as the piston is forced down. The pressure from combustion remains approximately constant throughout the power stroke.

Diesels operate efficiently because of the injection and autoignition of the fuel. More engineering has been devoted to control of injection and autoignition than on any other factor in a diesel engine.

COMPRESSION RATIOS

CI engines have compression ratios ranging from 18:1 to 23:1. The volume of air drawn into the cylinder is compressed to a small fraction depending on the ratio. When air is compressed its temperature increases. In CI engines the heat of compression can range from 1100 to 1500 degrees Fahrenheit. This high temperature in the cylinder ignites the diesel fuel when it is injected.

INJECTION TIME

With this elevated temperature, the injection system must be able to inject the fuel into the cylinder within 0.001 to 0.006 second. The liquid fuel sprayed into the cylinder quickly vaporizes, helped by the intense heat.

At the beginning of the injection cycle, the fuel instantly vaporizes then autoignites, resulting in audible knock. With this combustion started, fuel injection continues to carry on combustion as the piston travels down. At the beginning of combustion, temperatures of 3500 degrees Fahrenheit are reached. This further vaporizes and burns the injected fuel. The shape of the fuel spray, turbulence in the combustion chamber, beginning and duration of injection, and the chemical properties of the diesel fuel all affect power output.

ADVANTAGES AND DISADVANTAGES

The basic advantages of the diesel engine are low fuel consumption (greater thermal efficiency), reduced fire hazard, and lower emission levels.

- *Low Fuel Consumption.* The three primary factors for the diesel engine's lower fuel consumption are air/fuel ratio, compression ratio (Figure 3–100), and low pumping losses.
- *Reduced Fire Hazard.* While not as great a fire hazard as gasoline, there is no reason to handle diesel fuel any less carefully than gasoline.
- *Lower Emission Levels.* The high air/fuel ratio lowers HC and CO emissions. The diesel also produces less NO_x.

The following disadvantages of the diesel engine can be attributed to the diesel's characteristics or to the subjective opinion of the owner:

- *Diesel Engine Construction Costs Are Higher.* Because the diesel puts more stress on its compounds, the engine must be constructed from special materials. The quality of the material must be exact, and the parts must be fitted together with very little tolerance for error. This means assembly costs are higher.
- *Different Maintenance Procedures.* Because the diesel engine's design, service, and maintenance procedures are different, the technician must be more precise in making repairs.
- *Cold Weather Starting.* Diesels use the heat of compression to ignite the fuel. The colder the air temperature, the harder it is to build up enough heat for ignition. To aid cold weather starting, manufacturers have added special starting aid packages.
- *Engine Noise.* Diesel engines produce a knock, particularly at idle, that is very noticeable.
- *Exhaust Smoke and Odor.* Anyone who has followed a diesel, particularly one with a malfunction, knows that the diesel produces smoke and an odor all its own.

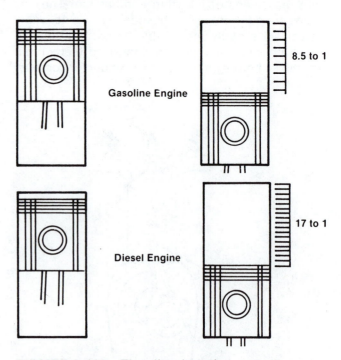

FIGURE 3-100 The diesel engine operates on a much higher compression ratio.

REVIEW QUESTIONS

1. Which of the following statements about engines is not true?
   a. The engine provides the rotating power to drive the wheels through the transmission and driving axle.
   b. Only gasoline engines, not diesel, are classified as internal combustion engines.
   c. The combustion chamber is the space between the top of the piston and the cylinder head.
   d. For the explosion in the cylinder to take place completely and efficiently, air and fuel must be combined in the right proportions.

2. Which stroke in the four-stroke cycle begins as the compressed fuel mixture is ignited in the combustion chamber?
   a. power stroke
   b. exhaust stroke
   c. intake stroke
   d. compression stroke

3. What is compression ratio?
   a. the volume the cylinder holds between top dead center and bottom dead center
   b. diameter of the cylinder
   c. cylinder arrangement
   d. the amount the air/fuel mixture is squeezed

4. When a customer talks about the engine component that is the mounting point for the carburetor or fuel injection, Counterperson A believes the customer is referring to the camshaft. Counterperson B believes the customer is referring to the intake manifold. Who is right?
   a. Counterperson A
   b. Counterperson B
   c. Both A and B
   d. Neither A nor B

5. When referring to a camshaft, what is flank?
   a. camshaft base
   b. ramp between the cam base and its nose
   c. highest point of the cam lobe
   d. distance the lobe will move the lifter

6. Which of the following is a type of lifter?
   a. solid
   b. mechanical
   c. hydraulic
   d. all of the above

7. What is piston slap?
   a. grooves on the side of the piston
   b. force applied to the piston
   c. noise made by the piston when it contacts the cylinder wall
   d. a piston ring

8. What is the function of the connecting rod?
   a. transmits the pressure applied on the piston to the crankshaft
   b. controls oil flow
   c. cools the engine
   d. transmits pressure to the piston

9. Which engine system removes burned gases and limits noise produced by the engine?
   a. exhaust system
   b. emission control system
   c. ignition system
   d. fuel system

10. What engine component meters, atomizes, and distributes the fuel throughout the air flowing into the engine?
    a. intake manifold
    b. carburetor
    c. fuel pump
    d. metering valve

11. Counterperson A looks at the top of the oil can to note oil viscosity. Counterperson B looks at the top of the oil can to note the design use for the oil. Who is right?
    a. Counterperson A
    b. Counterperson B
    c. Both A and B
    d. Neither A nor B

12. What cooling system component blocks off circulation in the system until a preset temperature is reached to speed engine warm-up?
    a. radiator
    b. water jacket
    c. temperature indicator
    d. thermostat

13. What component of both the exhaust and emission control systems works as a gas reactor to speed up the heat-producing chemical reaction between the exhaust gas components to reduce pollutants in the exhaust?
    a. catalytic converter
    b. air injector
    c. crankcase ventilation
    d. exhaust gas recirculation

14. Which of the following is an advantage of electronic engine controls?
    a. fewer moving parts
    b. flexibility
    c. capability to meet government standards
    d. all of the above

15. Which engine system ensures that the engine gets the right amount of both air and fuel needed for efficient operation?
    a. ignition system
    b. fuel system
    c. exhaust system
    d. emission control system

CHAPTER FOUR

PRODUCT KNOWLEDGE: OTHER AUTOMOTIVE SYSTEMS

Objectives

After reading this chapter, you should be able to:

- Explain the electrical systems of an automobile, including its subsystems and their components.
- Explain the operation of an air conditioning system in an automobile and the various related items a counterperson might sell to someone servicing the system.
- Name the components of the steering system and how a counterperson might help a customer understand its operation.
- Name the parts of a suspension system.
- Explain powertrain and transmission systems, differentiating between manual and automatic transmissions.
- Demonstrate how product knowledge helps a counterperson make related sales.
- Name the components of a brake system.
- Name some of the body and paint equipment a counterperson might have to work with.
- Explain how product knowledge can help in the sale of substitute items.
- Name four sources of the most helpful information for a counterperson.

Product knowledge of the various other automotive systems is as important for counter personnel as is knowledge of the engine and its systems. Customers will have questions and need advice on parts for systems such as the following:

- Electrical system
- Air conditioning and heating systems
- Steering and suspension systems
- Drivetrain and transmission systems
- Brake system
- Tires and wheels
- Body finishing systems

ELECTRICAL SYSTEM

Automobiles have many circuits that carry electrical current from the battery to individual components. The total electrical system (Figure 4–1) includes such major subsystems as:

- Starting system
- Charging system
- Lighting circuit
- Other electrical circuits, including horn, windshield wipers, and flashers
- Ignition system

The latter system is covered in Chapter 3.

THE STARTING SYSTEM

The starting system (Figure 4–2) operates as follows: When the ignition key is turned to the START position, electric current is sent to the solenoid and battery voltage is supplied directly to the starter motor. The starter motor then turns a flywheel mounted on the rear of the crankshaft that starts all engine parts in motion. The ignition system provides a spark to the spark plugs that ignites the air/fuel mixture from the carburetor. If all components are in good working condition, the engine

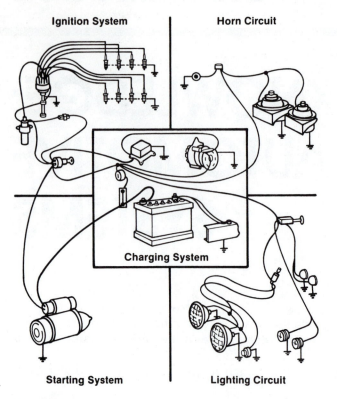

Ignition System

Horn Circuit

Charging System

Starting System

Lighting Circuit

FIGURE 4-1 Simplified electrical system

should start immediately. The essential components of the starting system are as follows:

1. *Battery.* In this system the battery uses stored electrical energy to supply the starter with its requirements. A parts person will be doing himself and the customer a big favor by making sure the customer has

a heavy-duty battery. Most people replace the original with a cheaper, less powerful battery. As a result, alternator, starter, or other electrical problems can develop. The battery is an integral part of the charging system, and as the car gets older and the resistance in the wiring increases, the replacement battery should be heavier rather than lighter.

2. *Neutral Safety Switch.* This switch allows the starting system to be operated only when the automatic transmission gearshift lever is in the NEUTRAL or PARK position. This switch is also called a starting safety switch.

3. *Ignition Switch.* The ignition switch has five positions: accessories, lock, off, on, and start. The first four positions of the ignition switch will automatically stay in position when the key is turned there. The START position, however, is like a momentary contact switch; it has to be held there to crank the engine.

4. *Solenoid.* The solenoid magnetic relay switch connects the battery to the starter by principles of magnetism. Electrical windings are wrapped around a hollow core and an iron plunger is placed partially in the area. When the key is turned to the START position, a magnetic field is created that pulls the plunger into the core. When the plunger touches two contacts, it connects the circuit between the battery and starter. Some solenoids, such as those

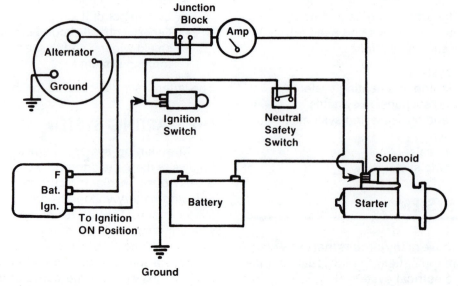

FIGURE 4-2 Starting system

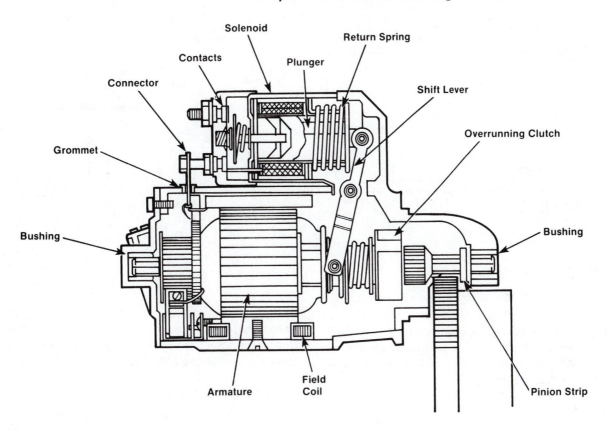

FIGURE 4-3 Starter motor

used on General Motors, Chrysler, and some late-model Ford products, are also used to engage the drive mechanism. The plunger is connected to a shift lever that engages the starter drive gear with the flywheel.

5. *Starter Motor.* The starter or "cranking" motor is a small but powerful electric unit that turns the engine crankshaft so that the engine can start and operate under its own power. It is usually located in the lower rear portion of the cylinder block. It drives the car's engine through a gear mounted on the starter motor's armature shaft (Figure 4-3). It is designed to operate under great overload and produce a high horsepower for its size. It can do this for a short period of time only, because a high current must be used to produce such power. If cranking motor operation is continued for any length of time, heat will cause serious damage. For this reason, the cranking motor must never be used for more than 30 seconds at any one time and cranking should not be repeated without waiting for a pause of at least 2 minutes to permit the heat to escape.

6. *Electrical Wiring.* The wiring in a car is of different types and gauges designed to do specific jobs. When replacing wire, use a piece of the same type, length, and gauge. If the new wire is longer or its gauge too small, it will increase the resistance or give poor conductivity, which can have a detrimental effect on some parts of the system.

THE CHARGING SYSTEM

The charging system performs two basic tasks:

1. Maintains the battery's state of charge
2. Provides electrical power for the ignition system, air conditioner, heater, lights, radio, and all electrical accessories.

In addition to the battery, the charging system includes (Figure 4-4):

- Alternator or generator
- Voltage regulator
- Indicator light
- Necessary wiring

Turned by a drive belt driven by the engine crankshaft, the charging unit—either the generator or alternator—converts mechanical energy into

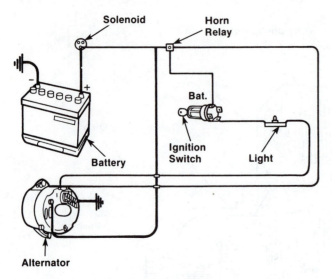

FIGURE 4-4 Charging system

electrical energy. When the electrical current flows into the battery, the battery is said to be charging. When the current flows out of the battery, the battery is said to be discharging.

Generator

The generator—direct current type—is driven by a V-belt from the engine crankshaft pulley (Figure 4-5). This drive belt must be properly adjusted and in good condition in order to charge the battery. If the belt slips or breaks, the generator does not turn and no current is produced.

Alternator

On all late-model vehicles, the alternator—alternating current type—has replaced the generator (Figure 4-6). Although the alternator operates in much the same way as the generator, it has one big

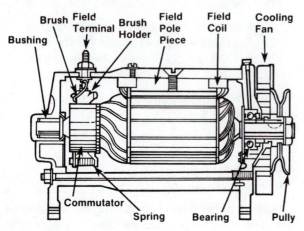

FIGURE 4-5 Generator

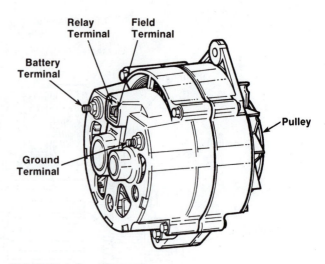

FIGURE 4-6 Alternator

advantage: Even when the engine is idling, the alternator charges to some degree. The conventional generator does not begin to charge the battery until the engine is above idling speed.

Voltage Regulator

An alternator has a self-limiting feature that controls current flow in the charging circuit. It does not, however, control voltage output. The function of the voltage regulator is to limit the output voltage to a preset value by controlling the strength of the alternator field current. Without the regulator, the voltage generated at high engine speeds would overcharge the battery.

Most parts operations carry three types of voltage regulators:

- Electromechanical
- Transistor electronic
- Integral electronic

Check the service manual to find the proper type of regulator for the customer's vehicle.

Charging Indicator

Like votlage regulators, there are three charging indicators in automobiles:

- Indicator light
- Ammeter
- Voltmeter

LIGHTING SYSTEMS

Automotive lighting systems have become increasingly more sophisticated. Headlights and tail lights have grown into multiple-light systems, turn signal indicators have changed from optional ac-

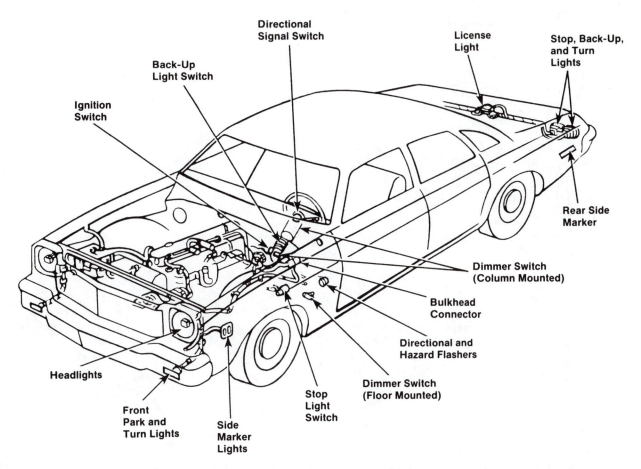

Directional
Signal Switch

Back-Up
Light Switch

Ignition
Switch

License
Light

Stop, Back-Up,
and Turn
Lights

Rear Side
Marker

Dimmer Switch
(Column Mounted)

Bulkhead
Connector

Directional and
Hazard Flashers

Dimmer Switch
(Floor Mounted)

Stop
Light
Switch

Side
Marker
Lights

Front
Park and
Turn Lights

Headlights

FIGURE 4-7 Typical lighting systems

cessories into standard equipment and evolved further into emergency and hazard warning systems. Indicator lights on the dashboard commonly warn of failure or improper operation of the charging system, seat belts, brake system, parking brakes, door latches, and other items on the vehicle (Figure 4-7).

Headlights have both a high and a low beam. The two-light systems have one headlight on each side, each of which has a high-beam and a low-beam filament. The four light systems have two lights on each side. Two of these should be lit for low beam and all four for high beam. The four-light system uses two distinctly different headlights (Figure 4-8). The headlights used in the two-light system and the four-light system are different in size. Until recently, all headlights were round, but recent styling developments created a need for a rectangular light, which is also supposed to be more efficient.

Halogen lights also may be used as headlights. A halogen light contains a small quartz-glass bulb, inside of which is a filament surrounded by halogen gas. The small, gas-filled bulb fits within a larger reflector and lens element. Halogen lights produce a whiter, brighter light than conventional sealed-beam headlights (Figure 4-9).

 SALES TALK _____

There are various methods of identifying bulbs, such as #1, #2, and the Halogen marking molded into the front lens. Type #1 is high beam with two connections on the back; type #2 is both high/low beam with three connections. A counterperson should be aware of the possible locations of the #1 and/or #2 applications. Halogen can be substituted for regular sealed beams, but it is suggested that this be done in pairs because Halogen is a whiter, brighter light pattern. The regular bulbs have a universal part numbering system. This means if a bulb is marked 1157 Tungsol, it would be the same basic part number with other bulb manufacturers such as G.E., Sylvania, or Toshiba.

Bulbs used in most other lighting fixtures fit into sockets and are held by spring tension or mechanical force (Figure 4-10). Bulbs are coded with num-

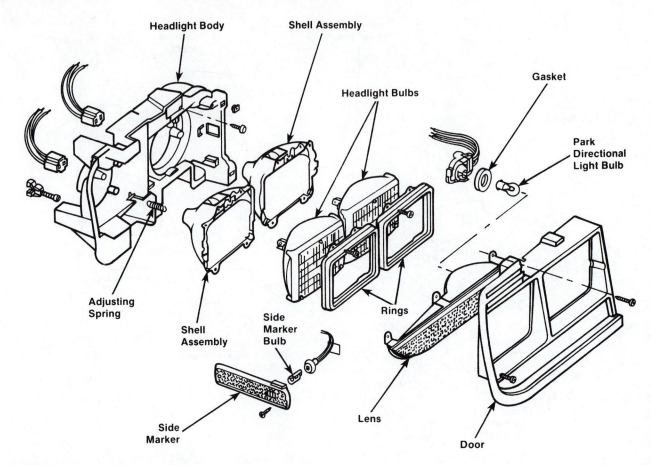

FIGURE 4-8   Typical headlight mounting components

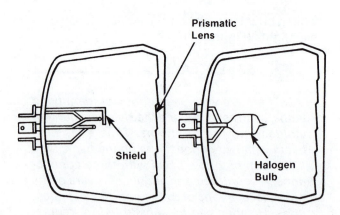

FIGURE 4-9   Halogen bulb headlight

bers for replacement purposes. Bulbs with different code numbers might appear physically similar but have different wattage ratings. Smaller bulbs can simply be pushed into a socket. Most bulbs must be pushed inward and turned counterclockwise to be removed. Replacement is accomplished by pushing in and turning clockwise.

Light systems normally use one wire to the light, making use of the car body or frame to provide the

ground back to the battery. Since many of the manufacturers have gone to plastic sockets and mounting plates to reduce weight, many lights must now use two wires to provide the ground connection. Some double-filament lamps use two "hot" wires and a third ground wire.

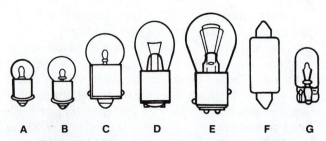

A, B—Miniature Bayonet for Indicator and Instrument Lights
C—Single-Contact Bayonet for License and Courtesy Lights
D—Double-Contact Bayonet for Trunk and Underhood Lights
E—Double-Contact Bayonet with Staggered Indexing Lugs for Stop, Turn Signals, and Brake Lights
F—Cartridge Type for Dome Lights
G—Wedge Base for Instrument Lights

FIGURE 4-10   Common automotive bulbs

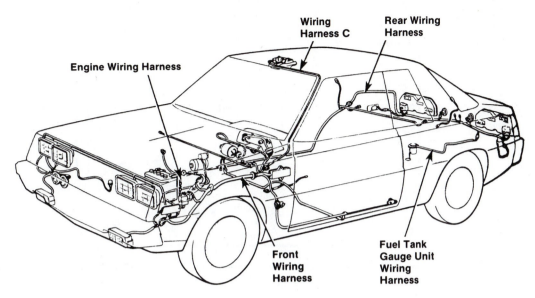

FIGURE 4-11 Wiring harness groupings

Wiring

Wire used in cars comes in an assortment of sizes and colors (for identification purposes). The sizes are referred to as gauges. The smaller the gauge number, the heavier the wire and the greater the load it can carry. The largest wires, used between the battery and starter or to ground on the block, the wires between the alternator and regulator, to and from the ammeter (if so equipped), and to the fuse block and the headlight switch, are normally 8-gauge or 10-gauge. Accessories that draw a lot of power use 12-gauge; 14-gauge is used for most other accessories. Smaller lights usually use 16-gauge or 18-gauge. The more power the accessory or light draws, the heavier the wire should be. If the circuit uses an unusually long length of wire (10 feet or more), then the next larger gauge size (next lower number) than indicated above is used.

Wires of different sizes run throughout a vehicle. Wires are grouped together in harnesses (Figure 4-11). A harness is a bundled group of wires. At specific points, wires lead from a harness to electrical units. Wiring harnesses can contain many wires, especially in the areas of the dash and fire wall.

A wiring diagram (Figure 4-12) always indicates the size and color of the wire used in the circuit. For example, 12R means the wire is 12-gauge and colored red; 16R/Y means the wire is 16-gauge and red with a yellow stripe or tracer. A wiring diagram can be found in the vehicle's service manual.

Many newer models use printed circuits (Figure 4-13), which are thin sheets of nonconductive plastic material on which conductive metal has been deposited. Parts of the metal are etched, or eaten away, by acid. The remaining metal lines form conductors for separate circuits. A wiring connector can be plugged into a common point. Power or ground circuits are completed through connected wires.

Circuit Protection

The vehicle must be protected from fire hazards that occur if a powered circuit is accidentally shorted or grounded. Fuses, circuit breakers, and fusible links can be used to protect circuits.

All the fuses are contained in a device called a *fuse block* (Figure 4-14). The fuse block is located on the inside of the fire wall, usually on the driver's side of the car. It is marked with the name of the circuit or circuits each fuse protects and usually the amp rating of the correct fuse. Inline fuses are used in some circuits, generally, for installed accessories such as stereo equipment, fog/road lights, and CB radios.

In addition to fuses, some circuits are protected by circuit breakers (Figure 4-15), which perform the very function that their name implies. The circuit breaker opens the circuit briefly during periods of overload. The headlight circuit is normally protected by a circuit breaker. When a short occurs, the breaker repeatedly cycles on and off. The circuit breaker does not leave one without lights as a fuse would. Because an overload does not destroy a circuit breaker, they are rarely in need of replacement.

Manufacturers often use a "fusible link" as a type of circuit protection. This technique, however, is used in circuits that are not normally fused. Most of these are in the engine compartment. These links are part of the wire. They are usually the same color and approximate size as the wire they protect. In

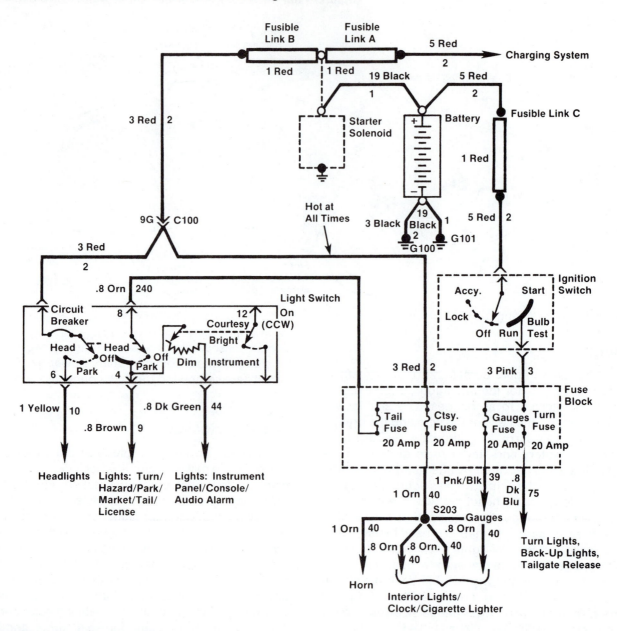

FIGURE 4-12 Wiring diagram

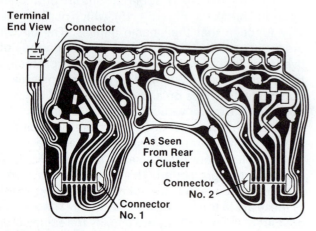

FIGURE 4-13 Printed circuit

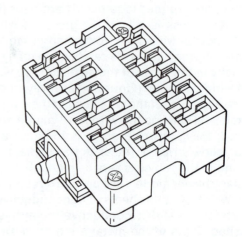

FIGURE 4-14 Typical fuse block

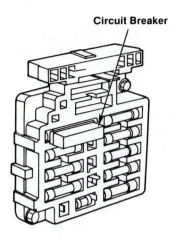

FIGURE 4-15 Typical circuit breaker in fuse block

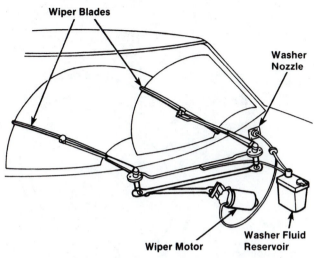

FIGURE 4-17 Windshield wipers and washers

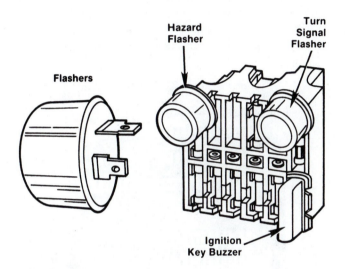

FIGURE 4-16 Flasher and where it is installed

some cases, however, they might be several sizes smaller and are designed to disintegrate when overloaded. Check the appropriate service manual for location, testing, and replacement procedures.

As shown in Figure 4-16 flasher turn signals or hazard warning lights are usually mounted under the dash on the driver's side; sometimes they are mounted to the fuse block or are taped to a wiring harness.

OTHER ELECTRICAL CIRCUITS

There are other electrical circuits in modern vehicles that operate such accessories as

- Power seat positioners
- Power window and locks
- Power positioned outside mirrors
- Automatic headlight dimmers
- Cruise control

Other electrical devices include radios, cassette players, speaker systems, chimes, buzzers, graphic displays, analog instruments, and computer commands. Speedometers, odometers, and various vacuum/pressure gauges are usually operated mechanically rather than electrically. There are two electric circuits in every automobile: windshield wipers and washers and horn.

Windshield Wipers and Washers

A typical windshield wiper operates on a small single- or multi-speed electric motor. A switch on the steering wheel assembly or dashboard activates the motor. The spray washer generally has its own motor, plastic container, or reservoir and pump, which forces liquid through tubing to a nozzle (Figure 4-17). The nozzles spray the liquid washer on the windshield.

Horn

Most horn systems are controlled by relays. When the horn button, ring, or padded unit is depressed, electricity flows from the battery through a horn lead, into an electromagnetic coil in the horn relay to the ground (Figure 4-18). A small flow of electric current through the coil energizes the electromagnet, pulling a movable arm. Electrical contacts on the arm touch, closing the primary circuit and causing the horn to sound.

AIR CONDITIONING AND HEATER SYSTEMS

An air conditioning unit works on the simple principle that when a liquid is converted to a gas, it

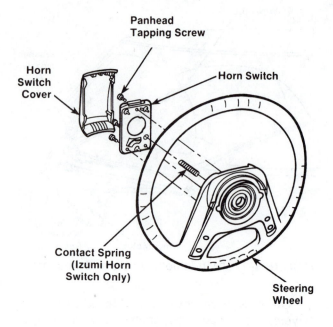

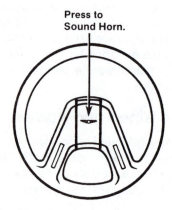

FIGURE 4–18 Horn system

3. Moisture and contaminants are removed by the filter/dryer, where the cleaned refrigerant is stored until it is needed.
4. The expansion valve converts the high-pressure liquid changes into a low-pressure liquid by controlling its flow into the evaporator.
5. Heat is absorbed from the air inside the passenger compartment by the low-pressure, low-temperature refrigerant, causing the liquid to vaporize.
6. The refrigerant returns to the compressor as a low-pressure, higher temperature vapor.

Refrigerants are the chemicals that transfer heat by absorbing it during evaporation and releasing it during condensation. Refrigerant 12, more commonly known as Freon®, is used in cooling applica-

absorbs heat from its surroundings. When it is reconverted from this gaseous state back to a liquid, it gives up this heat. The air conditioning system consists of the following components:

- Compressor
- Condenser
- Receiver/dryer
- Refrigerant control
- Evaporator

In the basic five-part air conditioner system, the heat is absorbed and transferred in these steps (Figure 4–19).

1. Refrigerant leaves the compressor as a high-pressure, high-temperature vapor.
2. By removing heat via the condenser, the vapor becomes a high-pressure, lower temperature liquid.

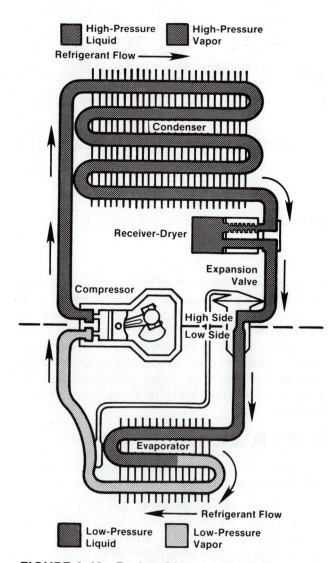

FIGURE 4–19 Basic refrigerant (R-12) flow cycle

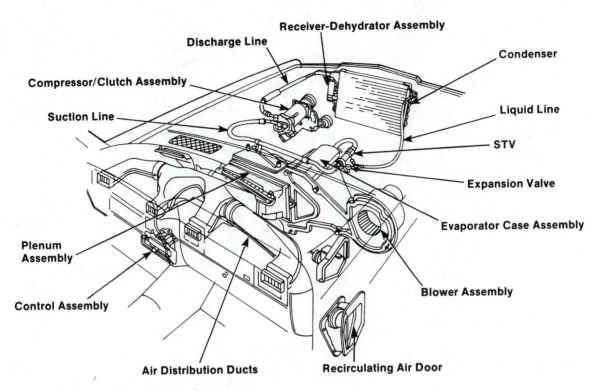

Receiver-Dehydrator Assembly
Discharge Line
Compressor/Clutch Assembly
Suction Line
Condenser
Liquid Line
STV
Expansion Valve
Evaporator Case Assembly
Plenum Assembly
Control Assembly
Blower Assembly
Air Distribution Ducts
Recirculating Air Door

FIGURE 4-20 Air conditioner components

tions because of its unique properties. It is a relatively stable chemical with a low boiling point (-21.7 degrees Fahrenheit). Refrigerant 12 or R-12 is available in 30- and 60-pound cylinders and in 14-ounce cans. In its container, under pressure, it is in liquid form. Refrigerants should not be stored at temperatures above 120 degrees Fahrenheit.

AIR CONDITIONER OPERATION

The compressor is nothing more than a pump (Figure 4-20) driven by the crankshaft via a V-belt. It picks up a gaseous refrigerant from the evaporator inside the car and compresses it. The compressor uses an electromagnetic clutch to permit it to be turned off when not needed.

The R-12 is metered into a cooling coil (evaporator), normally located inside the car on the fire wall, at about 30 psi. The R-12 is in liquid form at this point. Its boiling point at 30 psi is just above the freezing point of water. The R-12 therefore tends to boil, absorbing heat from the coil.

A blower forces either inside or outside air, depending on the setting, through the evaporator. The air then passes into the passenger compartment. As the air passes through the evaporator, heat and moisture are removed.

The R-12 boils inside the evaporator then passes into the compressor as a gas, where its pressure is increased. R-12 pressure as it leaves the compressor is usually 200 psi. The R-12 then enters the condenser, a heat-exchanging coil resembling a radiator and usually located in front of the car's radiator. The high pressure caused by the compressor is put to work at this point and raises the boiling point of the R-12 to over 150 degrees Fahrenheit. When outside air is passed over the thin tubes and fins of the condenser, it cools and changes the R-12 back to a liquid, losing the heat it picked up from the interior of the car.

The liquefied R-12 then enters the receiver (dryer), a small black tank located next to the condenser or on one of the fender wells. This unit has the job of separating liquid refrigerant from any gas that might have left the condenser, and also filters the refrigerant and absorbs any moisture it might contain. It incorporates a sight glass, in most systems, that allows the R-12 to be visually checked for the presence of bubbles. Remember this sight glass—it is invaluable in troubleshooting air conditioner problems.

R-12 then flows through a liquid line to the expansion valve. This valve is usually located near the evaporator, on or near the fire wall. The valve, shaped like a mushroom on some systems, controls the flow of R-12 to the evaporator. It provides only the amount of flow that the evaporator can handle.

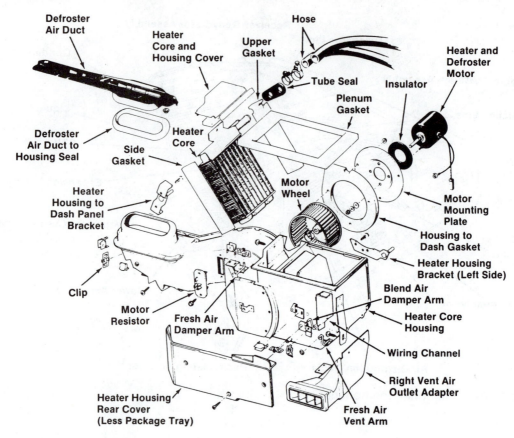

FIGURE 4-21 Components of a car's heating system

 SALES TALK _____

In addition to selling refrigerant, the counterperson should keep in mind three other products.

1. *Refrigerant Oil. This very thin oil lubricates and helps seal the air conditioning system. During the refining process, all of the solid and liquid contamination is removed. Refrigerant oil is available in quart and gallon bottles (for adding oil to a system while it is apart), and in 4- and 14-ounce oil charges. The oil charge is a combination of oil and refrigerant that is used to add oil to a system that is already charged.*
2. *Flush Solvent. Flush solvent is used with a flush gun to clean contamination out of the system. This chemical is used for flushing the air conditioning system because it evaporates quickly.*
3. *Leak Detection Chemicals. Leak detection chemicals make it possible to trace air conditioning leaks visually.*

CAUTION: Handle R-12 with care.

HEATER OPERATION

The heater is a comfort control item especially in colder climates. It may be part of the air conditioning system or a separate item. Actually, the heater core could be considered a miniature version of the radiator (Figure 4-21). That is, as hot coolant flows through the heater, a fan blows air over the tubes, warming it and delivering it to the passenger compartment. The blower fan, located in the heater housing, forces air through the heater core and into the passenger compartment.

The air heating distributor system is a duct system. Outside air enters the system through a grill, usually located directly in front of the windshield, and goes into a plenum chamber where rain, snow, and some dirt is separated from it. The air from the plenum is directed through the car's heater core, through the air conditioner evaporator, or into a duct that runs across the fire wall of the car. Outlets in the duct direct the airflow into the passenger compartment (Figure 4-22). Doors inside the system either recirculate the air inside the compartment or circulate outside air, according to the control settings. They also route the air inside the system through or around the heater or air conditioner or to the windshield defroster.

Some cars have an additional distributor system that brings air into the vehicle through intakes located in the engine compartment and carries it through ducts to side vents located in the sidewall of the passenger compartment near the front seat passenger's feet. The airflow through these vents can be controlled mechanically, by a vacuum system, or electronically.

STEERING AND SUSPENSION SYSTEMS

The suspension system keeps the vehicle's wheels and tires in solid contact with the road while it lets them move to cushion bumps and jolts, or road shocks. The steering system keeps the wheels and tires pointed in the right direction and works with the suspension system to let the wheels be steered with precise directional control.

The functions of steering and suspension are different. Steering is an operation performed by the driver to direct the vehicle. Suspension is a structural arrangement at each of the four wheels, including the rear wheels. In short, even though steering and suspension do different things, they do them together. The suspension system includes the shock absorbers, springs, and control arms, which keep the wheels and axles in their correct positions. The steering system connects the steering wheel to the front wheels to provide the directional control. The steering system is supported by the vehicle frame. The relationship of the wheels to the frame determines the type of suspension system.

STEERING SYSTEM

The typical steering system (Figure 4–23) consists of the following:

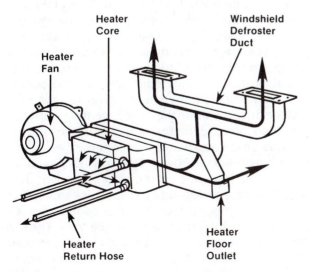

FIGURE 4-22 Blower fan and motor

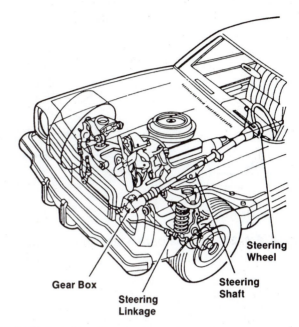

FIGURE 4-23 Conventional steering system

- *Steering Wheel.* Allows the driver to control the direction of the vehicle—the motion of the steering wheel goes through the steering system to the wheels.
- *Steering Shaft.* Connects the steering wheel to the steering linkage. It is attached to the steering wheel on one end and the gear box on the other.
- *Gear Box.* Allows the driver to turn the wheels by mechanical gear reduction. There are two types of gear boxes—conventional and rack and pinion. Each of these systems is available in either manual or power steering.
- *Steering Linkage.* Connects the steering system to the wheels. Steering motion goes from the steering wheel and steering shaft through the steering gear into the steering linkage, which sends movement to the wheels.

There are five types of steering systems: cross steer, Haltenberer linkage, center arm steer, parallelogram linkage (Figure 4–24), and rack and pinion. Steering systems are further divided into manual steering and power assisted. For the most part, the average counterperson will be dealing only with parallelogram linkage, which is also called traditional or conventional, and with rack and pinion.

Conventional or Parallelogram Steering Linkage

Parallelogram linkage uses two tie-rod assemblies connected to the steering arms to support a

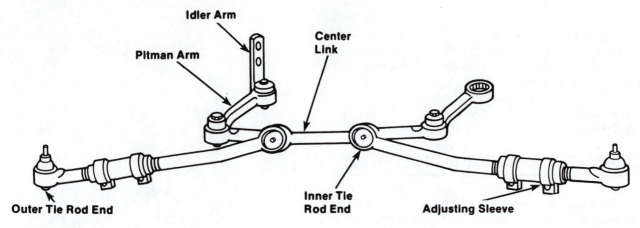

FIGURE 4-24 Parallelogram steering linkages

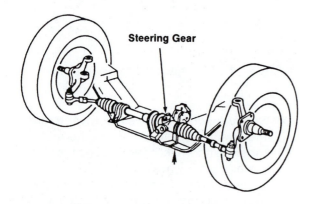

FIGURE 4-25 Rack-and-pinion steering

long center link. The center link holds the tie-rods in position. An idler arm supports the center link on one end; the other end of the center link is attached to the pitman arm. The arrangement forms a parallelogram shape (each end moves equal and parallel to the other end). On some systems a steering damper is attached to the frame and the steering linkage to help absorb road shocks and wheel shimmy.

Rack-and-Pinion Steering

The rack-and-pinion design (Figure 4-25) uses a simplified steering system that weighs less and takes up less space. For these reasons, it is used on many domestic and import vehicles.

Sometimes a customer might mistakenly use other terms to describe rack-and-pinion steering For instance he or she might say it is a MacPherson strut or front-wheel system. This is a key that it might be rack-and-pinion steering, but look it up in the catalog to be sure.

The rack-and-pinion type of steering unit takes the place of the idler arm, pitman arm, center link, and gear box used in conventional systems.

The rack-and-pinion unit consists of a pinion gear at the end of the steering shaft, which meshes with a toothed shaft (a bar with a row of teeth cut into one edge) in the steering gear box called the rack. When the steering wheel is turned, the pinion rotates in a circle and moves the rack sideways—left or right—to turn the wheels.

Two inner tie-rod ends are threaded onto the end of the rack and are covered by rubber bellows boots. These boots protect the rack from contamination by dirt, salt, and other road particles. The inner tie-rods are threaded onto the outer tie-rod ends, which connect to the steering arms. Rack and pinion has fewer friction points than a traditional steering system, so more energy and movement from road forces get through to the steering wheel. This gives the driver a more positive feel of the road.

Power Steering

The power steering unit is designed to reduce the amount of effort required to turn the steering wheel (Figure 4-26). It also reduces driver fatigue on long drives and makes it easier to steer the vehicle at slow road speeds, particularly during parking.

Power steering can be broken down into two design arrangements: conventional and nonconventional or electronically controlled. In the conventional arrangement, hydraulic power is used to assist the driver, while in the nonconventional arrangement an electric motor and electronic controls provide power assistance in steering.

Conventional Power Steering. Conventional power steering systems can also be broken down into two categories: integral and linkage (Figure 4-27). The integral type has the spool valve and power piston integrated with the gear box. The linkage type has an external power piston and spool valve and uses a manual gear box. Regardless of the linkage design, a gear box is necessary to give the

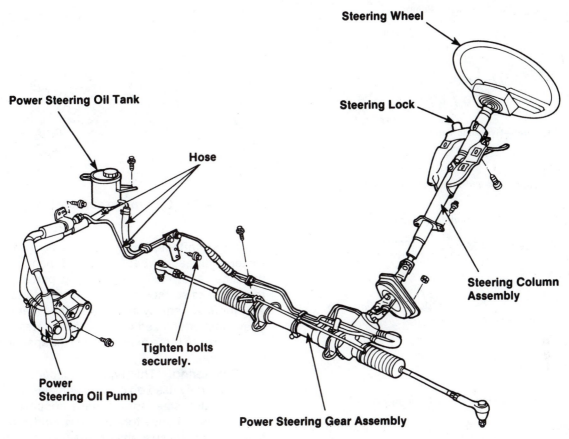

FIGURE 4-26 Typical power steering system

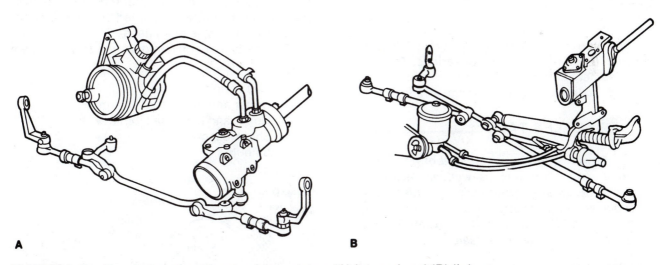

A

B

FIGURE 4-27 Two types of power steering systems: (A) integral and (B) linkage

vehicle operator the mechanical advantage necessary so that the wheels will turn without great difficulty. The mechanical advantage is supplied by simple gear reduction.

Electronically Controlled Power Steering. In this arrangement, an electric/electronic rack-and-pinion unit replaces the hydraulic pump, hoses, and fluid

associated with conventional power steering systems with electronic controls and an electric motor located concentrically to the rack itself (Figure 4-28). The design features a DC motor armature with a hollow shaft to allow passage of the rack through it. The outboard housing and rack are designed so that the rotary motion of the armature can

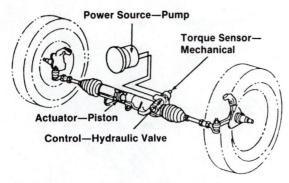

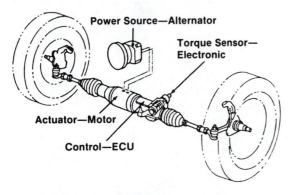

Electric Rack-and-Pinion Steering System

FIGURE 4-28 In an electrically driven power rack-and-pinion system the alternator becomes the power source, replacing the conventional system's hydraulic pump.

be transferred to linear movement of the rack through a ball nut with thrust bearings. The armature is mechanically connected to the ball nut through an internal/external spline arrangement.

 SALES TALK _____

Power steering systems—both conventional and electronic as well as rack-and-pinion designs—are generally available as new replacement parts and remanufactured or rebuilt units. The latter are less expensive and often can be included in a sale.

SUSPENSION SYSTEM

The suspension system suspends the vehicle's frame, body, and driveline (engine, drive shaft, and differential) above the wheels. Its function is to cushion the vehicle from road shocks to the wheels.

Front Suspension

There are two main types of front suspension: solid axle and independent. With solid axle, both

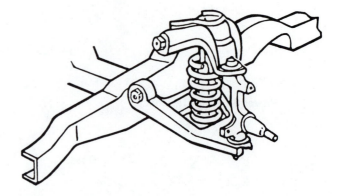

FIGURE 4-29 Coil spring suspension

wheels are affected by road shocks. Independent systems let each wheel move independently from the other and have more moving parts to absorb road shocks so they give a more comfortable ride. The independent systems described here are the ones most often used on passenger cars.

There are three main types of independent front suspension.

Coil Spring. This type of suspension (Figure 4-29) is widely used on current vehicles. In this system, a coiled steel spring helps support the car's weight. It is actually a torsion bar wound into a coil to make a compact suspension design that will absorb load.

The spring can be on either the upper or lower control arm. The spring mounted on the lower control arm, or A-arm, is often used.

The parts in a conventional coil spring system are:

- Coil spring
- Upper and lower control arms
- Upper and lower ball joints
- Shock absorber
- Spindle

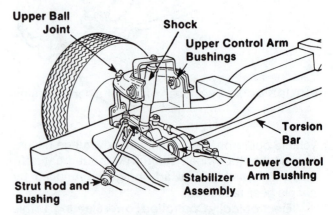

FIGURE 4-30 Conventional torsion bar suspension

- Strut rod (if applicable)
- Stabilizer bar (if applicable)

In this system, the coil spring is the front end's central point. It helps support vehicle weight and maintains chassis height (distance between the ground and the frame). When the coil spring weakens or sags, it forces steering and suspension parts out of position and makes every alignment angle incorrect.

 SALES TALK _____

Coil springs should always be sold in pairs. It is important that the ride height (which affects appearance) and the spring's compression rate (which affects handling) are equal on both sides of the car. A new spring can have a different compression rate from an old spring.

Torsion Bar System. In this system, a steel bar that twists lengthwise is used to provide the spring action (instead of a coil spring). As this bar twists, it resists up-and-down movement. One end of the bar attaches to the frame; the other end attaches to the lower control arm. When the wheel moves up and down, the lower control arm raises and lowers. This twists the torsion bar, which causes it to absorb road shocks.

In conventional torsion bar suspension (Figure 4-30), the torsion bar runs from front to rear. Either an A-arm or a single inner bushing control arm with a strut rod is used for the lower control arm.

In a transverse torsion bar system—the latest in torsion bar suspension—the bar is mounted from side to side and runs across the width of the chassis back to the control arm (Figure 4-31). This eliminates the need for a strut rod or A-arm.

MacPherson Strut. The MacPherson strut system reduces both the weight and the space of the suspension system because it eliminates the upper control arm, upper bushings, and upper ball joint (Figure 4-32).

In this system, the shock absorber, strut cartridge, and spindle are built as a combined unit. The lower control arm (also called transverse link or track control arm) supports it at the bottom. A follower ball joint (not a load carrier) is attached to the lower part of the spindle.

Rear Suspension

The most typical types of rear suspension systems are:

1. *Coil (Figure 4-33A).* Coil springs are the most often used type of rear spring. Rear

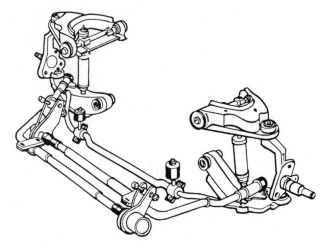

FIGURE 4-31 Transverse torsion bar suspension

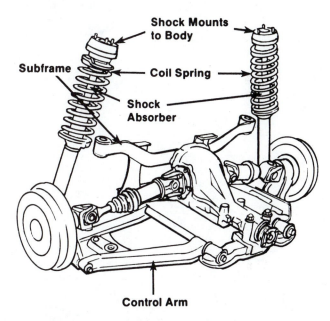

FIGURE 4-32 MacPherson strut suspension

control arms are always used, because coil suspensions cannot control the location of the axle.

2. *Leaf (Figure 4-33B).* Leaf springs can be single or multileaf, depending on vehicle weight and load. In the multileaf design, long, narrow strips of steel are bolted together; they bend with weight but spring back to shape when weight is reduced.

3. *MacPherson Strut (Figure 4-33C).* A MacPherson strut is attached to a lower control arm or axle to isolate each wheel.

Shock Absorbers

Without shock absorbers (Figure 4-34) the continuing jounce and rebound of the spring is uncon-

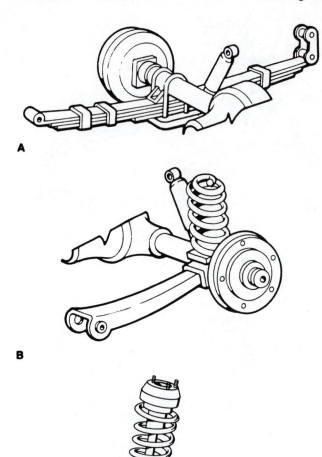

A

B

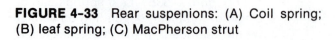

C

FIGURE 4-33 Rear suspenions: (A) Coil spring; (B) leaf spring; (C) MacPherson strut

trolled. This would be hard on the steering and suspension system and would provide a rough and unstable riding vehicle as well. Shock absorbers reduce the number and severity of spring oscillations. Faulty shock absorbers can cause spring breakage.

Shock absorbers are mounted between the frame and the suspension by means of brackets and rubber bushings. The most common shock absorber is the direct acting, telescopic, hydraulic, double-

acting type. Both standard and heavy-duty shock absorbers are available. Gas-types are also available (Figure 4-35).

The shock absorber operates on the principle of forcing fluid through restricted openings (orifices) on both jounce (compression) and rebound. Fluid is forced from one compartment to another in the shock absorber by piston and cylinder movement. Shock absorbers are mounted vertically or at an angle. Angle mounting of shock absorbers is used to improve vehicle stability and to absorb accelerating and braking torque.

POWERTRAIN AND TRANSMISSION SYSTEMS

The powertrain of an automotive vehicle transfers the power developed by the engine to drive the car's wheels. The components of the powertrain include

- Clutch
- Driveline
- Axle
- CV joints and front-wheel drive
- Transmission

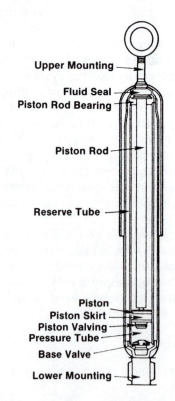

Upper Mounting

Fluid Seal
Piston Rod Bearing

Piston Rod

Reserve Tube

Piston
Piston Skirt
Piston Valving
Pressure Tube
Base Valve

Lower Mounting

FIGURE 4-34 Cross section of a conventional shock absorber

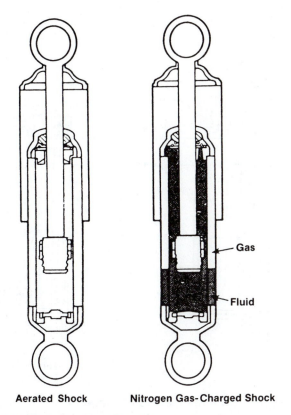

FIGURE 4–35 Gas-filled shock absorbers

- Universal joints
- Differential

Note that all powertrains do not contain all these components (Figure 4–36).

An internal combustion engine cannot develop appreciable torque at low speeds to drive the vehicle in direct drive. Also, the crankshaft of an engine must always rotate in the same direction. The transmission provides the mechanical advantage that enables the engine to propel the vehicle under various loads. It also furnishes the driver with a selection of vehicle speeds while the engine is held at speeds within the effective torque range. It allows disengaging and reversing the power flow from the engine to the wheels. The transmission provides the operator with a selection of gear ratios between the engine and wheels so that the vehicle can operate most efficiently under a variety of driving conditions and loads. It also provides reverse action and allows the engine to brake.

Gear ratios vary considerably depending on vehicle and engine size, rear axle ratio, and wheel and tire size. When a small gear drives a larger gear, the result is gear reduction. Gear reduction provides an increase in torque and a decrease in speed. When a small gear is driven by a larger gear, the result is a decrease in torque and an increase in speed.

If a gear with 15 teeth is used to drive a gear with 30 teeth, the ratio is 2:1. This is calculated by dividing the number of teeth on the drive gear into the number of teeth on the driven gear: driven/drive. In this example, $\frac{30}{15}$:1 or 2:1.

When power flows through a series of gears, the ratio can be calculated in a similar manner. For example, if a 20-tooth drive gear drives a 24-tooth cluster gear and the second speed cluster gear has 16 teeth driving a 20-tooth second speed driven gear, the result would be a 1.5:1 ratio. This is calculated as follows:

$$\frac{\text{Driven}}{\text{Drive}} \times \frac{\text{Driven}}{\text{Drive}} \text{ or } \frac{24}{20} \times \frac{20}{16} = 1.5{:}1$$

which is an acceptable second gear ratio.

The transmission—manual or automatic—provides the driver with a selection of gears to permit the car to operate under a variety of conditions and engine loads. The majority of passenger cars in use today are equipped with an automatic transmission that performs all clutching and gearshifting operations with minimum assistance from the driver. However, many cars equipped with manual transmissions are sold each year. These cars must have a clutch, which is foot pedal-operated, and a transmission that shifts from gear to gear by hand.

MANUAL TRANSMISSION

The clutch in a manual transmisison is the mechanical connection or link between the engine and the gears that move the car. It is designed to connect or disconnect the transmission of power from one working part to another—in this case, from the engine to the transmission. The clutch, a friction-type

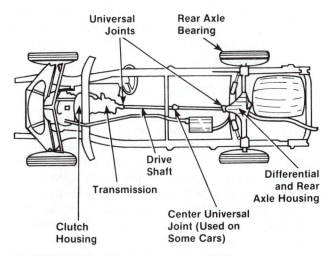

FIGURE 4–36 Typical car powertrain

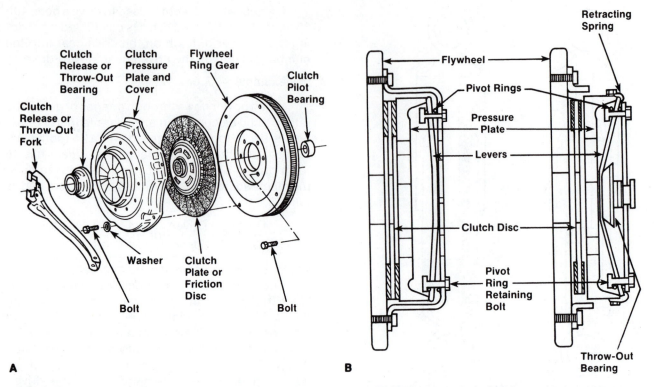

FIGURE 4-37 (A) Main components of a typical assembly; (B) how a manual clutch works

device, is linked to a clutch pedal in the driver's compartment.

The clutch assembly is divided into four main components (Figure 4-37A).

1. *Flywheel.* Bolted to the engine's crankshaft.
2. *Friction disc or clutch plate.* Splined to the transmission input shaft.
3. *Pressure plate assembly.* Installed over the friction disc and bolted around its outside edge to the flywheel.
4. *Clutch or throw-out control and mechanical or hydraulic linkage.* Allows the driver to engage or disengage the clutch as required.

Manual Transmission Operation

When the clutch is engaged in the UP position, the spring pressure plate forces the disc hard against the face of the flywheel which, when the engine is running, causes the flywheel, disc, and pressure plate to rotate together (Figure 4-37B). When the clutch is disengaged (pedal depressed), the pressure plate is pulled away from the disc, allowing the flywheel and pressure plate to turn without rotating the disc (Figure 4-37B). The pressure plate is operated by a clutch release bearing, which is controlled

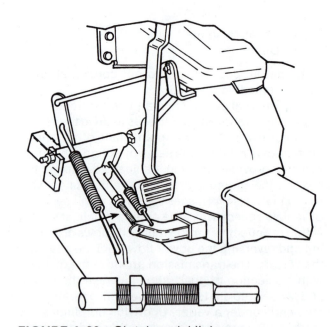

FIGURE 4-38 Clutch pedal linkage

by a series of linkages attached to the clutch pedal (Figure 4-38). After placing the transmission in gear, the driver gradually engages, or raises, the clutch pedal and at the same time depresses the gas pedal. The flywheel then transfers its rotary energy to the disc which, in turn, is connected to the transmission

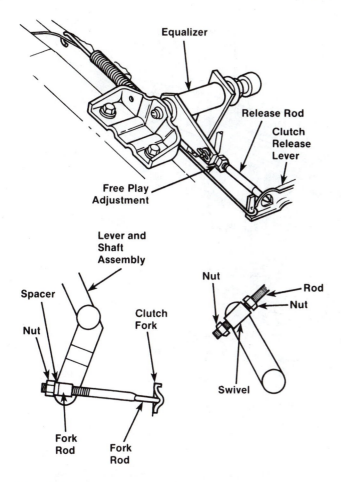

FIGURE 4-39 Linkage adjustments

threaded and has two nuts on the thread; this is the adjustment (Figure 4-39). If the car has a cable linkage, the adjustment is where the cable end attaches to the clutch fork. Generally, shortening the length of the linkage will give more free play. Change the length and then check the pedal to determine whether or not you are increasing the play. When the correct play is obtained, tighten the locknut at the adjustment. Although adjusting the free play is easy to do, more than half the clutches that are replaced fail as a result of this oversight.

All mechanical clutches will wear out in time. The friction material facings inside the assembly wear off as they are squeezed between the pressure plate and the whirling flywheel. Many dealers, wholesalers, and specialty shops offer a clutch rebuilding service.

Standard Manual Transmission

The purpose of the transmission is to provide the operator with a selection of gear ratios between the engine and wheels so that the vehicle can operate at best efficiency under a variety of driving conditions and loads. The typical domestic car fitted with standard transmission has four gear selections, or ratios. They are first (or low) gear, second gear, third (or high) gear, and reverse gear (Figure 4-40). Some domestic sports models, as well as many imports with less powerful engines, offer four or five forward-gear selections, which provide greater utilization of the engine's power.

AUTOMATIC TRANSMISSIONS

Modern automatic transmissions have eliminated the necessity of depressing the conventional clutch pedal to shift gears by replacing the mechan-

by a steel shaft, and the vehicle is set into motion. When the car is moving, each time the gears are shifted, the engine's driving force is disconnected from the transmission by disengaging the clutch; if it were not, the transmission gears would be destroyed.

A clutch must have free play, which means the pedal should move down toward the floor with little effort for a distance of between 3/4 and 1 inch. It should then require much more pressure to push it all the way down to the floor. Without this free play, the clutch is partly disengaged, even when the foot is not touching the clutch pedal, and ultimate failure soon occurs. The clutch release bearing might also be affected. It is meant to operate only when the clutch is disengaged, and if there is no free play, it revolves constantly, burns out, and makes a grinding noise. It might then be difficult to get the car into or out of gear.

The free play is adjusted by changing the length of the linkage. Trace the linkage from the pedal to the left side of the clutch housing where the clutch fork enters it. Look for a portion of the linkage that is

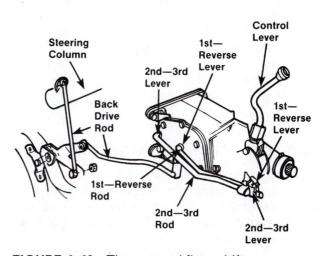

FIGURE 4-40 Three-speed floor shift

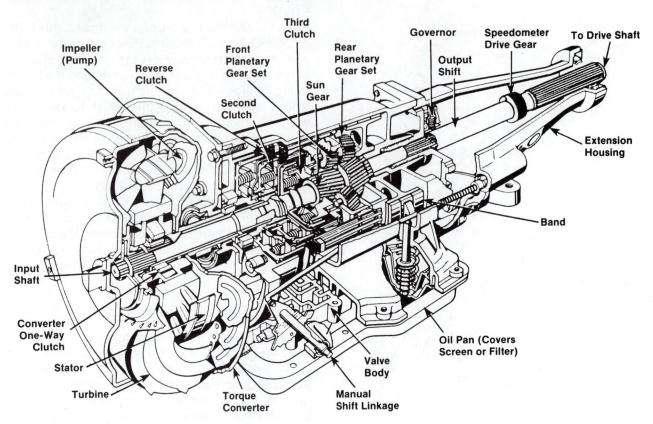

FIGURE 4-41   Main components of an automatic transmission

ical link with a fluid coupling that allows the engine to idle while the gear lever is in DRIVE. Actually, an automatic transmission is more than just a link. It is really a computer that matches the demand for acceleration with engine speed, wheel speed, and load conditions (Figure 4-41). It then chooses the proper gear ratio and initiates a gear change, if necessary, and does it quite accurately if it is in good operating condition.

There is no doubt that an automatic transmission is a complex and sophisticated bit of machinery. Specialists spend years learning the specifics and idiosyncrasies of the many types and varieties used today. But neither the principles behind the automatic nor the routine maintenance required to keep it working correctly are that mysterious or intricate.

Automatic Transmission Operation

Various types and styles of automatic shifters have been tried since the first semiautomatic was introduced in the 1930s. Today, however, most automatics are basically the same and consist of three hydromechanical devices: a torque converter to provide the fluid link, a set of valves operated by oil pressure to control transmission bands and initiate

shifts, and two planetary gear assemblies to transmit the power (Figure 4-42).

Torque Converter.   The torque converter provides the fluid coupling. It fits onto the front of the transmission housing and bolts into the engine. Inside its housing are three sets of blades known as the pump, stator, and turbine, immersed in light oil (transmission fluid). Energy passes from the engine through the fluid and the blades and is transmitted as mechanical torque to the drive wheels by the output shaft. All three blade sets rotate together at cruising speed, providing maximum output with little slippage. At idle, the torque converter allows maximum slippage and very little torque output.

Valves.   A small pump inside the gearbox initiates the big job of changing gear ratios. This pump supplies fluid under pressure to various piston-in-cylinder servos (Figure 4-43A). These servo units perform the hydromechanical functions of operating the clutches and brakes (inside the transmission), which affect gear changes.

Planetary Gears.   Actual gear changes are possible through two sets of planetary gears. These gears are so arranged that those not held by a brake (transmission band) will move. A transmission band is simply a circular strip of spring steel that fits around a drum or shaft, which controls a planetary

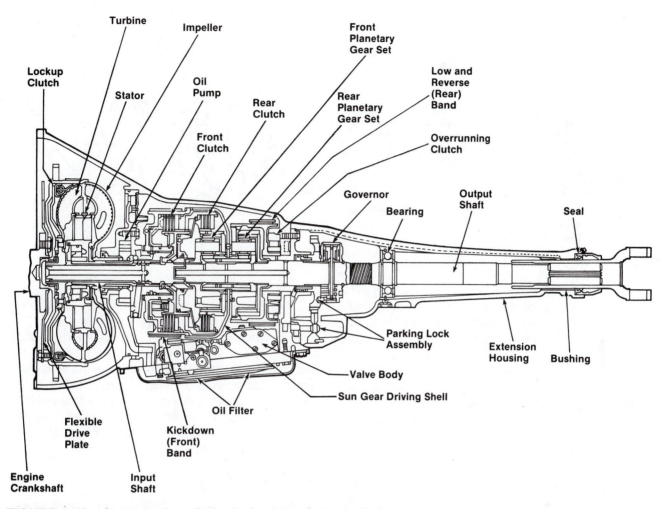

FIGURE 4-42   Cross section of a typical automatic transmission

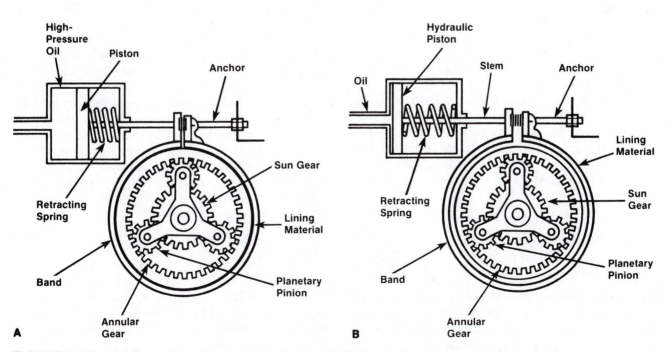

FIGURE 4-43   (A) Operation of a transmission band; (B) band closed properly around a drum

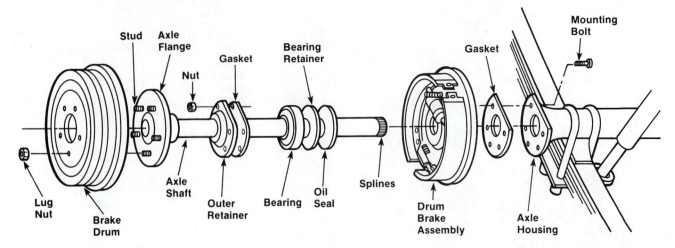

FIGURE 4-44 Semi-floating axle

gear or one of its members. When the servo unit piston closes the band properly around a drum or shaft to keep it from turning, a gear change occurs (Figure 4-43B).

The transmission bands, however, sometimes loosen or get out of adjustment and cannot clamp around the component tightly enough; slipping or erratic shifts can occur and need a band adjustment. But on most late-model cars, band adjustment is seldom necessary after the initial warranty work.

Fluid. Since the entire transmission is bathed in fluid and operates on hydraulic pressure, the best way to prevent transmission problems is to check the transmission dipstick frequently.

AXLES

Axles can be divided into two types:

1. *Dead Axle.* Remains stationary while the wheel rotates.
2. *Live Axle.* Both the axle shaft and the wheel rotate as a unit. The front wheels of most conventional automobiles are mounted on dead axles and the rear wheels are mounted on live axles. However, the reverse is true for cars with front-wheel drive.

There are three types of live axles:

1. Semi-floating
2. Three-quarter floating
3. Full floating

Each is identified by the manner in which the outer end of the axle is supported in the axle housing. The inner ends of all axles are attached to the differential side gears by means of straight splines.

The semi-floating axle is the type used in conventional rear-wheel drive cars (Figure 4-44). The outer or wheel end of the axle is supported in an axle housing by a single bearing mounted near the outer end of the axle. With this type of axle, the axle shaft not only transmits the driving torque, but also resists the bending movements caused by the forward motion of the car and the side thrusts imposed when the vehicle makes a turn. It also must carry its share of the weight of the vehicle.

DIFFERENTIAL

In the rear axle assembly, two small gears known as differential or side gears are splined to the inner ends of the axles (Figure 4-45). These two gears face each other and mesh with two smaller bevel gears, called *spider gears*, to form a square. The spider gears are mounted on a shaft that is mounted in the differential case. This case surrounds the spider and side gears and keeps these four gears in constant mesh. A large gear, called the *ring gear*, is attached to the outside of this case, which is supported in the differential housing by two bearings. These bearings permit the differential case to rotate freely inside its housing. The ring gear and differential case are driven by a pinion gear, which is connected by a universal joint to the drive shaft.

DRIVE SHAFT AND U-JOINTS

To connect the vehicle's transmission to the differential, a drive shaft must be used in a conventional rear-wheel drive automobile. The function of the various components of the drive shaft (Figure 4-46)—also known as a driveline and propeller shaft—are as follows:

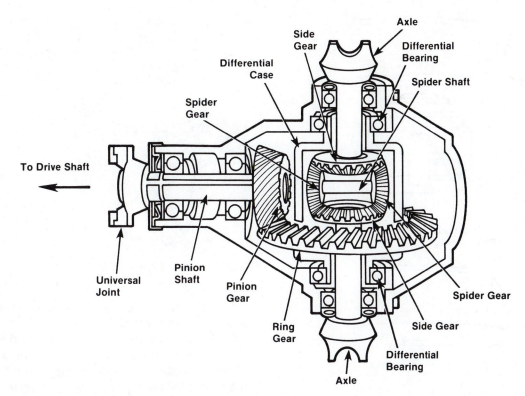

FIGURE 4-45 Differential allows wheels to rotate at different speeds on turns.

- *Slip Yoke.* The most common slip or sliding yoke design (Figure 4–47) features an internally splined externally machined bore that lets the yoke rotate at transmission output shaft speed and slide at the same time (hence the name slip yoke). While the need for rotation is obvious, without the linear flexibility, the drive shaft would bend like a bow the first time the suspension jounced.
- *Drive Shaft and Yokes.* The drive shaft is nothing more than an extension of the transmission output shaft. That is, the drive shaft, which is usually made from seamless steel tubing, transfers engine torque from the transmission to the rear driving axle on a rear-wheel drive vehicle. The yokes, which

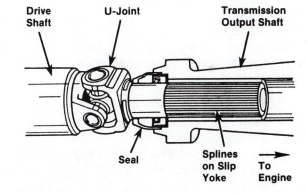

FIGURE 4-47 Slip yoke design

are either welded or pressed on the shaft, provide a means of connecting two or more shafts together. At the present time, a limited number of vehicles are equpped with fiber composite—reinforced fiberglass, graphite, and aluminum—drive shafts. The advantages of the fiber composite drive shaft—other than weight reduction and torsional strength—are fatigue resistance, easier and better balancing, and reduced interference from shock loading and torsional problems.
- *Differential Pinion Flange.* The pinion flange serves as a rear anchoring point for the drive

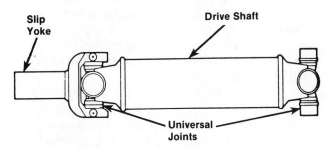

FIGURE 4-46 Drive shaft and components

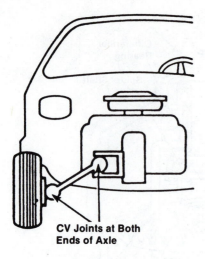

CV Joints at Both
Ends of Axle

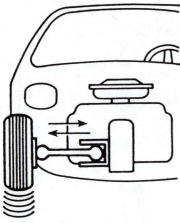

FIGURE 4-48 CV joints must permit axle to bend and allow the the length to change.

shaft. It is splined, bolted, or pressed to hold it securely in place.
- *Universal Joint.* The U-joint allows two rotating shafts to operate at a slight angle to each other.

FRONT-WHEEL DRIVE AND CV JOINTS

Front-wheel drive cars differ radically from conventional rear-wheel drive cars because their powertrains do not have a drive shaft, U-joints, or rear drivetrain. Because their drive wheels are in the front of the car where the engine is located, they have a transaxle, which consists of a clutch and manual transmission (or automatic transmission), front differential drive shafts, and shaft bearings. The transaxle, which is fastened to the engine, has constant velocity joints that act as steering and suspension members because they take part in all front-end movements. They convey power from the transmission to the wheels and serve as axle, suspension, and steering members.

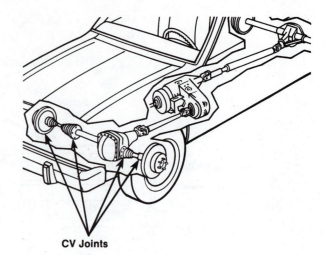

CV Joints

FIGURE 4-49 Typical CV joints

The constant velocity joint is the driveline on front-wheel drive vehicles. Most front-wheel drive cars have two drive shafts—one on each wheel. Each shaft has a CV joint on each end. CV joints next to the transmission are inboard joints, or plunging joints. Most inboard joints, which ride in a tulip-shaped housing, use roller bearings on the end of the shaft. This allows the inboard joint to plunge when the shaft changes in length as the wheel follows the road (Figure 4-48). The CV joint at the wheel hub is an outboard joint. It is fixed and does not change in length. Typical types of CV joints are shown in Figure 4-49.

 SALES TALK ─────────

Constant velocity joint repair and replacement is a new market—even to the professional installer. For this reason, the counterperson might have to help the customer with some CV joint problem-solving.

For example, if the installer asks for a boot kit, the counterperson should ask, "Is the boot cracked or torn?" If the boot is cracked or torn and has been leaking, in most cases, the joint will be pitted and worn and should be replaced.

If both the boot and joint have to be replaced, the counterperson should recommend a CV joint service kit, which includes CV joints, boot, correct grease for lubrication, and all attachment hardware.

If the shaft is bent, the counterperson should explain that the entire drive shaft assembly has to be replaced. Bent shafts cannot be repaired. The counterperson should remind the customer to be sure to use all the boot lubricant that comes prepackaged with the kit.

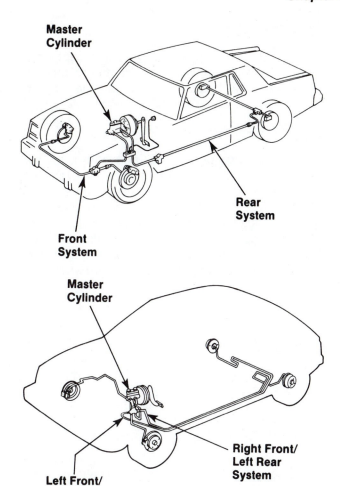

FIGURE 4–50 (A) Front/rear split hydraulic system; (B) dual diagonal split hydraulic system

BRAKE SYSTEM

In today's front-wheel drive automobiles, the front brakes are now doing as much as 80 percent of the work in stopping the car. Thus, the conventional front/rear split hydraulic system (Figure 4–50A) has given way to a dual diagonal split system. This combines a front brake with its opposite rear brake (Figure 4–50B). This system allows straight stopping and provides 50 percent of the braking capacity in case of failure in either of the two hydraulic systems.

The actuating system of today's cars employs hydraulic pressure or power-assisted hydraulic pressure. When the driver steps on the brake pedal, the force from the foot creates a hydraulic pressure in the master cylinder. This pressure is transmitted through lines and hoses to the wheel cylinders where it turns into force again to act upon the drums or discs that stop the car.

The components of a vehicle's brake system with which a counterperson should be concerned are

- Master cylinder
- Brake lines and hoses
- Wheel cylinders
- Brake safety switches and hydraulic valves
- Brake fluid
- Drum brakes and their hardware
- Disc brakes and their hardware
- Parking brakes
- Power brake and antiskid systems

MASTER CYLINDER

The master cylinder is the heart of the hydraulic system. It is located in the engine compartment usually on the driver's side and connected to the brake pedal by a special rod. The master cylinder initiates braking when the brake pedal is depressed by pushing out a piston inside the cylinder, exerting pressure that is transferred through the system.

To protect against a total failure of the system, all cars are now required to have two hydraulic systems. The old single system had all four wheel cylinders using one brake reservoir and one master cylinder piston (Figure 4–51). The dual master cylinder (Figure 4–52) provides for either a front/rear split or a diagonally split system.

 SALES TALK _____

Many late-model cars use aluminum cylinders to save weight. Although overhaul kits are available for these units, warn customers to proceed with caution. The bore should not be sanded or honed as would normally be done with a cast iron master cylinder. Aluminum is a porous material, especially under high pressure, so the inside of the bore is anodized at the factory to prevent seepage of brake fluid through the casting. If the seal coating is removed by honing, the overhauled master cylinder will probably leak. If the bore is worn to the point where anodizing has rubbed through, the master cylinder should be replaced rather than overhauled.

Another warning to pass along to would-be rebuilders is that some late-model master cylinders have plastic fluid reservoirs. The reservoirs can be easily broken or cracked during disassembly and reassembly, and if that happens, a customer must buy a whole new replacement master cylinder because the reservoirs are not yet available as a separate repair item.

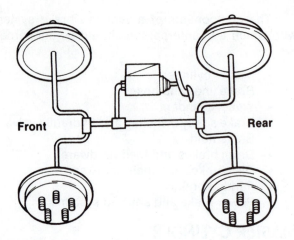

FIGURE 4-51 Single master cylinder system used in older cars

BRAKE LINES

The brake lines are usually steel, except where they have to flex—between the chassis and the front wheels, and the chassis and the rear axle. At these places, flexible hoses are used.

Steel Tubing

Most tubing used in a braking system is double-walled welded steel coated with a rust resisting material. The most common diameters are 3/16, 1/4, and 5/16 inch. Tubing is available in various lengths with the ends already flared and provided with fittings or it can be cut to length and flared. Before flaring steel tubing, slip the tube nuts onto the tubing (and make sure they are pointed the right way). Then

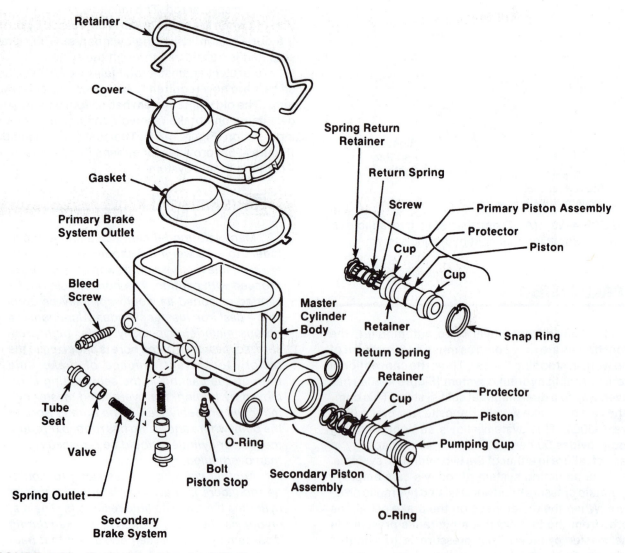

FIGURE 4-52 Typical dual master cylinder

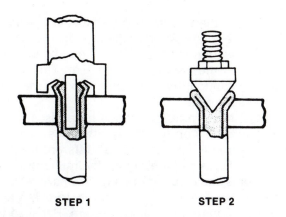

STEP 1   **STEP 2**

FIGURE 4-53 Steps in making a flare connection

select the type of flare based on the type of fittings used. In all cases it will be a male fitting on the tubing itself, and the flaring tool should be the double flaring type (Figure 4-53). Steel tubing is easily bent.

Hydraulic Hoses

Hoses must be strong enough to withstand high fluid pressures without expanding yet be free to flex during motion of the steering and suspension systems. When replacing a hydraulic hose, always use a new one of exactly the proper length. A hose that is too long will chafe against the chassis (or some other metal), eventually wearing through and leading to hydraulic failure. A hose that is too short will break in tension. Most hoses have a male fitting on one end, a female on the other.

WHEEL CYLINDER

The wheel cylinder (Figure 4-54) performs the opposite function of that of the master cylinder: It converts hydraulic pressure to a mechanical force. In a disc brake system, the wheel cylinder is usually integral with the caliper housing instead of being a separate piece, but both types of systems, disc and drum, work the same way hydraulically: Fluid pressure is transmitted to the bore by a flexible hose. As the pressure increases inside the bore, the pistons are forced outward to bear against the friction elements.

In a drum brake system, the wheel cylinder has two pistons (Figure 4-55). The hydraulic principle is

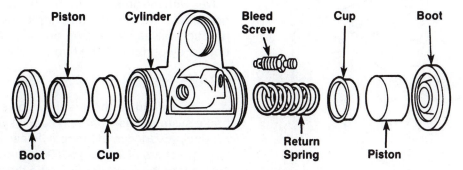

FIGURE 4-54 Wheel cylinder used with most disc systems

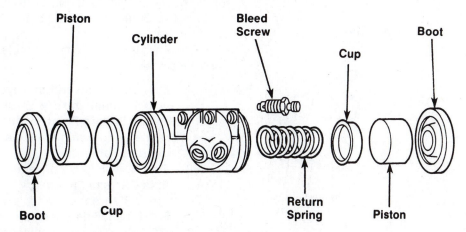

FIGURE 4-55 Wheel cylinder used on most drum brakes

the same, only instead of one end of the cylinder being blanked off, there are pistons at both ends, opposing each other.

Disc brakes, and usually front brakes in general, have larger bores. Disc and drum systems differ in the method of returning the pistons to rest after an application. With drum brakes, the return is done with springs. Disc brakes have no return springs. On some models, the pads and pistons are retracted by the natural runout of the spinning rotor. On newer models, a special type of piston seal is used that pulls the piston and pads back upon brake release.

BRAKE SAFETY SWITCHES AND HYDRAULIC VALVES

Switches and valves installed in the vehicle's brake system are the principal safety devices (Figure 4-56). For example, if failure should occur in either brake system, the pressure loss differential activates a brake failure warning switch to light the brake warning light in the instrument panel and alert the driver to the failure. Apart from the hydraulic system, the brake warning light serves a second purpose. It signals that the parking brake is set. This occurs when the ignition switch is turned on and the parking brake is applied.

Most modern auto systems have discs on the front and drums on the rear. Both types of brakes work differently and activate at different levels of pressure. Discs require higher hydraulic pressure than drums and respond quickly. Drums must overcome the tension of return springs before they activate, but require lower line pressure. Because the two types of brakes differ in action and hydraulic requirements, metering (hold-off) valves are used to balance pressure in the system. A metering valve momentarily delays front disc braking action until the rear drum brake shoes are forced against the brake drums. The proportioning valve (or P-valve) balances front-to-rear braking action as the car comes to quick stops; and, it prevents premature locking of the rear wheels.

Most newer cars have a combination valve in their hydraulic system. This valve is simply a single unit that combines the metering and proportioning valves with the pressure differential valve. Combination valves are described as three-function or two-function valves, depending on the number of functions they perform in the hydraulic system.

BRAKE FLUID

Brake fluid is the lifeblood of any hydraulic brake system. It is what makes the system operate properly. Every can of brake fluid carries the identification letters of SAE and DOT. These letters (and corresponding numbers) indicate the nature, blend, and performance characteristics of that particular brand of brake fluid. SAE comes from the Society of Automotive Engineers; DOT stands for the Department of Transportation.

There are three basic types or classifications of hydraulic brake fluids.

1. *DOT 3.* A conventional type brake fluid with a minimum dry ERBP (Equilibrium Reflux Boiling Point) of 401 degrees Fahrenheit and a minimum wet ERBP of 284 degrees Fahrenheit
2. *DOT 4.* This is a conventional type brake fluid with a minimum dry ERBP of 446 degrees Fahrenheit and a minimum wet ERBP of 356 degrees Fahrenheit.
3. *DOT 5.* A unique silicone-based brake fluid with a minimum dry ERBP of 500 degrees Fahrenheit and a minimum wet ERBP of 356 degrees Fahrenheit.

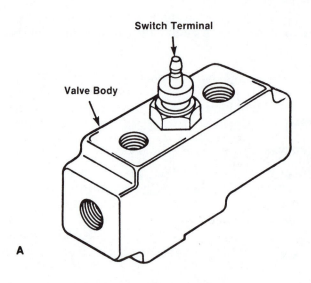

Switch Terminal

Valve Body

A

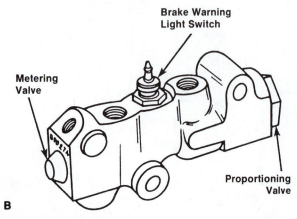

Brake Warning Light Switch

Metering Valve

Proportioning Valve

B

FIGURE 4-56 Brake system's safety devices: (A) Typical brake warning light switch; (B) three-function combination valve

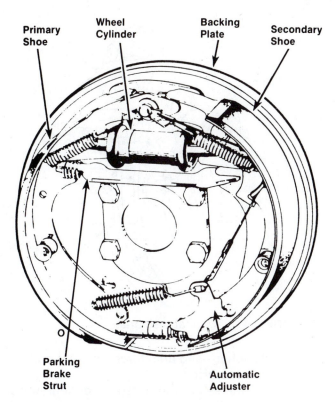

FIGURE 4-57   Typical drum brake

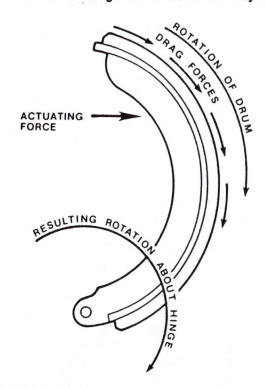

FIGURE 4-58   How a shoe became self-energized

CAUTION: Because of its ability to absorb moisture, care should be used when handling brake fluid.

TYPES OF HYDRAULIC BRAKES

As already mentioned, there are two basic types of hydraulic brakes: drum and disc.

Drum Brakes

A drum brake assembly consists of a cast-iron drum, which is bolted to and rotates with the vehicle wheel, and a fixed backing plate to which are attached the shoes and other components—brake cylinders, return springs, linkages, and so on (Figure 4-57). Additionally, there might be some extra hardware for parking brakes. The shoes are surfaced with frictional linings, which contact the inside of the drum when the brakes are applied. The shoes are forced outward, against the action of the return springs, by pistons which are actuated by hydraulic pressure. Just how this pressure is brought to the pistons will be covered later. As the drum rubs against the shoes, the energy of the moving vehicle is transformed into heat, and this heat energy is passed into the atmosphere.

When the brake shoe is engaged, the frictional drag acting around its circumference tends to rotate it about its hinge point, the brake anchor. If the rotation of the drum corresponds to an outward rotation of the shoe, then the drag will pull the shoe tighter against the inside of the drum, and the shoe will be self-energizing (Figure 4-58).

Nonservo Drum Brakes. In most brake designs, shoe energization is usually supplemented by another feature called "servo" action, but for now let us consider a simple nonservo drum brake and how the energization figures in (Figure 4-59). Each shoe is separately anchored at its lower end, the heel; the upper end, the toe, is where the actuating force is

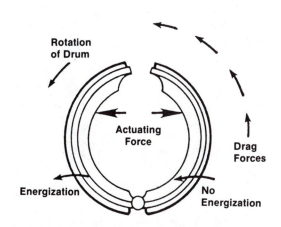

FIGURE 4-59   Typical nonservo brake

applied. When the brakes are applied in normal vehicle motion, the forward, or leading, shoe is energized, while the trailing shoe receives no energization. If the brakes are applied when the car is in reverse—remember, the drums are rotating the other way now—the leading and trailing shoes exchange roles; that is, what was previously the trailing shoe now becomes energized. Since most of the time it is the front shoes that receive the energization in a non-servo brake system, they will tend to wear faster.

Servo Type Drum Brakes. A servo brake is termed such because the action of one shoe is governed by an input from the other. There is an anchor, but it only serves as an anchor for the secondary shoe, while the secondary shoe serves as an anchor for the primary shoe.

When the brakes are applied, the primary shoe reacts first—it has a weaker return spring. The shoe lifts off the anchor (which at this point is only acting as a stop) and contacts the drum surface. Since the shoe is pivoted at the bottom, the drag forces tend to rotate it deeper into the drum; it is energized. Now take a look at what happens to the secondary shoe (Figure 4-60): it is being acted upon at its heel, via a connecting link, by the primary shoe. This forces the secondary shoe in the direction of the drum rotation—but it cannot rotate, because its upper end is butted against the fixed anchor. Now, since the drag forces (from contact with the drum) try to rotate it about the anchor, it too becomes energized. In the applied position, the servo brake behaves as if it were one continuous shoe—the actuating force is pushing on one end, and the other end is held fast by

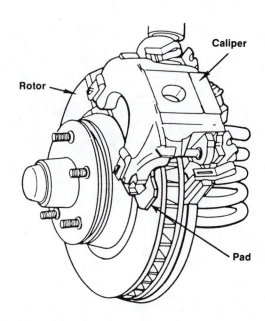

FIGURE 4-61   Typical disc brake

the fixed anchor—and drag forces cause the whole brake to be energized.

Servo brakes exist in different configurations. There is usually some form of piston pressure against the secondary shoe, but this is incidental; the principal of operation remains the same.

Disc Brakes

Disc brakes resemble the brakes on a bicycle: the friction elements are in the form of pads, which are squeezed or clamped about the edge of a rotating wheel. With automotive disc brakes, this wheel is a separate unit, called a *rotor,* inboard of the vehicle wheel (Figure 4-61). The rotor is made of cast iron, and since the pads clamp against both sides of it, both sides are machined smooth. Usually the two surfaces are separated by a finned center section for better cooling. The pads are attached to metal shoes, which are actuated by pistons, the same as with drum brakes. The pistons are contained within a caliper assembly, a housing that wraps around the edge of the rotor. The caliper is kept from rotating by way of bolts holding it to the car's suspension framework.

The caliper is a housing containing the pistons and related seals, springs, and boots, as well as the cylinder(s) and fluid passages, necessary to force the friction linings or pads against the rotor. The caliper resembles a hand in the way it wraps around the edge of the rotor. It is attached to the steering knuckle. Some models employ light spring pressure to keep the pads close against the rotor; in other caliper designs this is achieved by a unique type of

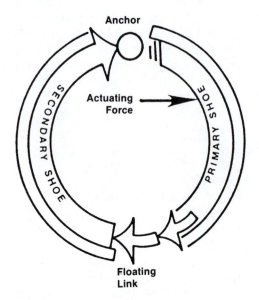

FIGURE 4-60   Typical servo type brake action

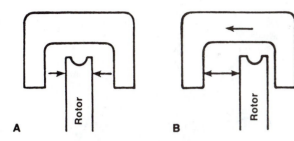

FIGURE 4-62 Disc brake caliper designs: (A) fixed and (B) moving. Moving caliper includes the sliding and floating types.

seal that allows the piston to be pushed out as far as is necessary, then retracts it just enough to pull the pad off the rotor.

Unlike shoes in a drum brake, the pads act perpendicularly to the rotation of the disc when the brakes are applied. This effect is different from that produced in a brake drum, where frictional drag actually pulls the shoe into the drum. Disc brakes are said to be nonenergized, and so require more force to achieve the same braking effort. For this reason, they are ordinarily used in conjunction with a power brake unit.

Disc brake calipers fall into two categories (Figure 4-62): fixed and moving designs. The latter includes both sliding and floating caliper types.

Disc brakes offer four major advantages over conventional drum brakes.

1. *Resistance to Heat Fade.* Disc brakes are generally more resistant to heat fade during high-speed brake stops or repeated stops. The design of the disc brake rotor exposes more surface to the air and thus dissipates heat more efficiently.
2. *Resistance to Water Fade.* Disc brakes also are affected very little by water fade because the rotation of the rotor tends to throw off moisture. That is, the squeeze of the sharp edges of brake pads clears the surface of water.
3. *More Straight-Line Stops.* Due to their clamping action, disc brakes are less apt to pull and generally produce more uniform straight-line stops.
4. *Automatic Adjusting.* Disc brakes automatically adjust as pads wear.

Because of the advantages that the disc brake offers over the drum brake, most cars manufactured in the United States since the 1960s employ a combination brake system. In this type of system, disc brakes are used on the front wheels and drum brakes on the rear wheels.

One of the major reasons for using a combination brake system is the increased popularity of front-wheel drive vehicles. Front engine/rear drive vehicles have their weight load reasonably distributed from front to rear while front-wheel drive vehicles exhibit extreme forward weight bias. The engine, transmission, and other drivetrain components are placed directly over the front wheels and contribute to the forward weight bias. The actual distribution varies from model to model, but in general, it is distributed forward of center so as to give the vehicle good handling and road-holding characteristics. (Light trucks, of course, have more than 80 percent of their weight forward of center when unloaded.) In addition, during braking as much as 75 percent of the vehicle weight can be transferred forward. Because of this, more braking force must be applied to the front wheels than to the rear. (If under hard-braking conditions, the brakes were applied equally to the front and rear, the rear wheels might lock up prematurely, causing skidding.)

SALES TALK _____

Switching from one brand of disc semimetallic pad to another can help reduce squealing on problem cars since different manufacturers use different compounds in their pads. Some might squeal less than others on a given application, but there might be trade-offs in terms of pad wear, performance, or price. Since there is no easy answer to this question, rely on feedback from customers as to which pads seem to work best on certain applications. Do not be tempted to substitute less expensive asbestos pads for semimetallics. Asbestos pads will sometimes cure a noise problem, but the risks far outweigh any benefits they offer. Pad wear will be a major problem and severe brake fade can result under hard use.

One of the reasons carmakers switched to semimetallic compounds is because of today's downsized braking systems. Brakes generate heat by turning the energy of motion into heat through friction—that is what stops the car. The brake systems on many front-wheel drive cars today are physically much smaller than those of the older rear-wheel drive cars. In spite of the fact that the downsized cars are lighter, the brakes work harder because of the proportionately higher loads, and that means more heat. Front-wheel drive has increased the load on front brakes to the point where the front wheels now handle 80 to 85 percent of the braking. This means that the front brakes run a lot hotter on a FWD car than on a typical RWD car.

Ordinary asbestos pads start running into trouble once brake temperatures get up around 350 to 400 degrees Fahrenheit. Since brake temperatures in many downsized FWD cars can easily exceed those limits, a more heat-resistant brake lining is required. If asbestos pads are substituted for semi-metallics, two things can happen: Pad wear will be unacceptable. Because of the higher operating temperatures, pad wear will be three to five times normal. Second, under hard braking, temperature can soar to the point where heat fade would prevent the brakes from stopping the car within a safe distance. Therefore, stay with the brake pad material specified by the lining manufacturers in their catalogs.

To help prevent new replacement pads from squealing, recommend the following to the customers:

- Remind customers to make sure any tabs or ears on the outer pads are tightly crimped to hold the pad firmly to the caliper. Any looseness here will guarantee a squeal. Clips and antirattle springs must likewise be installed following manufacturer's instructions.

- Recommend the application of an antisqueal compound to the backs of both pads. The material should be applied to the back of the outer pad where it contacts the caliper and to the back of the inner pad where it butts up against the piston. The material forms a flexible layer to soak up vibrations and reduce the possibility of squealing. Follow manufacturer's directions as to the application thickness and drying time.

- Always recommend new mounting hardware for any vehicle with floating calipers. If the guide pins are corroded, the caliper will not be able to center itself over the rotor. This can cause the outer pad to drag.

- When a customer brings in a rotor to turn on the company's lathe, it is important to always check the rotor thickness and condition because it is a waste of time to attempt machining rotors that are worn out or damaged. Many jobber machine shops offer rebuilding disc and drum brake service.

OTHER BRAKE SYSTEMS

There are three other systems for which the counterperson might be asked to furnish components: power, antilock, and parking brakes.

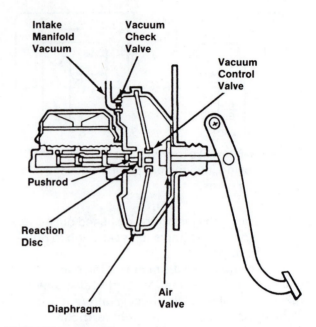

FIGURE 4-63 Vacuum booster components

1. *Power Brakes.* Power brakes on passenger cars are used in conjunction with a basic hydraulic braking system. All of the systems produced today employ some sort of a booster unit (Figure 4-63), which acts like an extra strong foot on the brake pedal. The booster unit can derive its force from vacuum or from the power steering pump. Either way, it is located between the brake pedal and the master cylinder. When the brake pedal is depressed, the booster unit multiplies the pedal force to the master cylinder. From here on, the system is identical to a basic hydraulic system.

2. *Antilock Brakes.* This system is another control arrangement that is used in conjunction with a basic hydraulic braking operation. During hard braking conditions, it is possible for the wheels of a vehicle to lock, resulting in reduced steering as well as braking. On vehicles equipped with the antilock brake system, however, an electronic sensor constantly monitors wheel rotation (Figure 4-64). If one or more of the wheels begins to lock, the system opens and closes solenoid valves, cycling up to ten times per second. This applies and releases the brakes rapidly and repeatedly, so that the front wheels alternately steer and brake. This makes it possible for vehicles equipped with the antilock brake system to avoid skidding under conditions

that might cause vehicles not so equipped to handle differently. The antilock or anti-skid brake system has a controller that senses rotation at each of the wheels through wheel sensors. It can apply the antilock brake system to each of the front wheels independently, to the rear wheels as a pair, or to any combination of these three, as the need arises.

3. *Parking Brakes.* The rear wheel brakes act to hold the car stationary for parking (Figure 4-65). Parking brakes, however, are not actually a part of the hydraulic brake service system. They are actuated mechanically, rather than by hydraulic pressure.

TIRES AND WHEELS

Most automobile parts and accessory stores and some dealerships sell tires and custom wheels. There are sales organizations that just sell these important vehicle items.

TIRES

The main function of tires is to provide traction. Tires also help the suspension to absorb road shocks, but this is a side benefit. They must do their work under a variety of conditions. The road might be wet or dry; paved with asphalt, concrete, or gravel; or there might be no road at all. The car might be traveling slowly on a straight road, or moving quick-

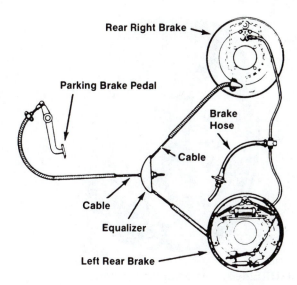

FIGURE 4-65 Typical integral parking brake

ly through curves and over hills. All of these conditions call for special requirements that must be present, at least to some degree, in all tires. Passenger car tires spend most of their miles on paved, dry roads, however.

Pneumatic tires are of two basic types: those that use inner tubes and those that do not. The latter are called tubeless tires and are about the only type used for passenger cars today. The inner tube tire was used exclusively before the 1960s. With this type, both the tube and the tire are mounted on the wheel rim with the tube being inside the casing. The tube is inflated with air, which causes the tire casing

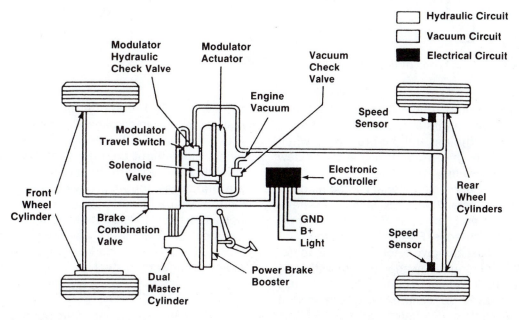

FIGURE 4-64 Control circuits of a typical antilock system

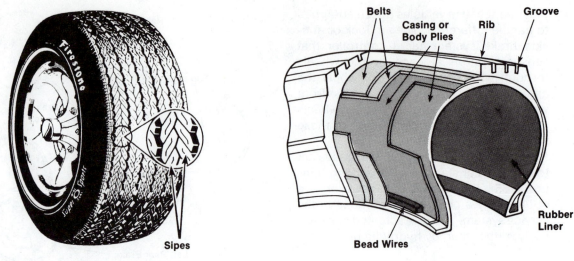

FIGURE 4-66 Typical tubeless tire

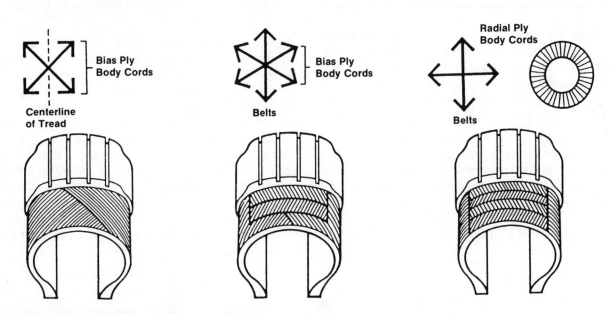

FIGURE 4-67 Three types of tire construction

to resist any change of shape. The tubeless tire is mounted on a special rim in such a way that air is retained between the rim and the tire casing when the tire is inflated. Figure 4-66 shows a cutaway view of a typical tubeless tire.

The tire body and belt material can be made of rayon, nylon, polyester, fiberglass, steel, or the newest synthetics—amarid or kevlar. Each has its advantages and disadvantages. For instance, rayon cord tires are low in cost and give a good ride, but do not have the inherent strength needed to cope with long high-speed runs or extended periods of abusive use on rough, unpaved, or semi-paved roads. Nylon cord tires generally give a slightly harder ride than rayon—especially for the first few miles after the car

has been parked—but offer greater toughness and resistance to road damage. Polyester and fiberglass tires offer many of the best qualities of rayon and nylon, but without the disadvantages. They run as smoothly as rayon tires but are much tougher. They are almost as tough as nylon, but give a much smoother ride. Steel is tougher than fiberglass or polyester, but it gives a slightly rougher ride because the steel cord does not give under impact as do fabric plies. Amarid and kevlar for their weight are much stronger than steel.

Types of Tire Construction

There are three types of tire construction in use today (Figure 4-67):

1. *Bias Ply Tires.* The oldest type currently in use, bias ply tires have a body of fabric plies that run alternately at opposite angles to form a crisscross design. The angle varies from 30 to 38 degrees with the centerline of the tire and has an effect on the characteristics of the tire: high-speed stability, ride harshness, and handling. Generally speaking, the lower the cord angle, the better the high-speed stability, but also the harsher the ride. Bias ply tires usually are available in 2-ply or 4-ply.

2. *Belted Bias Ply Tires.* This type is similar to the bias ply, except that two or more belts run the circumference of the tire under the tread. This construction gives strength to the sidewall and greater stability to the tread. The belts reduce tread motion during contact with the road, thus improving tread life. Plies and belts of various combinations of rayon, nylon, polyester, fiberglass, and steel are used with belted bias construction. Belted bias ply tires generally cost more than conventional bias ply tires, but give up to 40 percent more mileage.

3. *Radial Ply Tires.* These have body cords that extend from bead to bead at an angle of about 90 degrees—"radial" to the tire circumferential centerline—plus two or more layers of relatively inflexible belts under the tread. This construction of various combinations of rayon, nylon, fiberglass, or steel gives greater strength to the tread area and flexibility to the sidewall (Figure 4-68). The belts restrict tread motion during contact with the road, thus improving tread life and traction. Radial ply tires also offer greater fuel economy, increased skid resistance, and more positive braking.

Although the newer synthetics are being used more in radial tires, steel is still the most popular belt material in their manufacture. Bias ply and belted bias are available in all cord materials mentioned earlier, except amarid and kevlar. Nonradial belts are usually of the same material as the sidewalls.

Specialty Tires. Besides standard tires, specialty tires reflect the advances made in the conventional tire field. Special snow and mud tires are available in all three construction types. Studded tires, which provide superior traction on ice, are slowly disappearing from the tire market because their performance in dry weather is poor. In addition,

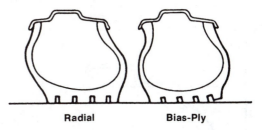

Radial   Bias-Ply

FIGURE 4-68   Radial tires have highly flexible sidewalls.

many states, in the last few years, have outlawed their use because they damage roads.

The present trend in specialty tire manufacturing is toward all-season or all-weather. Their use does eliminate the twice-a-year tire change for many northern motorists. The all-season tire, however, is a compromise and might not perform as well as specialty tires under certain circumstances.

 SALES TALK ————

Advise the customer that tire chains are not recommended by most tire manufacturers. In case of emergency, or where required by law, chains can be used if they are the proper size for the tires and are installed tightly (no slack) with ends securely restrained. Follow the chain manufacturer's instructions. Drive slowly. If any contact of the chains against the vehicle is heard, stop and retighten the chains. Use only SAE Class "S" tire chains. Use of other chains can damage the vehicle.

Tire Ratings and Designations

Tires are rated by their profile, ratio, size, and load range. The tire's profile is the relation of its cross-section height—from tread to bead—compared to the cross-section width—from sidewall to sidewall. Today, this ratio is also known as series.

For many years, the accepted profile ratio for standard bias ply passenger car tires was approximately 83. This meant the tire was 83 percent as high as it was wide. With the coming of bias belted and radial ply construction, lower profile tires with ratios of 78, 70, 60, and even 50 have become popular. The lower the number, the squatter the tire. For instance, a 50-series tire is quite low and fat, being only 50 percent as high as it is wide. Most new cars are equipped with 80-, 78-, 75-, or 70-series tires. That is, "70" indicates the aspect ratio, or the ratio of width to height.

$$\text{Aspect Ratio} = \frac{\text{Section Height}}{\text{Section Width}}$$

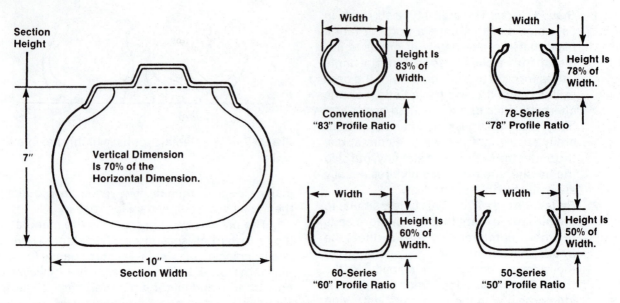

FIGURE 4–69 Aspect ratio of a tire is its cross-sectional height compared to its cross-sectional width expressed in percentage figures

In this example, the aspect ratio is 70; the section height is 70 percent of the cross-section width as illustrated in Figure 4–69.

Prior to 1967, tire sizes were designated by a series of numbers such as 7.75-14 or 9.50-15. The first number (9.50) referred to the cross-section width in inches of an inflated tire, and the second number (15) was the rim diameter. Today, three sizing systems are used: all United States; half United States, half metric; and all metric.

In the standard United States tire size, the width/load letter scale starts at "V" and runs through "Z," then takes up with "A" and runs through "N." The "V" through "Z" ratings are uncommon. The wider the tire, the more weight it can carry. A V-rated tire can handle 650 pounds, while an N-rated one can take 1880 pounds. Table 4–1 shows the maximum weight each tire size can carry when inflated to

24 PSI (as inflation is increased so is the tire's carrying capacity up to a usual maximum of 32 PSI).

In the all United States system, the tire letter size is followed by a number to indicate the tire's approximate profile ratio, followed by the rim diameter. For example, with a tire designated as F78-15, the 78 means that it belongs to the 78 series and will fit a 15-inch rim. Radial ply tires use the same designation but have an "R" inserted in the number, such as FR78-14.

The half United States, half metric system gives the tire's width in millimeters, but gives its rim diameter in inches, such as 195R-15. Although the profile ratio is usually omitted in part-metric designations, a few manufacturers designate the tire series, such as 185-70-14. The 70 stands for a 70-series tire. In 1977, "P metric" tires were introduced. A P195/75R14, for example, means the following:

- "P" identifies passenger car tire. (If this designation is followed by "M," "S," or "MS," the tire's tread is rated for mud and snow use.)
- 195 is the width in millimeters.
- 75 is the height-to-width ratio.
- R identifies radial construction.
- 14 is the rim diameter.

All metric system measurements are now being given along with a standard translation. A typical metric tire shows its width in millimeters, its inflation pressure in kilopascals (kPa), and its load capacity in kilograms (kg). One kilopascal equals 6.895 PSI. A typical all-metric radial size is 190/65R-390. It fits a 390-mm-diameter wheel.

TABLE 4-1: WIDTH/LOAD LETTER STANDARDS	
V—650 pounds	F—1280 pounds
W—710 pounds	G—1380 pounds
Y—770 pounds	H—1510 pounds
Z—830 pounds	J—1580 pounds
A—900 pounds	K—1620 pounds
B—980 pounds	L—1680 pounds
C—1050 pounds	M—1780 pounds
D—1120 pounds	N—1880 pounds
E—1190 pounds	

Additionally, the letters "B," "C," or "D" might appear on the sidewall, separate from the tire designation coding. These are holdovers from the old load-rating system, which replaced the still older ply system. Load Range "B" is the lightest design, suited to passenger car use. Load Range "C" is in between, while Load Range "D" is heavy-duty, best for trucks and off-road vehicles.

The newer tires have abandoned the load range system replacing it with "SL" and "XL" for standard load and extra load. Standard load falls between load ranges "B" and "C." "XL" is slightly more heavy duty than Load Range "D."

Federal law requires that all tires carry designations (Figure 4–70) indicating size, load range, maximum load, maximum inflation pressure in pounds per square inch (PSI), number of plies under the tread and in the sidewalls, manufacturer's name, tubeless or tube construction, radial construction (if a radial tire), and DOT (United States Department of Transportation) symbol indicating conformation to applicable federal standards. Adjacent to the DOT symbol is a tire identification number, the first two characters of which identify the tire manufacturer. The remaining characters identify size, type, date of

FIGURE 4–71   Uniform Tire Quality Grading

manufacture, and whether the tires are tubeless or tube-type.

In addition, DOT requirements mandate that tires must have the Uniform Tire Quality Grading System molded into their sidewalls. As shown in Figure 4–71, this consumer comparison information includes:

- *Treadwear.* The treadwear grade is a comparative rating based on the wear rate of the tire when tested under controlled conditions on a specified government test course. For example, a tire grade 160 would wear twice (two times) as well on the government course as a tire graded 80. The relative performance of tires depends upon the actual conditions of their use, however, and might depart sig-

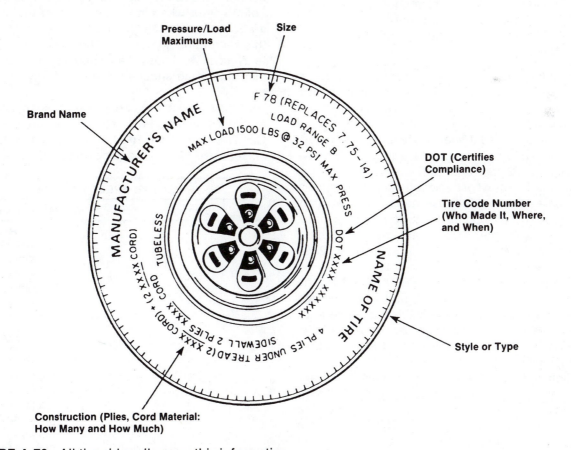

FIGURE 4–70   All tire sidewalls carry this information

nificantly from the norm due to variations in driving habits, service practices, and differences in road characteristics and climate.

• *Traction.* The wet-weather traction grades, from highest to lowest, are "A," "B," and "C," and they represent the tire's ability to stop on wet pavement as measured under controlled conditions on specified government test surfaces of asphalt and concrete. The federal government warns consumers not to choose tires that earn only a "C" traction rating, no matter what other ratings they have. They have poor traction on wet roads.

CAUTION: The traction grade assigned to this tire is based on braking (straight-ahead) traction tests and does not include cornering (turning) traction.

• *Temperature.* The heat dissipation or resistance to high-temperature grading system also uses an "A," "B," and "C" rating, but the "C" grade indicates that the tire meets DOT's present standards for temperature resistance and is acceptable. Tires that earn "A" and "B" temperature resistance ratings are still better.

High-Performance (Speed-Rated) Tires. In recent years, the so-called "high-performance" tire has made an increasing appearance on cars driven in America. Auto industry figures indicate that in the next few years, at least one out of every five new cars sold in the United States will be a sports or sporty-type vehicle. Speed-rated radial performance tires will be found on most of these vehicles. Many drivers do not care if they actually use the speed capabilities of the tires, just as long as the performance is there. In fact, many of their cars are not capable of these higher speeds. A comparison of the Porsche 928S4 and Renault Alliance GTA points this out.

The Porsche has a top speed of 165 MPH; the GTA tops out at 110 MPH. Yet, both are equipped with V-rated tires (up to 149 MPH; see Table 4-2). While the Porsche can certainly use the V-rated tire, the GTA is using it strictly for handling, and, of course, image.

High-performance tires, like all tires, will eventually have to be replaced. In some European countries, the replacement tire must have, by law, the same speed rating as the OE tire. Although it is not the law in the United States, trading down in speed ratings would probably not be a good idea. Optimum

performance and, perhaps, even safe handling might be sacrificed or reduced. If the customer insists on doing this, the undercar shop owner should note his/her recommendation on a receipt just for liability protection. However, low-profile, high-performance radials have been common in Europe for some time. Aspect ratios of 50, 45, and even 35 are not uncommon on European sport cars.

To make a tire safe at high speeds, manufacturers concentrate on the bead package and belting—the two most demanding points at high speed. The addition of bead fillers and chafers are used to reinforce the bead area, while different belting systems are used to stiffen the tread area.

Steel belts are commonly used because of their strength. But, steel expands as speed and heat increase. Most manufacturers overlay the steel belts with nylon so that the carcass is not pulled apart by centrifugal force. Nylon, which contracts as it is heated, holds the carcass together, providing greater stability, improved handling at high speeds, and greater durability.

Three belting systems are available for today's tires: cut belts, folded belts, and woven belts. Cut belts are the most popular, but some weakness has been shown at high speeds with edge separation. Folded belts, wherein the belts closer to the wheel are folded back over the top belts, seem to reduce this separation somewhat. The woven belting system has no weakness at the shoulder and it seems to distribute road shocks more effectively than steel belting.

The speed ratings are currently set by the Economic Commission using the European Passenger Car Tire Regulation R-30. While United States tire makers use the testing procedures established by

TABLE 4-2: SPEED RATINGS	
Symbol	**Maximum Speed**
F	50 MPH
G	56 MPH
J	62 MPH
K	68 MPH
L	75 MPH
M	81 MPH
N	87 MPH
P	93 MPH
Q	100 MPH
R	106 MPH
S	112 MPH
T	118 MPH
U	124 MPH
H	130 MPH
V	149 MPH
Z	+149 MPH

this association, there are no minimum standards to which they must comply. Table 4–2 is a complete list of European speed symbols, indicating the speed at which a tire can carry a load corresponding with its load index under specified service conditions.

Another area of specialization in speed-rated, high-performance tires is tread design. Few drivers can pull into the garage when it starts raining, so while tire makers look for a tread that performs well on dry pavement, they also have to contend with hydroplaning. The most popular tread pattern on speed-rated tires is a block design. The blocks in the design offer good grip while, at the same time, channeling water to the side. The effectiveness of block shapes and spacing on lateral traction has been aided a great deal by computers. Computers can project the results of a tread design long before a prototype tire is made.

Combining Tire Types

Tire construction affects both dimensions and ride characteristics, creating differences that can seriously affect vehicle handling. As a general rule, tires should be replaced with the same size designation or an approved optional size as recommended by the auto or tire manufacturer. In addition to following the vehicle manufacturer's recommendations for tire size, type, inflation pressures, and rotation patterns, the customer should be made aware of the following safety points:

1. Never mix size or construction types on the same axle.
2. Tires on the same axle should be of approximately equal tread depth.
3. All tires on station wagons and all other vehicles used for trailer towing should be of the same size, type, and load rating.
4. There are some vehicles where the use of radial tires might hinder ride quality and control due to the design of the suspension system.
5. It is recommended that new tires be installed in pairs on the same axle. When replacing only one tire, it should be paired

TABLE 4–3: TIRE COMBINATIONS

Construction	Series (Profile)	Bias on Front (Read down for rear)			Belted Bias on Front (Read down for rear)			Radial on Front (Read down for rear)			
		78 Series	70 Series	60/50 Series	78 Series	70 Series	60/50 Series	Metric	78 Series	70 Series	60/50 Series
Bias on Rear (Read across for front)	Conventional (83 Series)	A	NO	NO	A	NO	NO	NO	NO	NO	NO
	78 Series	P	A	NO	A	NO	NO	NO	NO	NO	NO
	70 Series	A	P	NO	A	A	NO	NO	NO	NO	NO
	60/50 Series	A	A	P	A	A	A	NO	NO	NO	NO
Belted Bias on Rear (Read across for front)	78 Series	A	A	NO	P	A	NO	NO	NO	NO	NO
	70 Series	A	A	NO	A	P	NO	NO	NO	NO	NO
	60/50 Series	A	A	A	A	A	P	NO	NO	NO	NO
Radial on Rear (Read across for front)	Metric	A	A	NO	A	A	NO	P	A	A	NO
	78 Series	A	A	NO	A	A	NO	A	P	A	NO
	70 Series	A	A	NO	A	A	NO	A	A	P	NO
	60/50 Series	A	A	A	A	A	A	A	A	A	P

P: Preferred applications. For best all-around car handling performance tires of the same size and construction should be used on all wheel positions.
A: Acceptable but not preferred applications. Consult the car owner's manual and do not apply if vehicle manufacturer recommends against this application.
NO: Not recommended.

with the tire having the most tread to equalize braking traction.

6. If radial tires are used on the car, combine them with radial snow tires on the driving wheels.

7. If equipping a four-wheel drive vehicle and the owner wants to use radials, use them for all four wheels.

8. Snow tires should be of a size and type equivalent to the other tires on the vehicle. Otherwise the safety and handling of the vehicle might be adversely affected.

9. If any doubt exists on combining tires on the same vehicle, use Table 4–3 as a guide.

 SALES TALK _____

Be sure to warn the customer that the mini, space-saver, skin, or similar type of compact spare should be used only temporarily for an emergency. Do not use as a regular tire. Service or replace the regular tire as soon as possible. Any continuous road use of the temporary, emergency compact spare tire might result in tire failure, loss of vehicle control, and possibly injury to vehicle occupants.

WHEELS

The wheels themselves are made of either stamped or pressed steel discs riveted or welded to a circular shape or a die-cast or forged aluminum or magnesium rim or disc. The latter are commonly referred to as "mag" wheels, although they are more commonly an aluminum alloy. Aluminum wheels are lighter in weight when compared with the stamped steel type. As shown in Figure 4–72, there are many so-called custom wheel designs that are sold by parts and accessory dealers.

Mounting holes are provided in the rim or disc. The mounting holes are tapered to fit tapered mounting nuts that center the wheel over the hub. The rim has a hole for the valve stem and a drop center area is offset for ease of tire removal and installation. Wheel offset (Figure 4–73) is the vertical distance between the centerline of the rim and the mounting face of the disc.

Wheel size is designated as rim width and rim diameter. Rim width is determined by measuring across the rim between the rim flanges. Rim diameter is measured across the bead seating areas from top to bottom of the wheel.

PAINT AND BODY EQUIPMENT

The assortment of paints and other related body shop items that an auto store and jobber carries can be a lucrative sales source. Once customers have been attracted to the store to buy paint and body equipment, the counterperson has to be careful to acquire the right information from them. Just as there is certain information needed from a customer before selling a part, certain information is required from a painter before any paint can be sold. The mechanic or do-it-yourselfer who orders a carburetor kit without checking the tag number is exactly like the painter who orders a quart of paint without checking the paint code.

The counterperson should also be familiar with paint equipment, which includes air compressors, spray guns, air regulators, air hoses, and exhaust fans. Other items used in this work include: orbital sanders, air dusting guns, apron taper or masking machines, putty knives, spray masks, dust masks, sanding blocks, squeegees (putty application), single-edge razor blades, sponges, and shop cloths.

UNDERCOAT PRODUCTS

Proper undercoating is part of the foundation for an attractive, durable topcoat. If the undercoat—or combination of undercoats—is not correct, the topcoat appearance will suffer and might even crack or peel.

Undercoats can be compared to a sandwich filler that holds two slices of bread together. The bottom slice of bread is called the substrate (or surface of the vehicle). That surface can be bare metal or plastic, or a painted or preprimed surface. The undercoat is the sandwich filler applied to the substrate. It makes the substrate smooth and provides a bond for the topcoat. The upper slice is the topcoat, the final color coat that the customer sees.

Undercoats contain pigment, binder, and solvent. There are four general or basic types of undercoats:

- Primer
- Primer-surfacer
- Primer-sealer
- Sealer

Most surfaces must be undercoated (Table 4–4) before refinishing for several reasons: to fill scratches to provide a good base for applications, to promote adhesion of the topcoat to the substrate, and to assure corrosion resistance. A primer alone, however, will not fill sand scratches or other surface

FIGURE 4-72 Custom wheel designs *(Courtesy of Motor Sport Wheel, Inc.)*

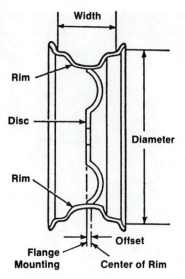

FIGURE 4-73 Wheel dimensions are important to tire replacement.

TABLE 4-4: FUNCTIONS OF UNDERCOATS				
Undercoat Function	Primer	Primer-Surfacer	Primer-Sealer	Sealer
Resists rust and corrosion	Yes	Yes	Yes	No
Makes topcoat adhere better	Yes	Yes	Yes	Yes
Fills scratches and nicks	No	Yes	No	No
Provides uniform hold out of the topcoat	No	No	Yes	Yes
Prevents show through of sand scratches	No	No	Yes	Yes

flaws. Primer-surfacers are used to provide both priming and filling in one step. Primer-sealers are applied to prevent solvents in the topcoat from being absorbed into the porous primer-surfacer. These three undercoats—primer, primer-surfacer, and primer-sealer—can be used together, singularly, or in various combinations, depending on the surface condition and size of the job (Table 4–5). Sealers are employed to improve adhesion between the old and new finishes. To provide good adhesion, a sealer should always be used over an old lacquer finish when the new finish is to be enamel. Under other conditions, a sealer can be desirable but not absolutely necessary.

With a great number of different kinds of undercoat products on the market, the refinisher is often faced with the problem of what one to use. No matter which type of undercoat product is used, the golden rule for selecting surface preparation and all other refinishing products is the same: Never mix manufacturers' products.

 SALES TALK _____

Some metal replacement parts are supplied by the car manufacturers with a primed surface. The car manufacturer might recommend that this primer not be removed because it is an anticorrosive as well as for paint adhesion. It is necessary to apply a primer over the factory primers.

Putties and Body Fillers

Putties and body fillers, while not precisely defined as undercoats, might be termed solid undercoats since they are frequently used in conjunction with one or more of the four liquid undercoats and perform many of the same functions. There are several types of putty that are called by different names depending on the manufacturer, but normally putty is classified as body filler, polyester putty, or lacquer putty (Table 4–6).

TOPCOATS

Automotive topcoat finishes range from paints that have been available for 50 years to new multi-component systems that provide the ultimate in durability. No single finish is the best for all applications; it all depends on a careful matching of finish capabilities and characteristics with the requirements of the application.

The type of paint used for the topcoat ultimately determines the attractiveness of the color, gloss, and finish. Table 4–7 is a general summary of the properties of paint used for repainting.

Table 4–8 illustrates the relative durability of the various popular topcoats. But remind the customer

TABLE 4-5: SURFACES FOR UNDERCOATS				
Undercoat Surface	Primer	Primer-Surfacer	Primer-Sealer	Sealer
Bare substrate (metal, fiberglass, or plastic)	Yes	Yes	Yes	No
Sanded old finish	No	Yes	Yes	Yes

TABLE 4-6: USE OF THREE MAJOR TYPES OF PUTTIES AND BODY FILLERS

	Body Filler	Polyester Putty	Lacquer Putty
Primary Use	Used to smooth out large depressions and fill in scratches.	Used to fill holes in body filler and sandpaper scratches in the metal.	Used to cover pinholes and small scratches after application of primer-surfacer, and to fill in small scratches in the old paint film.
Maximum Film Thickness per Application	Below 1/4"	Below 1/8"	Below 1/16"

TABLE 4-7: SUMMARY OF TOPCOAT PAINT FEATURES

Nomenclature	One-Component Type			Two-Component Type	
	Alkyd Enamels	Acrylic Lacquer	Acrylic Enamel	Polyurethane	Acrylic Urethane Enamel
Spray characteristics	Excellent	Excellent	Good	Good	Good
Possible thickness per application	Fair	Fair	Good	Excellent	Excellent
Gloss — without polishing	Fair	Good	Good	Excellent	Excellent
Gloss — after polishing	Good	Good	Good	—	Good
Hardness	Good	Good	Good	Excellent	Excellent
Weather resistance (frosting, yellowing)	Fair	Fair	Good	Excellent	Excellent
Gasoline resistance	Fair	Fair	Fair	Excellent	Good
Adhesion	Good	Good	Fair	Excellent	Excellent
Pollutant resistance	Fair	Fair	Fair	Excellent	Excellent
Drying time — to touch	68° F 5–10 minutes	68° F 10 minutes	68° F 10 minutes	68° F 20–30 minutes	68° F 10–20 minutes
Drying time — for surface repair	68° F 6 hours 140° F 40 minutes	68° F 8 hours 158° F 30 minutes	68° F 8 hours 158° F 30 minutes	—	68° F 4 hours 158° F 15 minutes
Drying time — to let stand outside	68° F 24 hours 140° F 40 minutes	68° F 24 hours 158° F 40 minutes	68° F 24 hours 158° F 40 minutes	68° F 48 hours 158° F 1 hour	68° F 16 hours 158° F 30 minutes

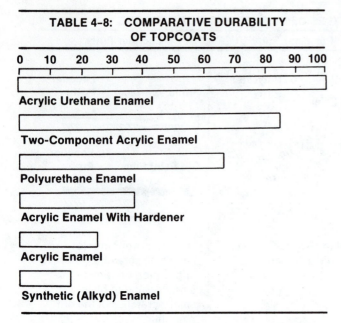

TABLE 4-8: COMPARATIVE DURABILITY OF TOPCOATS

0	10	20	30	40	50	60	70	80	90	100

Acrylic Urethane Enamel

Two-Component Acrylic Enamel

Polyurethane Enamel

Acrylic Enamel With Hardener

Acrylic Enamel

Synthetic (Alkyd) Enamel

TABLE 4-9: PAINT CODE LOCATION
(See Figure 4-76)

Manufacturer	Position
AMC	10
	2
ARROW	5
CHALLENGER 1978–82	5
1983	4
CHAMP	3
CHRYSLER	3
	4
	5
CONQUEST	1
COLT 1974–82	3
1983–84	5
COLT VISTA	4
COURIER	2
	5
	8
DATSUN/NISSAN	6
	8
	9
DODGE D50	5
FORD	2
HONDA	2
ISUZU	6
LUV 1972–80	1
1981–82	6
MAZDA	5
	8
	9
MITSUBISHI	
Starion	1
Montero/Pickup	5
Cordia/Tredia	7
OPEL	
SAPPORO 1978–82	5
1983	4
SUBARU	6
TOYOTA	
Passenger	1
Truck	7

that costs can be deceiving, particularly with the polyurethane and acrylic urethane colors, both of which are substantially more expensive in their unreduced prices per gallon. When costs are viewed in terms of life cycle—for example, the length of time the finish will continue to present an acceptable appearance before requiring refinishing—acrylic urethane enamels prove to be the most economical.

Automotive Colors

There are two basic automobile general color finish types: solid and metallic.

* *Solid Colors.* For many years all cars were solid colors, such as black, white, tan, blue, green, maroon, and so on. These colors are composed of a high volume of opaque type pigments. Opaque pigments block the rays of sun and absorb light in accordance with the type of color they are. That is, the darker the solid color is, the more light it absorbs and the less it reflects. Black will absorb more light and will reflect less; white absorbs less light but reflects a great deal more. When polished, solid colors reflect light in only one direction. Solid colors are still used by the refinisher, but to a lesser degree when compared with a few years ago.
* *Metallic Colors.* Metallic (or polychrome) paint contains small flakes of metal suspended in liquid. The metal particles combine with the pigment to impart varying color effects. The effect depends on the position the flakes assume within the paint film. The position of the metal flakes and the thickness

of the paint affect the overall color of the painted surface. The flakes reflect light, but some light is absorbed by the paint. The thicker the layer of paint, the greater the light absorption.

In the past few years, OEMs around the world have increasingly adopted basecoat/clear coat systems as the finish of choice for new cars rolling off assembly lines. The technology for basecoat/clear coat finishes was developed in Europe. The durability and popularity of these finishes prompted Japanese and American automobile manufacturers to begin offering them, too. In fact, most automotive finishing experts agree that the basecoat/clear coat system will be used on the vast majority of refinished vehicles before the turn of century. When the system was first used, it utilized either an acrylic lacquer clear or a polyurethane enamel clear over an acrylic lacquer basecoat. Early in 1985, the first acrylic enamel basecoat/clear coat refinish system to actually simulate the OEM basecoat/clear coat finish was introduced. Like OEM finishes, the new system loads more transparent pigments into the basecoat and locks the metallic flakes into a flat arrangement, enabling it to match the brightness, the color intensity, and the travel of an OEM basecoat/clear coat finish. This makes it easy to get the best color match available.

Color Selection

If spot or panel repair is planned, it is important to purchase the topcoat color that will accurately match the old color. When planning an overall re-painting, the customer might wish to match an old finish or might want a completely new one.

To order a matching topcoat color, first locate the vehicle identification plate (VIP). Write down the car manufacturer's paint code shown on the plate. Use Figure 4-74 and Table 4-9 to find the position of the paint code number on almost all vehicles, except General Motors. Location of paint code numbers for General Motors is shown in Figures 4-75 and 4-76.

Most auto paint shops have a color book (Figure 4-77). This book contains color chips and color information for almost all makes and models worldwide. First locate the car manufacturer's code number. This permits the refinisher and/or counter-person to identify the color chip next to it. As a double check, it is wise to compare the color chip with the car color, for there is always the chance that the car has been repainted with a different color.

If the color match is correct, order the topcoat from a local supplier by color stock number. Refinish jobbers supply topcoat colors in two ways:

- If it is a recent model or a popular color, chances are they will have it ready-mixed in pint, quart, and occasionally gallon cans. These ready-mixed colors are called factory packaged.
- If it is an older color, they might have to mix it in pint, quart, or gallon quantities. Paint manufacturers work extensively to develop OEM matches with mixing color formulas for all top qualities. Custom-mixed colors are those colors that are mixed to order at the jobber supplier. Today, approximately 50 percent of PBE jobbers mix customer paints.

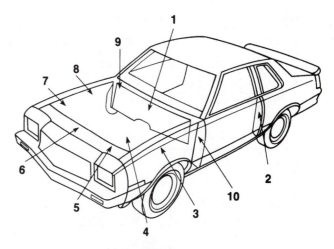

1. **Engine Compartment (center)**
2. **Lock Pillar (left side)**
3. **Fender Apron (left side)**
4. **Hood (left underside)**
5. **Radiator Support (left side)**
6. **Radiator Support (center)**
7. **Fender Apron (right side)**
8. **Fender Apron (right side)**
9. **Engine Compartment (right side)**
10. **Front Door Body Hinge Pillar**

FIGURE 4-74 Locating paint code number on various car makes. *(Courtesy of DuPont Co.)*

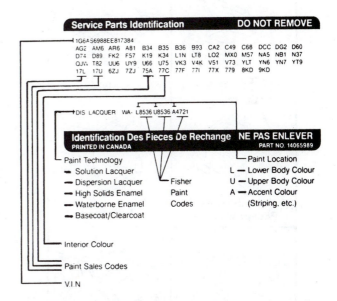

FIGURE 4-75 Typical General Motors' service parts identification label. *(Courtesy of DuPont Co.)*

BODY TYPES

In the mid to late 1970s, American car manufacturers began to produce unibody vehicles. Compared to Europe, where unibody technology has been used for more than twelve years, the U.S.

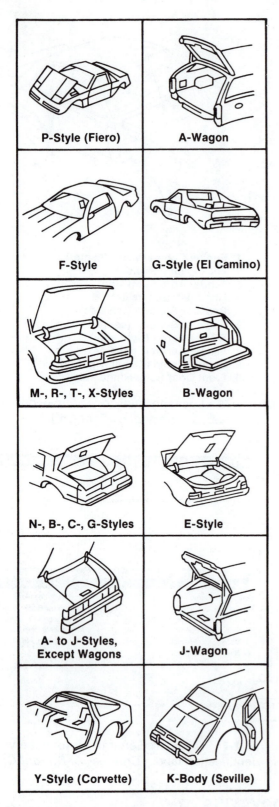

FIGURE 4-76 Since 1985, General Motors has added identification colors used on various parts of the car. These labels or tags define the type of paint used. *(Courtesy of DuPont Co.)*

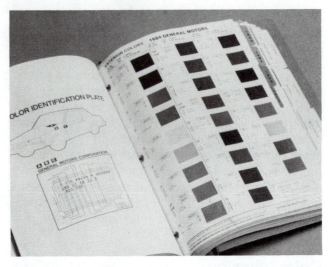

FIGURE 4-77 Refer to the shops's color book containing color chips that identify make and model year. *(Courtesy of DuPont Co.)*

market is still in its infancy. Actually, the unibody is the most important automotive design change since the invention of the automobile itself, but the industry was not ready for change.

There are many shops—including many dealership PBE operations—that perform improper repairs on unibody vehicles. Some of them do not even realize the need to change repair procedures.

What can the counterperson do to improve the situation? First, learn the difference between traditional frame construction and unibody design, and why the unibody has become so popular in recent years. Then, learn to encourge customers to use recommended repair procedures when working on a unibody.

The difference between traditional and unibody construction is simple. Traditionally, a car was built upon a frame (Figure 4-78). The engine, drivetrain, suspension, and body were added to the base structure. The unibody design (Figure 4-79) has no separate frame; the body is constructed in such a manner that the body parts themselves supply the rigidity and strength required to maintain the structural integrity of the car. The unibody design significantly lowers the base weight of the car, and that in turn increases gas mileage capabilities. But, do not assume that unibody vehicles, because they are lighter, are unsafe. Unibodies are built with passenger safety considerations as a top priority. In fact, occupants of a unibody vehicle are often safer than those in a frame car because the exterior sheet metal is designed to absorb the impact of a crash and shift it away from the passenger compartment.

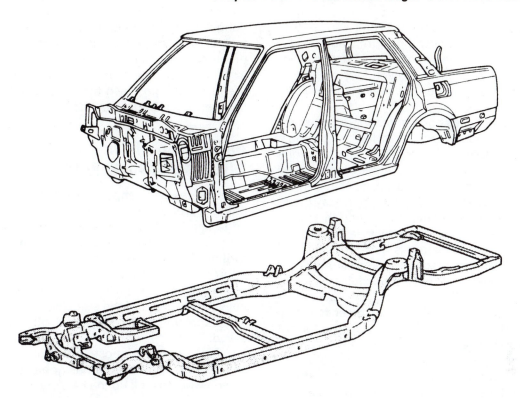

FIGURE 4-78 Typical conventional perimeter frame construction *(Courtesy of Toyota Motor Corp.)*

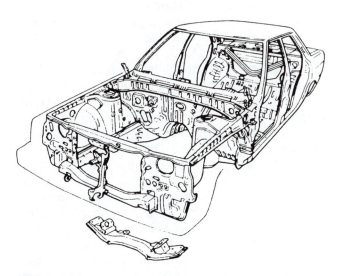

FIGURE 4-79 Typical unibody construction *(Courtesy of Toyota Motor Corp.)*

OTHER ITEMS

Many other items are sold by a jobber or auto parts organization. Auto belts and hoses are subsystems that have to be considered. Fortunately, most aftermarket manufacturers of these subsystem parts provide excellent catalogs to help the counterperson provide the customer with the knowledge to become a good problem-solver.

Knowledge of tools and equipment is also important to give the customer the needed service. Actually this knowledge can range from the simple hand tools of the mechanic trade to such items as wheel balancers, dynamometers, hoists, exhaust gas analyzers, and welders. Often, customers will order special machine tools such as cutters for lathes, grinding wheels, routers, wire wheels, and even common drill bits. The more the counterperson becomes familiar with such tools—especially the ones the store carries—the easier the job becomes.

USE OF PRODUCT KNOWLEDGE

Some people maintain that it is impossible for counterpeople to know too much about the products they sell. This can be true, but only with respect to knowing these things about a product that are important to customers. This critical qualification brings into focus the "features and benefits" concept of selling. It means that to sell a product, a counterperson must know only those particulars about how a product is made—its composition or components, finishes, functions, and so on—that have some bearing on how it performs and what it does for a customer. Everything else about that product might be nice to know or useful for other purposes, but is of little value for selling the product effectively.

It follows that it is shortsighted for counterpeople to devote too much time and effort to learning a mass of particulars about a product—its functions and how it is manufactured or created—unless they also learn how each of those particulars affects performance. Always answer the customers' key question, "What will this do for me?" even though it might not be verbalized openly. Thus the measure of relevance of any product feature is how it benefits the customer. Absorbing this concept and learning how to apply it are vital to selling successfully.

Counterpeople sometimes confuse product features with their corresponding customer benefits. Or they are inclined to think it is not very important to distinguish between the two. Some say, in effect, "Why not just tell a prospective customer all you can remember about either product features or benefits?" Two pitfalls in this matter must be considered. The first to be avoided is the mistake of spending too much time in studying product features to the neglect of user benefits. The second is the mistake of talking too much about the product and too little about what it will do for the buyer.

Because he or she is apt to be attracted to and impressed by product features, the inexperienced counterperson often spends much time describing them to prospects. Nothing could be more unproductive than unduly stressing product features, because nothing is more certain to cause a "so what?" reaction from the prospect. Instead of selling product features, sell the product benefits that result from buying the item. For example, when selling a complete boot set for a CV joint, stress the function of the boot set—it includes new boot hardware and premeasured lubricant for replacing this critical part. This replacement will increase the life of the CV joint. Or, when selling a CV joint service kit, show the customer all that is included in the package: CV joints, boot, special grease, and attachments for plunging or fixed—which insures the customer will have all important parts on hand when it is time to replace the CV joint. Or, if selling a complete shaft replacement, show the fully assembled drive shaft for a front-wheel drive car, ready for installation with no need to tear down and rebuild the old shaft.

The conclusions seem to be that a person cannot know too much about the products he or she sells, and that there is no need to memorize every possible product fact. Recognize that the greatest

danger is overburdening sales talk with so much technical data that the customer is either confused or indifferent, whether he or she is a do-it-yourselfer or professional mechanic. And when that happens, the prospects go down the street and test another counterperson's product knowledge.

A NECESSITY FOR INTERCHANGING PARTS

Besides being important in a good sales presentation of products to your customers, product knowledge can help you increase sales in another way, also. Product knowledge combined with mechanical skill and imagination enable the counterperson to interchange parts for the varying needs of customers—parts that are not in stock, but have close cousins in other areas of the stocking room. In this way sales that might have slipped otherwise can be saved.

For example, suppose a customer needs a plain ball bearing but the store has none. If the counterperson were to look up the bearing codes, he or she could probably find a workable substitute. There might be a shielded bearing on the shelves—a different part number, but a bearing that will do the job. A bearing with a lock ring, again, a different part number, will fit if the lock ring is thrown away. One with a shield and a lock ring is still a different number, but, in all probability, will also fit.

Many times, in the case of alternators, the only difference between the unit in stock and the one the customer needs is the placement of the mounting plate. If the bolts are loosened and the plate is twisted from a 6 o'clock to a 3 o'clock position, the unit becomes identical to the one the customer must have.

Some rebuilders list two part numbers for starters: those with solenoids and those without. Naturally, the units are virtually the same, save the cost. While there is a cost factor in most substitutions, many customers will pay the price added if the part has been difficult to locate or if they need it badly enough. Often that is the case when dealing with construction crews needing an important part for some equipment. They are highly paid workers who cannot continue without that part. Obviously, cost of the interchange is of little trouble in the sale.

INCREASING PRODUCT KNOWLEDGE

Inevitably, there will be times when a customer requests a part or makes an observation that completely exceeds a counterperson's knowledge or experience; this with good reason when considering the vastness of information involved in the automotive aftermarket. No one can possibly hope to know the answer to every question. The important thing is that when this occurs, and the information is available, the counterperson should tell the customer he or she will look up the answer and get back to the customer with it.

The counterperson should not be afraid to swallow his or her pride and admit ignorance on the matter. He or she might be able to bluff long enough to gain a redeeming amount of information as the conversation develops. It might work, and maybe no one will be the wiser. But what if it backfires? In spite of every effort to prevent it, ignorance has an uncanny facility for shining through. Do not become its next transparent victim.

Knowing how to find the necessary information has always been a part of a counterperson's job, but never more than today, and more so in the future, as models proliferate quickly and the components supporting them become more diversified and integrated.

In addition to answering phones and selling parts, being a counterperson means developing and keeping track of as many information sources as possible. These sources fall into four categories:

1. Documents (books, catalogs, bulletins)
2. Organizations (booster clubs, industry association)
3. Events (classes, clinics, trade shows)
4. People (co-workers, salespeople, customers)

Of these four categories counterpeople are generally most comfortble with documents. After all, most working hours are spent using catalogs and price sheets. For more insights on making the most of these books, see Chapter 5.

When one catalog does not have the information, try the catalog of a competing line. There is a lot of difference in the information various manufacturers think is important.

Service manuals are also good sources of information (Figure 4–80). Naturally, not every store keeps a lot of these on file, but certainly the garage accounts will have a large collection. Knowing which customer has which manual can be a real asset.

Try to arrange meetings with mechanics for updates on technical advances. Ask them about technical bulletins, tools, and repair procedures. Later, if stumped by a customer, the counterperson will have a good idea of who might have the answer.

Besides catalogs, bulletins, and repair manuals, the automotive industry is served by the trade press

(Figure 4–81). Becoming familiar with all of the trade publications is a great way to expand one's information network.

Although it would be impossible for the counterperson to collect, organize, file, and maintain all the documents that relate to automotive topics—and still sell a few parts—he or she can keep tabs on where to find all those books.

Organizations are another good source of information, and there are many of them serving the automotive world. Local wholesaler's associations, automotive booster clubs, or garage associations are other potential sources of information, primarily as places to establish contacts for future reference.

Car clubs usually have libraries of factory service manuals, parts catalogs, and magazines that apply to their favorite autos. Many of their members are steeped in the lore of their particular cars. Contacting these people, attending a few of their meetings, and using them as expert information sources has the added advantage of generating new business.

Parts manufacturers themselves are organizations full of information. Many of them are suppliers to carmakers and have vast technical resources. In most cases, a call to the right manufacturer will yield all the needed help. Here are several tips for using manufacturers as information sources:

Check the manufacturer's catalog for a toll-free number. If one is not listed, see if a local rep can provide one, or if collect calls are accepted. If it takes a long-distance call to make contact, ask the switchboard operator to have the appropriate person return the call. Many organizations have WATS lines or other inexpensive long-distance services.

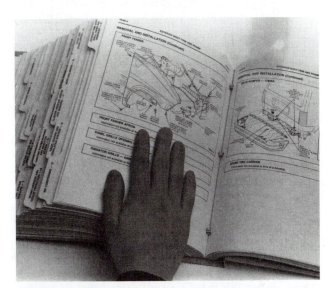

FIGURE 4–80 Typical service manual

FIGURE 4–81 Trade publications help expand a counterperson's knowledge

If calling from a different time zone, be sure to place the call during business hours at the manufacturer's location.

When the counterperson reaches someone who is helpful, get his or her name and ask if the person has a direct line. Note these for future use. There is no sense going through a switchboard more than once.

If the information is available in printed form, ask for a copy. There is a good chance that the publication will contain other information that will be useful later.

Events are another source of information, although not as immediate as documents, organizations, and people. You cannot ask the customer to wait for an answer until the next ignition clinic. However, clinics, classes, and trade shows deserve mention because they not only provide data, they are also good places to build a professional network of contacts.

Of all the outside sources of information, the "people network" is probably the most important. Throughout the local trading area, there are counterpeople, mechanics, salespeople, and service managers who constitute a vast human library.

For each unique problem that confronts the counterperson, there is probably someone in a nearby business who deals with it routinely. Finding and cultivating those sources is one of the most productive ways to enlarge one's network.

There are all sorts of people who can solve that unique problem, supply an obscure part, or at least get one headed in the right direction. Consider just a few examples.

Suppose a customer wants a new horn ring for a '57 Chevy or fresh window rubber for an old Cadillac. A quick call to the owner of a restoration shop could result in the names, addresses, and phone numbers for suppliers of each. An afternoon spent browsing through the restorer's bookshelves should convince a counterperson that he or she can still get just about anything for almost any car.

Often there will be requests for parts that have never made it to the aftermarket and never will. If handling these orders, the one to call is the counterperson at the vehicle's dealer.

Other times, customers need technical information that is not in the aftermarket repair books. In those cases, the dealer's service manager or shop foreman is a good contact. Dealerships have an information pipeline from the carmakers that the aftermarket can only envy. They definitely belong in the counterperson's network.

 SALES TALK _____

The marketing and selling techniques for automobile parts and materials described in Chapters 3 and 4 are covered in detail in Chapter 9.

REVIEW QUESTIONS

1. The electrical system carries electrical current to individual components from the
   _____ .
   a. ignition switch
   b. solenoid
   c. battery
   d. starter motor

2. In addition to the battery, the charging system includes the _____ .
   a. alternator
   b. indicator
   c. regulator
   d. all of the above

3. What protects circuits from fire hazards, accidental shorting, or grounding?
   a. circuit breakers
   b. fusible links
   c. fuses
   d. all of the above

4. What component of the air conditioning system absorbs heat during evaporation and releases it during condensation?
   a. refrigerant
   b. condenser
   c. compressor
   d. solvent

5. What component of the air conditioning system is a pump driven by the crankshaft?
   a. condenser
   b. compressor
   c. blower
   d. heater core

6. Shock absorbers, springs, and control arms are part of what system?
   a. suspension
   b. steering
   c. electrical
   d. transmission

7. What type of steering unit takes the place of the idler arm, pitman arm, center link, and gear box used in conventional systems?
   a. power steering
   b. parallelogram linkage
   c. rack and pinion
   d. cross steer

8. What type of front suspension system utilizes a steel bar that twists lengthwise instead of a spring?
   a. coil spring
   b. torsion bar
   c. stabilizer bar
   d. shock absorber

9. Which of the following is not a component of the powertrain?
   a. clutch
   b. transmission
   c. differential
   d. transverse torsion bar

10. What component of an automatic transmission provides the fluid coupling?
    a. torque converter
    b. planetary gears
    c. mechanical link
    d. stator

11. What component is the heart of the hydraulic brake system?
    a. tubing
    b. drum or disc
    c. water cylinder
    d. master cylinder

12. What type of brakes are actuated manually and are not actually part of the hydraulic brake system?
    a. antilock brakes
    b. parking brakes
    c. antiskid brakes
    d. all of the above

13. When a customer expresses an interest in a tire that offers greater fuel economy, greater strength and flexibility, and increased skid resistance, Counterperson A suggests bias ply tires. Counterperson B suggests radial ply tires. Who is right?
    a. Counterperson A
    b. Counterperson B
    c. Both A and B
    d. Neither A nor B

14. Which of the following is not a type of undercoat?

    a. primer-surfacer
    b. sealer-surfacer
    c. primer
    d. sealer

15. In order to stay abreast of recent automotive trends, Counterperson A refers often to service manuals. Counterperson B attends clinics. Who is right?
    a. Counterperson A
    b. Counterperson B
    c. Both A and B
    d. Neither A nor B

CHAPTER FIVE

CATALOG USE

Objectives

After reading this chapter, you should be able to:
- Describe the basic format components in most catalogs, including the cover, contents, applications section, illustrations, and specifications.
- Explain how to use the catalog for conducting parts and related items sales.
- Explain how to use footnotes in a catalog.
- Explain how to use supplements.
- Explain confusing aspects of some catalogs and what to do about them.
- Explain the importance of comparative cataloging.
- Explain the points to review before putting a new catalog in the rack: current publication; index, table of contents and aids; inaccurate and new numbers; coverage years; back section; corrections; and "sell" pages.
- Explain how to use bulletins for catalog maintenance.
- Describe the most important aspects of arranging the catalog rack, including the alphabetical approach, one-rack approach, Weatherly Index, superseded catalogs, specialty catalogs, and miscellaneous lists.

A good counterperson has a lot of tools at his or her disposal. The personality traits of charm, patience, and diplomacy come in handy in many situations. In addition, the telephone, pen, and invoice form are also necessary tools of the trade. But of all the tools a counterperson relies on for increased sales and improved profitability, none are more important than the catalogs.

Catalogs are the bibles of the auto parts trade. They help the counterperson and parts manager find the correct parts and solve customers' problems quickly. Almost every customer served requires some form of catalog consultation.

Since no two catalogs are exactly the same, a good counterperson should take the time to learn how to use each one. Because of a lack of an industry standard in catalog format, catalog listings vary between different parts manufacturers.

The goal of this chapter is not to advocate total retention but rather to teach where and how to locate and also how to understand the information necessary to solve each customer's problem. The catalog's greatest potential as a sales tool can be reached only when the counterperson becomes familiar with its content.

FORMATS

Catalog format is the arrangement of the information on each page. Formats vary from one type of product to another. The variety of product lines and interrelationship of parts dictate that certain parts groups should be cataloged together. For example, parts needed for electrical systems can be grouped together in one catalog. Likewise, suspension and steering components and engine parts, regardless of vehicle application, can have separate groupings. Even within one product line, each manufacturer's catalog has its own peculiarities. In addition, the major automobile manufacturers each have their own catalogs for dealers containing information specific to their vehicles and components.

COVER

Each catalog has in the upper right-hand corner the form number, the date of issue, and the number and date of the catalog it replaces or supplements (Figure 5-1). The cover also identifies the product line and manufacturer and often provides an area

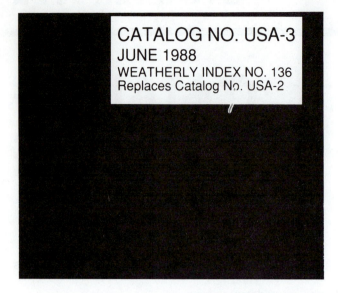

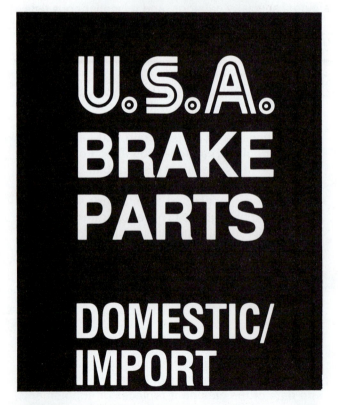

FIGURE 5-1 Most catalog covers include certain information in the upper right-hand corner.

where the jobber can stamp his or her name for the convenience of dealer customers.

Weatherly Index

Many catalogs also list the Weatherly Index number (Figure 5-1). The Weatherly Index is a list of more than 50 pages of three-digit codes to organize catalog categories in the fields of automotive, aeronautical, and marine products (see Appendix C). Because of the system's limitations and inconsistencies, it has been eliminated from some catalogs. For instance, many current producers of catalog material consolidate all of their products into a single publication, making it impossible to list a single product code on the cover. In addition, one product with many various applications—such as switches, which most manufacturers have combined in a single catalog—has different index numbers for each application. For example, switch/assembly lock has code number 592, while switch/oil pressure has 664.

CONTENTS

Every catalog starts with a table of contents that lists alphabetically all sections of the catalog and the make and model numbers (Figure 5-2). Sometimes the various sections are identified by a printed tab on the edge of each right-hand page.

Next, there is a section on how to use the catalog. This section is important because it explains the particular features of the specific catalog. For example, the page in Figure 5-3 uses illustrations to help the counterperson locate the correct parts numbers.

Another useful tool is the vehicle model cross-reference chart. This section lists all the nameplates, (that is, Caprice, Catalina, Centurion) and tells the section of the catalog in which they are listed. For example, Caprice will be under the "Chevrolet Full Size" heading.

Some catalogs also give a Vehicle Identification Number (VIN) chart. Encoded in the vehicle identification number are such things as make, model, body style, engine, model, year, plant of manufacture, and sequence of manufacture. For example, 3J37K4M100231 means:

(3) Oldsmobile
(J) Cutlass Supreme
(37) coupe
(K) 350" 4 bbl
(4) 1974 model
(M) made in Lansing, Mich.
(100231) It was the 231st car assembled

TABLE OF CONTENTS

GENERAL INFORMATION

APPLICATIONS LISTINGS

PRODUCT INFORMATION

NEW PARTS FOR 1988

94-0300	94-1370	94-2110	94-2881
94-0301	94-1510	94-2218	94-2911
94-0304	94-1537	94-2222	94-3018
94-0305	94-1611	94-2272	94-3019
94-1097	94-1621	94-2368	94-4022
94-1160	94-1640	94-2402	94-4073
94-1170	94-1720	94-2617	94-4075
94-1182	94-2090	94-2871	94-5003
94-1208	94-2100	94-2880	

SUPERSEDURES

DEPLETE STOCK	THEN USE	DEPLETE STOCK	THEN USE
94-1020	94-1060	*94-205X	94-2090
94-1181	94-1182	*94-206X	94-2100
94-1240	94-1230	*94-207X	94-2110
94-1451	94-1271	94-2220	94-2850
94-1600	94-1260		

*Part can be relabeled to surving number.

FIGURE 5–2 The table of contents lists all sections of the catalogs, sometimes by make and model, such as this one from a catalog for constant velocity joint parts.

HOW TO READ THIS CATALOG

(1) Find part number of carburetor. Part number will be located in one of seven places illustrated below.

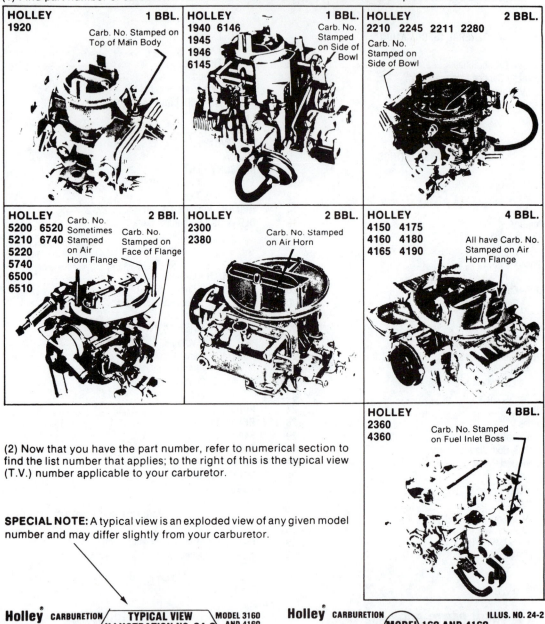

HOLLEY 1 BBL.
1920
Carb. No. Stamped on Top of Main Body

HOLLEY 1 BBL.
1940 6146
1945
1946
6145
Carb. No. Stamped on Side of Bowl

HOLLEY 2 BBL.
2210 2245 2211 2280
Carb. No. Stamped on Side of Bowl

HOLLEY 2 BBL.
5200 6520
5210 6740
5220
5740
6500
6510
Carb. No. Sometimes Stamped on Air Horn Flange
Carb. No. Stamped on Face of Flange

HOLLEY 2 BBL.
2300
2380
Carb. No. Stamped on Air Horn

HOLLEY 4 BBL.
4150 4175
4160 4180
4165 4190
All have Carb. No. Stamped on Air Horn Flange

HOLLEY 4 BBL.
2360
4360
Carb. No. Stamped on Fuel Inlet Boss

(2) Now that you have the part number, refer to numerical section to find the list number that applies; to the right of this is the typical view (T.V.) number applicable to your carburetor.

SPECIAL NOTE: A typical view is an exploded view of any given model number and may differ slightly from your carburetor.

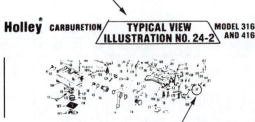

Holley CARBURETION | **TYPICAL VIEW ILLUSTRATION NO. 24-2** | MODEL 3160 AND 4160

By matching the part desired with the appropriate picture, determine the *"Call Out Number"*. This same number appears on the facing page. Reading to the right you will find a *"Letter Code"*. This indicates how the item is sold, L is "Loose", G is "Gasket Kit", R is "Renew", and M is "Master Repair Kit". Only those parts coded with an "L" can be ordered

Holley CARBURETION | **MODEL 160 AND 4160** | **ILLUS. NO. 24-2**
Refer to Numerical Listing Section for Variable Parts

28	L	Throt. Plate Screw - Secondary	5R-595	81	L	Air Vent Valve	X
29	L	Pump Level Adjusting Screw	5R-854	82	L-M	Pump Discharge Needle Valve	23R-256
30	L	Pump Discharge Nozzle Screw	5R-623	83	L-M-R	Power Valve Assy - Primary	X
31	L	Fast Idle Cam Plate Scr. & L.W.	5R-623	85	G-M-R	Fuel Line Tube 'O' Ring Seal	X
32	L	Choke Cont. Wire Brkt. Clamp Scr	5R-641	86	G-M-R	Balance Tube 'O' Ring Seal	X
33	L	Pump Cam Lock Screw	5R-682	87	G-M-R	Fuel Valve Seal 'O' Ring Seal	X
34	L	Fuel Pump Cov. Assy. Scr. & L.W.	X	88	G-M-R	Idle Needle Seal	X
35	L	Secondary Metering Body Screw	5R-691	89		Choke Rod Seal	27R-238
36	L	Throt. Body Screw - Special	5R-716	90	L-M	Diaphragm Housing Check Ball-Sec	X
37	L	Fuel Valve Seat Lock Screw	5R-765	91	L	Pump Inlet Check Ball	28R-35
37	L	Float Shaft Brkt. Scr. & L.W.	5R-1016	92	L	Throttle Lever Ball	28R-65
	L	Spark Hole Plug	7R-5	93	L	Pump Discharge Check Ball	28R-94

individually. All others must be ordered by the appropriate kit.

Continuing to read to the right you will find a *Part Description*, then either a *Part Number* or a *Code Letter* "X". The "X" indicates you must look in the first section of the catalog..."Numerical" for that specific part.

FIGURE 5-3 Most catalogs include a section explaining how to use the specific catalog, such as this page in a carburetor catalog and one in a catalytic converter catalog.

HOW TO USE THIS CATALOG

USING THE APPLICATION LISTING FOR MONOLITH 1™ AND AIR CONNECTORS:

1. Find the auto or truck model listing in the INDEX on the following page, and turn to the page number indicated for that model. Locate the model listing that page. Major vehicle heading are in bold letters on a tinted background for quick identification.

2. Locate the specific year, engine and model vehicle required. Be sure to check all information shown for correct application.

3. The proper MONOLITH 1™ 3-way catalytic converter.

4. Read across to the MONOLITH 1™ installation kit column for the correct part number.

5. Read across to the proper column for the 3-way Plus Oxidation catalytic converter, listed to the right of the kit.

6. Read across to the proper MONOLITH 1™ 2-way catalytic converter.

7. Locate the MONOLITH 1™ direct-fit catalytic converter if desired, no kit is then required.

8. The last column will list the Air Connector part number. Be sure to check installation instructions in kit. If N/R appears, no Air Connector is required.

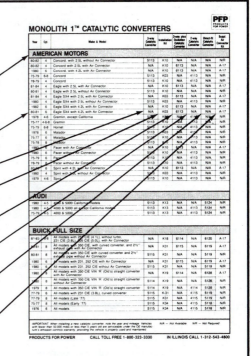

IMPORTANT NOTICE: The exhaust system components and catalytic converters manufactured by Products For Power are warranted to buyers for resale to be free of defects in material and workmanship. Liability is specifically limited to the value of mechandise sold or furnished, and does not include labor charges or any other claims incidental to replacing a defective item. Products For Power will replace any Products For Power exhaust component which is found defective by the Company after inspection at the factory. If a warranty claim relates to emission perfomance, proof of failure must be provided, i.e., test results. Furthermore, proof of the engine being tuned to manufacturers specifications and in proper condition is also required.

It is essential the engine is properly tuned to manufacturers specifications **BEFORE** installing a new catalytic converter. All engine emission control equipment must be operating within the vehicle manufacture specifications **PRIOR** to installation of the new catalytic converter. An improperly tuned engine may damage the catalytic converter. The manufacturer assumes no liability for catalytic converter failures caused by engines improperly maintained or the use of leaded fuel.

EXCLUSIONS: Warranty does not cover exhaust components that are blown out by backfire, or subjected to faulty conditions. Converters that have been installed on vehicles with modified exhaust or emission systems are not warranted. Converters that have been modified or installed on a vehicle for which they have not been cataloged are not warranted.

MONOLITH 1™ CATALYTIC CONVERTER LIMITED WARRANTY

Products for Power warrants its Catalytic Converter will meet Federal EPA emission reduction standards for a period of 25,000 miles. Products For Power further warrants the external converter shell for a period of 5 years or 50,000 miles, (whichever comes first) from the date of installation. Liability is specifically limited to the value of the merchandise sold or furnished, and does not include labor charge or any other claims incidental to replacing a defective item.

Products for Power will replace such defective Converter provided that it be returned to the original place of purchase along with the warranty claim card and proof of purchase. If a warranty claim relates to original emission performance, proof of failure of an emission test (test results) must be provided. Furthermore, proof of the engine being tuned to manufacturers' specifications and maintained in proper operating condition is also required. Products for Power reserves the right to make a determination as to the proper operating condition of such vehicle.

This warranty is conditioned upon the Converter being properly installed on the vehicle for which it is cataloged and maintained in a proper operating condition. The warranty does not cover the exhaust components that are blown out by backfire, damaged by accidents, or subjected to faulty conditions. The manufacturer assumes no liability for Converter failures caused by engines improperly maintained or the use of leaded fuel. Converters that have been installed on vehicles with modified exhaust or emission systems, are not warranted. Converters that have been modified or installed on a vehicle for which they have not been cataloged are not warranted.

This warranty gives you specific legal rights, you may have other rights which may vary from state to state.

PRODUCTS FOR POWER
1035 Republic Drive
Addison, IL 60101

FIGURE 5–3 continued

DISC BRAKE PARTS
IDENTIFICATION CHART

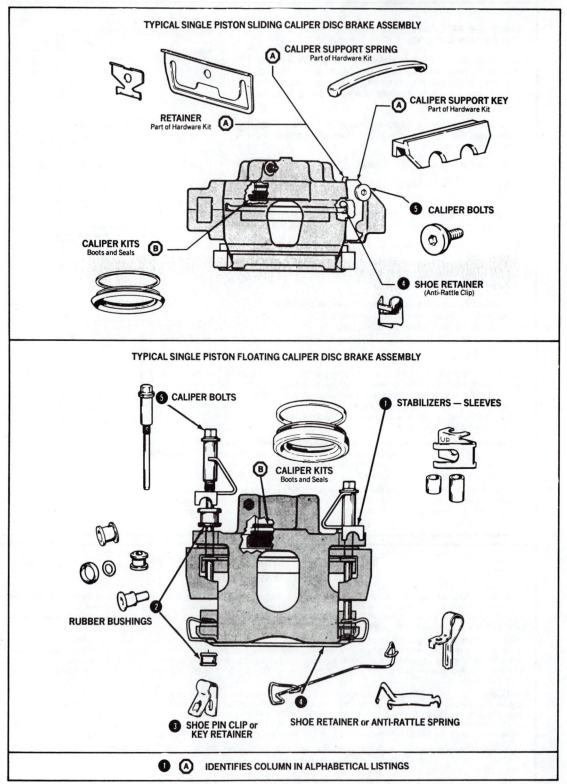

FIGURE 5-4 Illustrations help counter personnel identify components, such as brake parts in these identification charts.

REPLACEMENT CARTRIDGE

DIMENSIONING

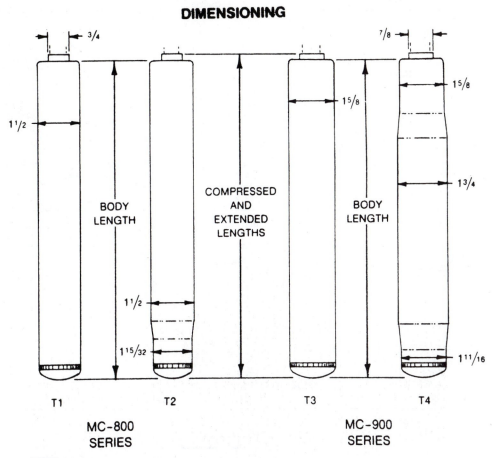

MC-800
SERIES

MC-900
SERIES

NOTE:
All Piston Rod Ends Are Designed Special For Each Application.

FIGURE 5-5   Illustrations can also be used to obtain dimensions and measurements, such as this diagram of a replacement cartridge in a shock and strut catalog.

Learning to read and interpret VINs can save the counterperson a great deal of time and problems.

APPLICATIONS SECTION

When it comes to standardization, the sequence of listings in the applications section is the most troublesome area. Various manufacturers arrange their listings and group vehicles in the sequence that makes the most sense to them. The end result is an inconsistency of listing between different parts manufacturers. For example, a Grand Prix may be listed under Grand Prix, Pontiac Full Size, or Pontiac Grand Prix.

ILLUSTRATIONS

A good catalog has illustrations that are useful and clearly labeled. Schematics, such as shown in figures 5-4 and 5-5, help to identify parts or dimensions of an assembly or system. Drawings or photographs of parts for core identification purposes are a must.

Some catalogs that offer only specific automotive parts include illustrations for every part. For example, Figure 5-6 shows heater illustrations and specifications that make up almost every page from a heater core catalog.

The usefulness of the illustrations found in a catalog can hardly be overestimated. Many catalogs contain very detailed pictures and/or illustrations with descriptions of parts. These illustrations or pictures sometimes contain measurements and instructions on how to measure and identify the parts correctly.

They also might contain exploded views of all components, designed to help the counterperson properly identify the part he or she is looking for.

HEATER ILLUSTRATIONS & SPECIFICATIONS

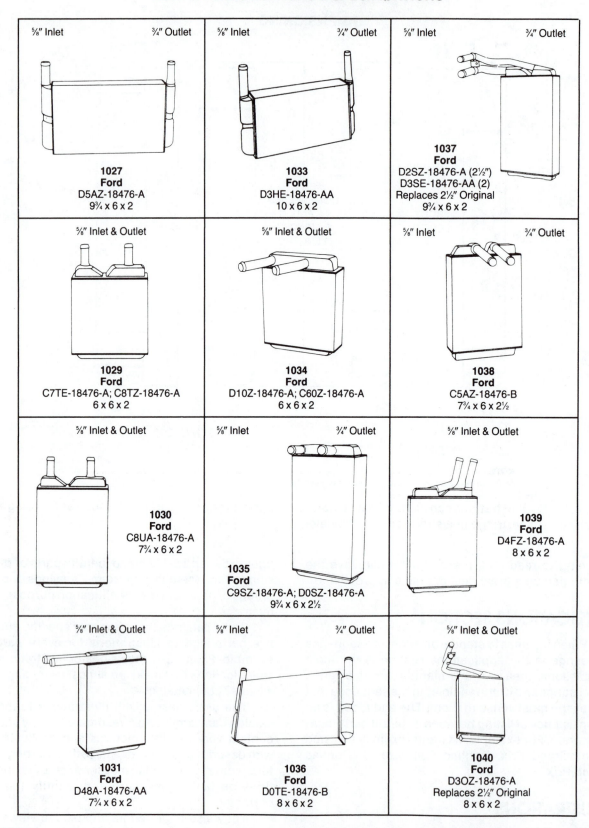

1027
Ford
D5AZ-18476-A
9¾ x 6 x 2

⅝" Inlet — ¾" Outlet

1033
Ford
D3HE-18476-AA
10 x 6 x 2

⅝" Inlet — ¾" Outlet

1037
Ford
D2SZ-18476-A (2½")
D3SE-18476-AA (2)
Replaces 2½" Original
9¾ x 6 x 2

⅝" Inlet — ¾" Outlet

1029
Ford
C7TE-18476-A; C8TZ-18476-A
6 x 6 x 2

⅝" Inlet & Outlet

1034
Ford
D10Z-18476-A; C60Z-18476-A
6 x 6 x 2

⅝" Inlet & Outlet

1038
Ford
C5AZ-18476-B
7¾ x 6 x 2½

⅝" Inlet — ¾" Outlet

1030
Ford
C8UA-18476-A
7¾ x 6 x 2

⅝" Inlet & Outlet

1035
Ford
C9SZ-18476-A; D0SZ-18476-A
9¾ x 6 x 2½

⅝" Inlet — ¾" Outlet

1039
Ford
D4FZ-18476-A
8 x 6 x 2

⅝" Inlet & Outlet

1031
Ford
D48A-18476-AA
7¾ x 6 x 2

⅝" Inlet & Outlet

1036
Ford
D0TE-18476-B
8 x 6 x 2

⅝" Inlet — ¾" Outlet

1040
Ford
D3OZ-18476-A
Replaces 2½" Original
8 x 6 x 2

⅝" Inlet & Outlet

FIGURE 5–6 Some catalogs that offer specific automotive parts include illustrations for almost every part such as these heater illustrations and specifications in a heater core catalog.

HEATER ILLUSTRATIONS & SPECIFICATIONS

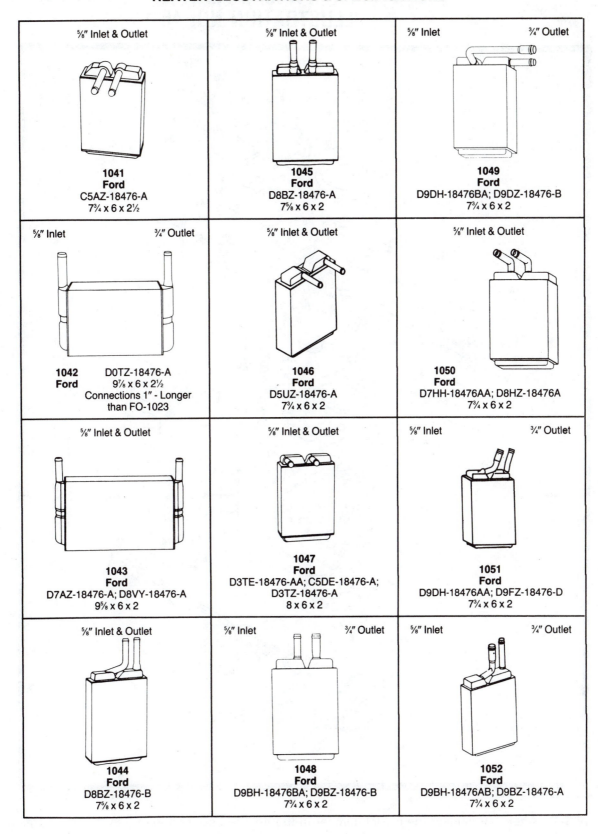

⅝" Inlet & Outlet	⅝" Inlet & Outlet	⅝" Inlet ¾" Outlet
1041 **Ford** C5AZ-18476-A 7¾ x 6 x 2½	**1045** **Ford** D8BZ-18476-A 7⅝ x 6 x 2	**1049** **Ford** D9DH-18476BA; D9DZ-18476-B 7¾ x 6 x 2
⅝" Inlet ¾" Outlet	⅝" Inlet & Outlet	⅝" Inlet & Outlet
1042 D0TZ-18476-A **Ford** 9⅞ x 6 x 2½ Connections 1" - Longer than FO-1023	**1046** **Ford** D5UZ-18476-A 7¾ x 6 x 2	**1050** **Ford** D7HH-18476AA; D8HZ-18476A 7¾ x 6 x 2
⅝" Inlet & Outlet	⅝" Inlet & Outlet	⅝" Inlet ¾" Outlet
1043 **Ford** D7AZ-18476-A; D8VY-18476-A 9⅝ x 6 x 2	**1047** **Ford** D3TE-18476-AA; C5DE-18476-A; D3TZ-18476-A 8 x 6 x 2	**1051** **Ford** D9DH-18476AA; D9FZ-18476-D 7¾ x 6 x 2
⅝" Inlet & Outlet	⅝" Inlet ¾" Outlet	⅝" Inlet ¾" Outlet
1044 **Ford** D8BZ-18476-B 7⅝ x 6 x 2	**1048** **Ford** D9BH-18476BA; D9BZ-18476-B 7¾ x 6 x 2	**1052** **Ford** D9BH-18476AB; D9BZ-18476-A 7¾ x 6 x 2

FIGURE 5-6 continued

TYPICAL VIEW
ILLUSTRATION NO. 46-1

MODEL 2280

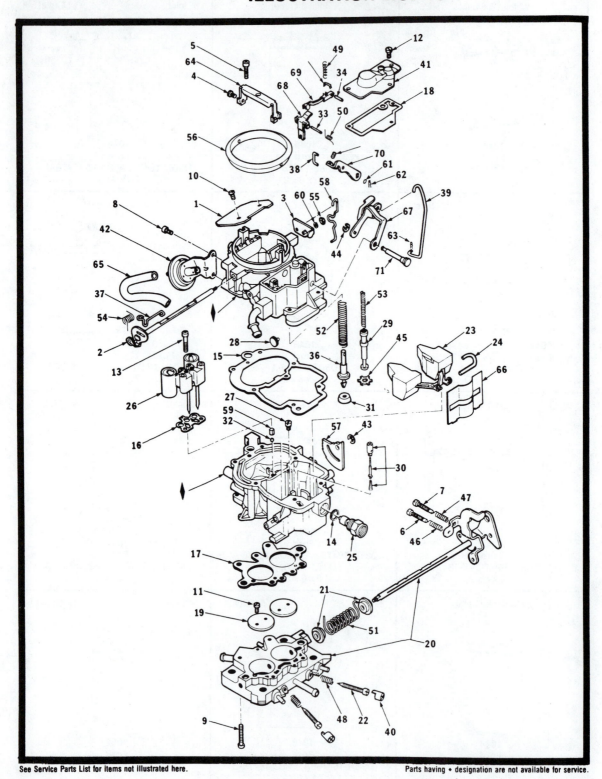

See Service Parts List for items not illustrated here.   Parts having • designation are not available for service.

FIGURE 5–7 After determining the model number in the how-to-use section of a carburetor catalog, the counterperson turns to the appropriate back pages, such as these, for a detailed illustration of components and a list of parts and numbers.

MODEL 2280

ILLUS. NO. 46-1

Refer to Numerical Listing Section for Variable Parts

Index Number	Letter	Part Name	Part Number or Reference Section	Index Number	Letter	Part Name	Part Number or Reference Section
1	L	Choke Plate	2R-488	43	L	Fast Idle Cam Retainer	36R-635
2	L	Choke Shaft & Lever Assembly	X	44	L	Pump Operating Lever Retainer	36R-635
3	L	Dechoke Lever	X	45	L	Power Valve Piston Retainer	36R-737
4	L	Air Cleaner Bracket Screw	5R-167	46	L	Throttle Stop Screw Spring	X
5	L	Air Horn to Main Body Screw & L.W.	5R-290	47	L	Fast Idle Adjusting Screw Spring	X
6	L	Throttle Stop Screw	X	48	L	Idle Adjusting Needle Spring	38R-320
7	L	Fast Idle Adjusting Screw	X	49	L	Vent Valve Operating Lever Spring	X
8	L	Bracket Retaining	5R-943	50	L	Vent Valve Spring	38R-1851
9	L	Throttle Body to Main Body Screw & L.W.	5R-1189	51	L	Throttle Return Spring	X
10	L	Choke Plate Screw	5R-1273	52	L	Pump Drive Spring	38R-1880
11	L	Throttle Plate Screw	X	53	L	Power Valve Piston Spring	38R-1909
12	L	Cover Screw	5R-1428	54	L	Choke Shaft Spring	38R-1926
13	L	Nozzle Bar Screw	5R-1519	55	L	Dechoke Lever Nut	39R-5
14	G.P.	Fuel Inlet Fitting Gasket	G-P	56	L	Air Cleaner Ring	40R-572
15	G.P.	Main Body Gasket	G-P	57	L	Fast Idle Cam	X
16	G.P.	Nozzle Bar Gasket	G-P	58	L	Fast Idle Rod	43R-677
17	G.P.	Throttle Body Gasket	G-P	59	L	Pump Discharge Weight	44R-173
18	G.P.	Pump Housing Cover Gasket	G-P	60	L	Dechoke Lever Nut L.W.	46R-265
19	L	Throttle Plate	9R-259	63	L	Pump Link Retainer	X
20	L	Throttle Body & Shaft Assembly	X	64	L	Air Cleaner Bracket	49R-1390
21	L	Throttle Return Spring Bushings	X	65	L	Vacuum Hose	52R-88
22	L	Idle Adjusting Needle	X	66	L	Float Baffle	59R-130
23	L	Float & Hinge Assembly	16A-510A	67	L	Pump Lever	63R-1217
24	L	Float Hinge Shaft & Retainer	17R-52	68	L	Vent Valve Lever	63R-1219
25	P.L.	Fuel Inlet Needle & Seat Assembly	6-139	69	L	Vent Valve Operating Lever	63R-1221
26	L	Nozzle Bar Assembly	X	70	L	Pump Operating Lever	63R-1244
27	L	Main Jet	X	71	L	Pump Operating Lever Shaft	67R-1480
28	L	Vent Valve	23R-699	72	L	Vent Valve Retainer	36R-746
29	L	Power Valve Piston	25R-693	73	L	T.P.T. Bracket Screw	X
30	P.L.	Power Valve Assembly	X	74	L	Power Valve Adjusting Screw	5R-1518
31	P.L.	Pump Cup	8-234	75	L	Power Valve Assembly (Staged)	X
32	L	Pump Discharge Ball	28A-6	76	L	Choke Diaphragm Link Pin	29R-743
33	L	Vent Valve Lever Pin	29R-854	77	L-R	Roll-Pin	X
34	L	Vent Valve Operating Lever Pin	29R-1268	78	L	Power Valve Stem Retainer	36R-169
35	L	Air Cleaner Ring Retaining Pin	29R-1184	79	L	Power Valve Spring	38R-1851
36	L	Pump Steam & Head Assembly	30R-644A	80	L	T.P.T. Lever Nut	X
37	L	Choke Diaphragm Link	X	81	L	T.P.T. Assembly Locknut	X
38	L	Pump Stem Link	33R-587	82	L	Mechanical Power Valve Stem	42R-528
39	L	Pump Link	33R-589	83	L	Mechanical Power Valve Stem Cap	42R-538
40	P.L.	Idle Needle Limiter	34R-6941-4	84	L	T.P.T. Lever Lockwasher	X
41	L	Accelerator Pump Housing Cover	34R-8692	85	L	T.P.T. Bracket	X
42	L	Choke Diaphragm Assembly Complete	X	86	L	T.P.T. Lever	X
				87	L	T.P.T. Assembly	X
				88	L	T.P.T. Wire Assembly	X

NOTE: "X" Symbol indicates "Refer to Numerical Listing Section for Specific Part Required"
G = Gasket Kit   L = Loose Parts   R = Renew Kit
*These items are not shown on the typical view.

FIGURE 5-7 continued

Matching dimensions to configurations can help determine the right part. Illustrations can be used to show measurement points as well as the specifications.

SPECIFICATIONS

The back section of the catalog contains reference materials, including product specifications in numerical and size sequences, interchanges or cross-reference to original equipment (OE) service parts or major competitors, and a numerical list of parts with a condensed applications listing.

A section on specialized tools and equipment designed to service the product line usually completes the catalog.

CONDUCTING SALES WITH THE CATALOG

Although important tools to the counterman, catalogs are a costly necessity for the automotive parts manufacturer. The manufacturers do not derive any direct sales from the distribution of their catalogs. Yet, these companies could not sell their products without them. Volumes such as interchange parts and reference catalogs can stimulate sales by allowing the purchaser to quickly determine what he or she should buy.

Since the ultimate objective of 99 percent of all catalogs is sales, catalogs are an important sales tool. Some catalogs build retail store trade while others provide valuable technical data resulting in sales.

To be a long-term sales builder, a successful catalog has to be prepared with the primary objective of expanding the customer base by generating sales in new markets. It should offer counter personnel, through design and planning, an assurance of product dependability and service.

To achieve the maximum sales potential, counter personnel must become familiar with the contents of each catalog. Without knowing what resources are available, it is impossible to select the best research strategy. Many counterpersons make it a habit to take new catalogs home for study before committing them to their counter racks.

RESEARCH STRATEGY

One of the most important catalog techniques involves developing a strategy for each order filled. The sequence in which the counterperson conducts research can have a dramatic effect on the time it

takes, how many questions are asked, and how much backtracking is done.

A counterperson should develop a strategy before beginning to file an order. For example, suppose the shop gets an order for tune-up parts, filters, belts, and hoses. Before opening the nearest book, the counterperson should take a few seconds to consider the best path to follow. If the filter catalog includes listings for PCV valves and spark plugs right along with the filters, there is not much sense in opening the spark plug catalog.

Also, much time can be saved by noting the details like the gap specification for spark plugs. The counterperson can write down the gap with the plug number to save time finding it again when the customer asks later.

If any of the old parts are on the counter, the counterperson can save steps by examining them for numbers before attempting to look them up.

Since different catalogs present information in different formats, there is no one method of researching material in a catalog. But no matter what part is needed for no matter what application, the best first step is to gather as much pertinent information about the vehicle and part as possible. Professional mechanics might be able to give counter personnel very detailed information about make, model, year, part needed, and perhaps even the number from it. However, walk-in customers and do-it-yourselfers might exclude information because they do not know it is important. It is up to the counterperson to ask as many questions as possible. For example, a customer might enter a shop and request a particular part for a 1984 Dodge Aries. The counterperson researches the part number, determines if it is in stock, finds it, and sells it to the customer. The next day, the customer, now angry, calls back claiming to have been sold the wrong part. Only then does the counterperson realize that the vehicle is a Dodge Aries station wagon.

To determine the particular method of usage for a catalog, the counterperson must refer to the how-to section in the front. For example, the how-to page from a carburetor catalog shown in Figure 5–3 uses seven illustrations to show the user where to find the part number on the carburetor that must be replaced or repaired. If the part number on the carburetor is 2280, the counterperson would turn to the pages shown in Figure 5–7 for Model 2280 to find a listing of 88 parts for that carburetor and a detailed illustration for the location of each part within the carburetor.

Another example of a how-to section in a catalog, this one for engine parts, lists what information is needed before researching the part number (Figure 5–8). The table of contents beneath the instruc-

CONTENTS

HOW TO USE THIS CATALOG

1.) Locate the cubic inch and configuration (In-line 4, V-6, etc.) of the engine.

2.) Locate the correct year and verify the application. Also confirm the original equipment camshaft part number if possible.

3.) Read across the pages to find the desired products.

LIMITED WARRANTY

Crane Cams Products are warranted to be free of defects in material and workmanship for a period of one year. This applies to the first retail purchaser only and excludes normal wear. Warranty is void if the products have been modified or not installed using our prescribed installation procedures.

IMPLIED WARRANTY OF MERCHANTABILITY IS LIMITED TO ONE (1) YEAR FROM DATE OF PURCHASE. Some states do not allow limitations on how long an implied warranty lasts, so the above restrictions may not apply to you.

THERE ARE NO WARRANTIES THAT EXTEND BEYOND THE DESCRIPTION ON THE FACE HEREOF.

Under any circumstance, liability for all warranties expressed or implied is limited to the repair or replacement of the defective item. Whether a defect in material or workmanship exists shall be determined solely by Crane Cams. Crane Cams specifically disclaims any and all responsibility for expense of diagnosis, removal and/or installation labor, and further specifically disclaims any liability for consequential or incidental damages incurred as a result of defects.

To request repair or replacement, the parts in question should be sent freight prepaid direct to Crane Cams, Incorporated, 530 Fentress Blvd., Daytona Beach, Florida 32014-1210, Attention: Warranty Department. Parts must be accompanied by a copy of your dealer's invoice.

This limited warranty gives you specific legal rights, and you may also have other rights which may vary from state to state.

FIGURE 5-8 The how-to-use section in some catalogs, such as this one for engine parts, includes instructions for obtaining all information needed from the customer before researching the part number

O.E. REPLACEMENT CAMSHAFTS

Crane Cams' new replacement camshaft program offers a quality line of popular replacement camshafts to handle your needs. Each camshaft comes packaged in an eye catching, multi-colored box and is covered by a one year limited warranty. The fastest movers are also available in attactive self-selling camshaft & lifter kit packages. Matching lifters and slipper followers are also conveniently shown for most applications.

AMERICAN MOTORS

C.I. DISP	CYL	YEARS	APPLICATION	REPLACES MFG. PART #	CAM PART #	LIFTER PART #	C & L PART #
151	L-4	1980-83	ALL (Pont.)	8132249	C671	C73-8	

BUICK

C.I. DISP	CYL	YEARS	APPLICATION	REPLACES MFG. PART #	CAM PART #	LIFTER PART #	C & L PART #
151	L-4	1980-84	ALL W/2BC OR FUEL INJECTION NOTE: IN REPLACEMENT APPLICATIONS FOR PART NO. 10029616, THE FUEL PUMP ECCENTRIC ON THE CAM IS NOT USED.	10006745	C671	C73-8	
181	V-6	1982-85	ALL W/2BC	1264395	C720	C73-12	
196	V-6	1978	ALL-EXC. PB ENG. (DO NOT USE WITH OEM PROD. CARB #17058188.)	1264395	C720	C73-12	
231	V-6	1984-85	ALL W/FUEL INJECTION	1264395	C720	C73-12	
		1978-83	ALL W/4BC, TURBOCHARGED	1264395	C720	C73-12	
		1979-84	ALL W/231-A ENG.	1264395	C720	C73-12	
		1979	ALL 2/231-2 ENG. & 2BC	1264395	C720	C73-12	
		1978	ALL 2/231-G ENG. & 2BC	1264395	C720	C73-12	
250	L-6	1968-75 1975	ALL-EXC. INTEGRATED CYL. HEAD, W/A.T. & CALIF. ALL 1975 W/INT. CYL. HEAD ALL	3864497	C663	C73-12	
		1974-75	1974 W/A.T. & CALIF.	6262809	C644	C73-12	
		1974	ALL W/M.T.	6262809	C644	C73-12	
252	V-6	1980-84	ALL W/EVEN FIRE ENG. & INTEGRAL DIST. DRIVE GEAR.	1264395	C720	C73-12	
267	V-8	1981-82	ALL	14014475	C707	C37	CK707
305	V-8	1979-85	ALL W/4BC	14014475	C707	C37	CK707
		1977-79	ALL W/2BC	361995	C628	C37	
307	V-8	1980-84	ALL W/4BC (Olds)	22506227	C590	C49	
350	V-8	1977-80	ALL W/VIN CODE NO. "R" ENGS. (Olds)	22506227	C590	C49	
		1977-79	ALL W/VIN CODE NO. "L" ENGS. (Chev.)	3896929	C274	C37	CK274
		1968-75	ALL-EXC. 1974 W/CALIF. (Buick)	1237736	C585	C73	
403	V-8	1977-79	ALL	22506227	C590	C49	

CADILLAC

C.I. DISP	CYL	YEARS	APPLICATION	REPLACES MFG. PART #	CAM PART #	LIFTER PART #	C & L PART #
252	V-6	1980-83	ALL W/EVEN FIRE ENG. & INTEGRAL DIST. DRIVE GEAR.	1264395	C720	C73-12	
350	V-8	1976-80	ALL W/VIN CODE NOS. "B", "R", & "8" ENGS. (Olds)	22506227	C590	C49	

NOTE: A letter or number in the application refers to the GM engine code displayed as the fifth digit (77-80) or eighth digit (81-85) of the vehicle identification number (VIN) located on the dash panel.

FIGURE 5-9 Pages 4 and 5 in the same engine parts catalog display the information needed for finding parts numbers for American Motors, Buick and Cadillac engines.

Manufactured to meet or exceed the highest quality control standards in the industry, these cams are produced in our new 145,000 square foot, Daytona Beach, Florida facility, to insure you of receiving the finest product available. With over 30 years of experience in producing camshafts for applications varying from heavy duty trucks to garden tractors, from diesel locomotives to motorcycles or from Indianapolis race cars to station wagons, Crane Cams offers you the utmost in design, tooling and manufacturing capabilities all under one roof.

VALVE SPRING	ROCKER ARM	PUSHRODS	TIMING SETS	BREAK IN LUBE
VS685-8		PR1029 (80-81) PR1027 (82-83)	TC2542	

VS685-8		PR1029 (80-81) PR1027 (82-83)	TC2542	
	RA542	PR1015	TC6100	
VS1071-12	RA542	PR1015	TC6100	
VS1071-12 (78-79E) VS1170-12	RA542	PR1015	TC6100	
VS886-12 (W/SEP INT) VS685-12	RK402	PR1002	TC2525	99003-1
VS1170-12	RA542	PR1015	TC6100	
VS880	RK401	PR1001	TC3900	
VS880 (EX 77) VS881 (77)	RK401	PR1001	TC3900	
VS1016		PR1018	TC3480	
VS1016	RK540 (77-79)	PR1010 (77-79) PR1018 (80)	TC3480	
VS881	RK401	PR1001	TC3900	
VS1001			TC3340	
VS1016	RK540	PR1010	TC3480	

VS1170-12	RA542	PR1015	TC6100	
VS1016	RK540 (76-79)	PR1010 (77-79) PR1018 (80)	TC3480	

FIGURE 5-9 continued

tions shows the correct page numbers to examine. For example, if a customer requested a valve spring for a 1981 Buick with a V-8 engine, the counterperson would turn to pages 4 and 5, (Figure 5–9), look down the left-hand columns to locate the correct Buick make, and then move toward the right until the correct number was found in the column marked valve spring, VS880.

PROBLEM: Using the two pages shown in Figure 5–9, find the correct number for a rocker arm for a 1980 Cadillac with a V-6 252 cubic inch engine.

ANSWER: RA 542

SELLING RELATED ITEMS

While looking for part numbers, the counterperson should watch for related items. Good catalogs are written to make related selling easy.

For example, when a customer comes in for an exhaust system, he or she probably is looking for the muffler and pipes. By using the catalog correctly, the counterperson can make additional sales since hangers and gaskets are used in the installation. Figure 5–10 shows a catalog containing brake parts that includes a section on brake hardware, complete with illustrations, dimensions, numbers, and applications.

Another opportunity for additional sales can be found in some of the technical data listed in the catalog. For example, exhaust systems could originally be one-piece systems consisting of an exhaust pipe, muffler, and tail pipe. The catalog should footnote these applications by informing the user that all pieces must be replaced. So, instead of selling a muffler, the counterperson can sell the entire system.

USING FOOTNOTES

Every counterperson will have occasion to refer to catalog footnotes at some point, either to reference additional information, determine exceptions, or maybe find alternative parts.

The basic definition of a footnote is: "A note of reference, explanation, or comment usually placed below the text on a printed page" (Figure 5–11). Sometimes footnotes are difficult to understand when a number of them are squeezed on a page. Nevertheless, the footnotes themselves are important. Footnotes should be referenced systematically in order to fully define those applications that have multiple possibilities.

Types

Generally, two types of footnotes are used in catalogs—alphanumeric and symbolic. Alphanumeric footnotes are noted by letters or numbers, symbolic footnotes are signified by symbols or shapes, such as an asterisk "*." These two types of footnotes cover almost everything from exceptions to alternate part numbers.

Common practice has been to utilize symbols for repetitious or on-going references such as "Available while stock lasts," while superior letters and numbers have been used in sequence for the constantly changing references like "1980 only" or "Use with high altitude or California emissions."

Trends

Within the recent past, there seemed to be a trend toward eliminating footnotes from catalog listings, where possible. Although the proliferation of models, optional equipment, and midyear changes encourages the growth of footnote usage, steps have been initiated by a sizeable group of manufacturers to suppress their usage. With fewer footnotes to read, the counterperson must pay closer attention to the listings because more lines are now present, representing the same applications as before in different form. Now, there might be three-line listings instead of one with two footnotes.

USING SUPPLEMENTS

Every counterperson will inevitably be confronted with catalog supplements due to the overwhelming number of updates that are continually made in the automotive aftermarket. Manufacturers are rarely consistent or uniform in publishing catalog supplements. For example, a few suppliers have regimented themselves into producing an update or catalog supplement for a certain date each year. However, certain trade-offs are made with this type of schedule. Pertinent model, engine, serial number, and build data codes information might not be known at a particular cutoff deadline. Of course without such valuable data, a publication will be less than complete. Some manufacturers choose to go to press in this manner while others try to accommodate all the latest changes.

After receiving a supplement, a counterperson should read and understand all accompanying cover letters and other pertinent data. The counterperson should determine if this supplement provides missing information or supersedes previous information and if previous supplements are still in effect.

DRUM BRAKE HARDWARE

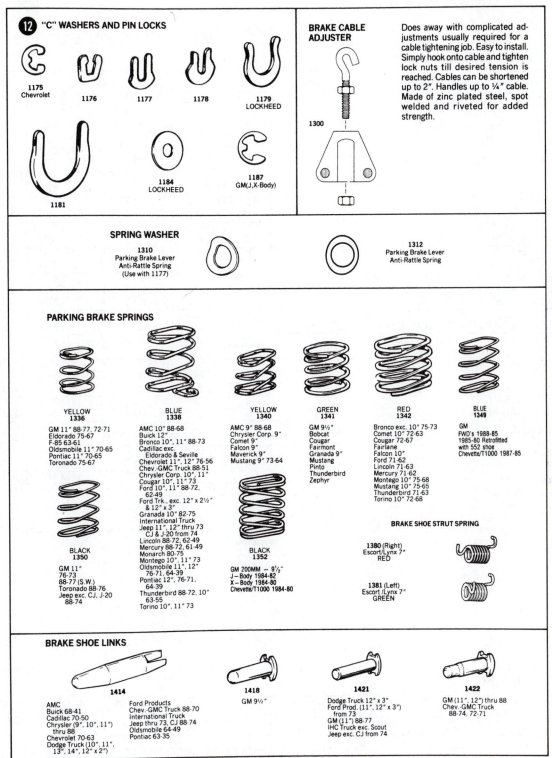

FIGURE 5-10 Most catalogs make it easy to sell related items, such as this brake part catalog, which includes a section on brake hardware.

PLYMOUTH   PONTIAC

D — Driver Side
P — Passenger Side

CONSTANT VELOCITY DRIVE AXLES & CV JOINT SERVICE KITS

Year	Model	Position Drivers Side	Comp. Shaft No.	Outer Joint w/Boot	Position Passenger Side	Comp. Shaft No.	Outer Joint w/Boot
	PLYMOUTH (Continued)						
83	Caravelle, Reliant (2.2) GKN Design (to 5/20/83)	D	*12-8071	14-2382J	P	*12-8070	14-2382J
84-86	Caravelle, Reliant (2.2 except Turbo) Citroen Design	D	*12-9740	14-4000J	P	*12-9741	14-4000J
83	Caravelle, Reliant (2.2 except Turbo) Citroen Design (from 5/21/83)	D	*12-8106	14-4006J	P	*12-8105	14-4006J
83	Caravelle, Reliant (Citroen Design) (2.2 except Turbo) (to 5/20/83)	D	*12-9760	14-4006J	P	*12-9761	14-4006J
81-82	Reliant (Citroen Design)	D	*12-9760	14-4006J	P	*12-9761	14-4006J
82	Reliant (GKN Design)	D	*12-8071	14-2382J	P	*12-8070	14-2382J
81	Reliant (GKN Design)	D	*12-9053	14-2382J	P	*12-9054	14-2382J
81	Reliant Citroen Design	D	*12-9760	14-4006J	P	*12-9761	14-4006J
78-82	Champ	D	*12-9540	14-3000J	P	*12-9541	14-3000J
85-87	Colt w/4 speed MT	D	12-9131		P	12-9132	
85-87	Colt w/AT&MT – Non-Turbo 5-Speed	D	12-9141		P	12-9142	
84-86	Colt Vista 2WD Wagon	D	12-9133		P	12-9134	
83-84	Colt – Hatchback – Non-Turbo	D	*12-9540	14-3000J	P	*12-9541	14-3000J
83-84	Colt – GTS – Turbo	D	12-9732		P	12-9733	
87	Voyager (exc. EFA) w/short wheel base (Citroen Design)	D	12-8472	14-4000J	P	12-8473	14-4000J
87	Grand Voyager (exc. EFA) GKN Design	D	12-8472	14-8608J	P	12-8473	14-8608J
87	Voyager (w/EFA) GKN Design	D	12-3000	14-8608J	P	12-3001	14-8608J
84-86	Voyager	D	*12-9307	14-4000J	P	*12-9308	14-4000J
87	Horizon, Turismo (2.2L w/AT) GKN Design	D	12-8480	14-8526J	P	12-8481	14-8526J
86-87	Horizon, Turismo (2.2L w/MT) GKN Design	D	12-8468	14-8526J	P	12-8469	14-8526J
85-86	Horizon, Turismo (2.2L) Turbo, All (Citroen Design)	D	12-8454	14-4000J		12-8455	14-4000J
83-86	Horizon, Turismo, Scamp, (1.6, 2.2) from 5/21/83 GKN	D	*12-8250	14-8526J	P	*12-8246	14-8526J
82-83	Horizon, Turismo, Scamp, TC3 (2.2) to 5/20/83	D	*12-8069	14-2382J	P	*12-8068	14-2382J
81	Horizon, Turismo, Scamp, TC3 (2.2)	D	*12-9047	14-2382J	P	*12-9048	14-2382J
83	Horizon 1.7	D	*12-8088	14-2381J	P	*12-8087	14-2381J
82	Horizon 1.7 (w/AT) or MT and splined type inner joint	D	*12-8067	14-2381J	P	*12-8066	14-2381J
81	Horizon 1.7 (w/AT or MT) w/splined type inner joint	D	*12-8035	14-2381J	P	*12-8034	14-2381J
81-82	Horizon 1.7 (w/MT) w/bolt-on type inner joint	D	*12-9041	14-2381J	P	*12-9042	14-2381J
79-80	Horizon (w/AT)	D	*12-9039	14-2380J	P	*12-9040	14-2380J
79-80	Horizon (w/MT) w/bolt-on type inner joint	D	*12-9037	14-2380J	P	*12-9038	14-2380J
78	Horizon (w/AT)	D	*12-9035	14-2379J	P	*12-9036	14-2379J
78	Horizon (w/MT) w/bolt-on type inner joint	D	*12-9033	14-2379J	P	*12-9034	14-2379J
86½-87	Maserati All Models w/2.2	D	12-9745		P	12-9746	

NOTE: Since 1983 GM has used both standard duty and heavy duty drivelines on Buick Century models. Driveshafts, Joints, Boots, and Tulips are different. To select the correct replacement part determine the driveline used by measuring the hub nut or fixed joint threaded stubshaft. Compare to the drawing below.

Standard Driveline   Heavy Duty Driveline

PONTIAC

Year	Model	Position Drivers Side	Comp. Shaft No.	Outer Joint w/Boot	Position Passenger Side	Comp. Shaft No.	Outer Joint w/Boot
85-87	Sunbird (w/AT and 2.0L Engine) except Turbo 2000 (20mm)	D	12-3170	14-8561J	P	12-3188	14-8561J
85-87	Sunbird (w/AT and 1.8L Engine) except Turbo 2000 (20mm)	D	12-3170	14-8561J	P	12-3162	14-8561J
85-87	Sunbird (w/MT) except Turbo 1.8L M.P.I. Engine 2.0P (20mm)	D	12-3173	14-8561J	P	12-3171	14-8561J

*Indicates axle is available in competitively priced E-line. **See Page 23 for details.**

FOR ORDERING: Wherever * appears delete 12- digit number in front of part number and add 24- digit number to that part number. For example: *12-8104 — use *24-8104.

FIGURE 5-11 Footnotes, usually located at the bottom of a page, contain additional information or exceptions.

Formats

Supplements are usually presented in various renditions, including self-cover, loose-leaf, and section by section. Not only does the outward physical appearance change, but also the actual material content within the supplements might occasionally be very different from the catalog it refers to.

Properly done, most supplements do provide additional data, typically presented in an identical format to the "mother" catalog.

CATALOG CONFUSION

After exhausting all the quick reference features of the catalog, a counterperson might begin to think that the information needed is not listed in the catalog.

Some companies have customer 800 phone numbers to call and verify applications. In some cases, the vehicle is too new to be covered in the catalog, or there might be some unique reason it is not listed. A counterperson should use all resources before telling a customer a part is not available.

Exceptions

Some applications have exceptions that should be noted by the counterperson. There can be different part numbers for the same application depending on the exceptions. For example, sometimes an application will list the part "with automatic transmission" and not list the part "without automatic transmission."

There are many types of exceptions within an application. For example, a counterperson looking for an OE-type replacement catalytic converter for a 1977 Dodge Aspen six-cylinder one-barrel engine without noise reduction might find:

```
          Except California & high altitudes
                  2 door,
        w/o air conditioning   12345
        w/air conditioning     12345
            4 door & station wagon,
        w/o air conditioning   45678
        w/air conditioning     48765
            Calif. & high altitudes
        2 door                 23456
        4 door & station wagon 42356
```

There are six possible part numbers in this situation.

The counterperson should research all the exceptions to determine if any of them apply to the customer's request. Some common exceptions are: before and after a certain date; with gas engine, with diesel engine; before and after chassis break; except heavy-duty, heavy-duty; except station wagon, station wagon; two-wheel drive, four-wheel drive; automatic transmission, standard transmission; front-wheel drive, and rear-wheel drive.

COMPARATIVE CATALOGING

A counterperson should realize the importance of making choices and decisions with the customer's best interest in mind. If a supplier's catalog does not have the part needed, other catalogs should be checked.

Also, with the recent emphasis on retailing, many decisions are coerced through the visual bombardment and vibrant suggestive merchandising encountered by the walk-in customer. The in-store decor, billboards, radio and TV advertising all influence customers. The counterperson's intervention might be required to assist in the decision-making process. Therefore, he or she must know the products within the store and back up or support his or her claims with reliable data, case histories, product stories, or even a physical demonstration of the product's capabilities and properties.

Many times, material other than the catalog rack publications might be required. Tool specification manuals can be an alternate source, for example, in hydraulics when the saddle and lifting height of a jack might be requested or possibly even greater details such as hydraulic fluid capacity, strokes per inch, or lighting capacity are required.

More than likely, the major areas of the counterperson's research will involve catalog rack publications. Comparing the information within these catalogs to the wants and needs of the customers will be the difference between a sale or a disgruntled customer.

CATALOG LITERACY

Before putting a new catalog away in the rack, a number of break-in procedures should be followed. The following lists some of the points to review.

CURRENT PUBLICATIONS

The counterperson should make sure this catalog replaces the store's current one and any supplements that have come out since the last catalog. This information can usually be found on the front cover (Figure 5-12). If there is any doubt, older publications should be kept.

SPARK PLUGS

1988
MARCH

M 200A
Replaces M 200Z
WEATHERLY 500

FIGURE 5-12 The counterperson must always be aware of which old catalog a new one replaces.

INDEX, TABLE OF CONTENTS, OTHER AIDS

The counterperson should scan the index and table of contents to get an overview of any new sections that have been added. A mistake often made by experienced counterpersons is to assume that these sections are only for beginners who are not familiar with a particular vendor. The table of contents and indexes are the road maps to guide the counterperson through the catalog.

The contents and index will show catalog sequence immediately (Figure 5-13). It will be quite evident if the trucks are listed with the passenger cars, or if the imports are integrated or listed separately. Product additions, possibly resulting from the consolidation of one or more formerly separate catalogs, can be quickly spotted. Instant recognition for several other catalog entities can also be found—numerical listings, buyer's guides, progressive size listings, illustrations, installation instructions, competitive interchanges, and so on. The list can be extensive.

Manufacturers usually include additional aids for using their publication in the form of cross reference lists, abbreviation lists, product descriptions, product identification, and product usage information. A counterperson who is familiar with these aids can make efficient use of the catalog. For example, abbreviations can often have more than one meaning. FWD can mean front-wheel drive or four-wheel drive, and OD can mean overdrive or outside diameter. The definition of these abbreviations and others, as well, can only be determined by checking the abbreviation list such as the one in Figure 5-14.

Aftermarket catalogs even have elaborate details on product/vehicle/parts identification on both the aftermarket side and the original equipment side. Quite informative are the VIN charts, engine ID, body and transmission type charts, and paint codes that frequent the pages of many of today's catalogs. These charts, in most cases, present the information in a very usable form, usually based on OEM data but without the confusion associated with OE charts of information.

Although expertise in the physical replacement of parts might not be in the counterperson's specific job description, he or she might often be asked for such instructions. That advice, although free, must be adequate to guide a do-it-yourselfer through a successful job. Installation instructions quite frequently are placed within the confines of the catalog for one specific purpose—to help customers. The instructions are generally well-defined and easy to understand, detailing the operations professionally.

INACCURATE NUMBERS

The counterperson should check for lists of superseded, renumbered, or obsolete numbers (Figure 5-15). This information is usually contained in the first few pages. Suppliers have only recently started putting this information in catalogs. In the past, this information was given to the warehouse distributors in bulletins and often did not reach jobbers.

At this time, the counterperson should check the inventory control system as well as inventory and take action on any supersedures, renumbers, or obsolete numbers listed in the catalog.

NEW NUMBERS

The counterperson should look for a list of new numbers to determine what new products have been added (see Figure 5-15). Often this list of new numbers is the only official notification that the counterperson receives. New numbers in the application section for which the store has had requests during the previous year can be highlighted with a colored outliner.

INDEX

IMPORTANT

The physical specifications of our parts may not be identical in every respect to those parts provided by other manufacturers. Our parts meet or exceed specifications established by the Society of Automotive Engineers. Refer any question you may have to our Customer Service Department.

The information contained in the catalog was compiled from reliable sources and carefully edited. WE CANNOT ASSUME ANY RESPONSIBILITY BEYOND OUR PART REPLACEMENT for any error.

FIGURE 5-13 Counter personnel should scan the table of contents and index of new catalogs to be aware of any changes that have been made from previous catalogs.

Abbreviations/Identification

ABBREVIATIONS

AC	- Air Conditioning	Man	- Manual	Qty.	- Quantity
AT	- Automatic Transmission	mm	- Millimeters	Veh.	- Vehicle
Auto	- Automatic	Max.	- Maximum	w/	- With
cc	- Cubic Centimeters	MT	- Manual Transmission	wo/	- Without
Comp.	- Competitor	NA	- Not Available	3AUT	- 3 Speed Automatic Transmission
C.V.	- Constant Velocity	No.	- Number	4AUT	- 4 Speed Automatic Transmission
D	- Driver's Side	NR	- Not Required	4MAN	- 4 Speed Manual Transmission
dia.	- Diameter	O.D.	- Outside Diameter	4WD	- Four Wheel Drive
exc.	- Except	O.E.	- Original Equipment	5MAN	- 5 Speed Manual Transmission
FWD	- Front Wheel Drive	P	- Passenger's Side	#	- Number
H.O.	- High Output				

FIGURE 5-14 The abbreviation list defines abbreviations used throughout the catalog, such as this list in a catalog for constant velocity products.

The counterperson should also make sure prices for these new numbers are in the latest price sheets. If not, the warehouse might have neglected to send the new price sheet or supplement. The counterperson should note new part numbers in the price sheet that are followed by a TBA (To Be Announced) or POA (Price On Availability). These "new" parts do not really exist at the time. Some suppliers add a part number to determine the demand before actually ordering the part themselves. They want the catalog coverage for competitive reasons, but have not completed their research into acquiring it before the catalog publication date rolls around. This practice can cause problems when the counterperson tries to fill an order for a part that really does not exist.

COVERAGE YEARS

The counterperson should check to see if any year's coverage was dropped. Some suppliers drop their earliest year's listing whenever they add a later year. The logic is that there is little need to catalog parts that pertain to vehicles over a certain age. Also, this is one way to offset the automatic growth of page count that happens each time another year's coverge is added.

If a supplier does eliminate a year, a counterperson might want to keep the old catalog in the back somewhere, especially if the store has regular customers who have a particular model of older vehicle.

BACK SECTION

The counterperson should check the back section of the catalog carefully. This section is often referred to as the unit section, where charts, interchanges, dimensional listings, or other helpful information might be found that, while not useful every day, could be a great help when no other way exists to confirm an application.

Another useful list often located in the back section is the interchange list (Figure 5-16). To interchange is to change places mutually or to put each of two things in the place of the other, permitting mutual substitution. Interchange lists can be valuable tools to the counterperson because they show what parts can be interchanged with another. For example, after a counterperson determines the part number for a customer, he or she checks the stock for availability. If the store is out of stock in one brand, the part number can be interchanged to a secondary line that might be in stock.

The counterperson should always use caution when interchanging. Parts that can be interchangeable might not be physically identical, or they might appear identical but differ in another characteristic such as heat resistance. In addition, many manufacturers place a disclaimer at the bottom of their interchange that states the interchanges are for reference only (see bottom of Figure 5-16). The counterperson should always verify the specific applications in the proper catalog.

CORRECTIONS

The counterperson should check any special markings or corrections in the old book before throwing it away. Since no catalog is perfect, a counterperson might have indicated where a correction is required, discovered through a customer complaint or possibly a correction bulletin from the factory.

CONSOLIDATION PART NUMBERS
RELABELING PROCEDURE CONTINUED

SPECIALTY

WHEN DEPLETED	THEN USE
14104	14116
14105	14115

GAS-MATIC® CARTRIDGE

WHEN DEPLETED	THEN USE
72853	72878

REPLACEMENT CARTRIDGE

WHEN DEPLETED	THEN USE
MC-853	MC-878

GAS-MAGNUM®

WHEN DEPLETED	THEN USE
34819	34958
34885	34960
34908	34958

MONRO-MAGNUM® 70

WHEN DEPLETED	THEN USE
74070	74410

NEW PART NUMBERS
NOT PREVIOUSLY CATALOGED

MONRO-MATIC PLUS™

31029	32211
31568	32212
32205	32213
32206	32214
32207	32215
32208	32217
32209	32617
32210	33116

MONRO-MAGNUM® 60

6966	6705
6969	6707
6701	6709
6702	

GAS-MAGNUM®

34967	34976
34968	34977
34971	34978
34974	34979
34975	34980

GAS-MATIC®

5931	5936

GAS-MATIC SPECIALTY

17201	17207
17202	17208
17204	17209
17205	17210
17206	

LOAD-HANDLER™

64058	64520
64059	

GAS-MATIC® CARTRIDGE

72878	73224
73213	

GAS-MATIC® STRUT

71774	71785
71775	71786
71776	71787
71779	71788
71783	71793
71784	71794

MONRO-MAGNUM® 70

74417	74418

REPLACEMENT CARTRIDGE

MC-878

MAX-AIR®

MA-763

FIGURE 5-15 Most catalogs will include lists of superseded, renumbered, or obsolete numbers, such as this page in a shock and strut catalog.

The use of trademarked vehicle brand names is for reference purposes only and not intended to imply that the products listed in this catalog are of their manufacture. The references to "O.E." and "O.E.M." are for the sole purpose of identifying products that are similar to, or can be used in place of, those that the vehicle was originally equipped with and not intended to suggest or imply that they were produced by the vehicle's manufacturer.

Popular Camshaft Interchanges

TRW P/N	CRANE P/N	SEALED POWER P/N	CRANE P/N	MICHIGAN P/N	CRANE P/N
TM274	C274	CS274	C274	274	C274
TM327	C327	CS327	C327	327	C327
TM443	C443	CS443	C443	443	C443
TM560	C669	CS560	C669	560	C669
TM572	C572	CS576	C576	579	C579
TM574	C574	CS579	C579	636	C574
TM576	C576	CS619	C619	641	C576
	C779	CS633	C663	641	C779
TM577	C577	CS636	C574	643	C572
TM578	C578	CS641	C576	700	C580
TM579	C579	CS643	C572	702	C580
TM580	C580	CS644	C578	709	C585
TM584	C643	CS645	C577	719	C643
TM585	C585	CS647	C585	723	C593
TM589	C589	CS648	C589	727	C596
TM590	C590	CS651	C590	773	C580
TM593	C593	CS654	C642	786	C671
TM596	C596	CS659	C593	805	C719
TM619	C619	CS660	C596	813	C720
TM628	C626	CS666	C664	814	C719
TM629	C707	CS667	C580	817	C707
TM642	C642	CS673	C628	819	C589
TM643	C643	CS674	C643	820	C577
TM644	C644	CS677	C644	820	C578
TM645	C720	CS696	C671	824	C720
TM663	C663	CS702	C720		
TM664	C664	CS710	C719		
TM667	C720	CS711	C707		
TM669	C669	CS740	C720		
TM671	C671				
TM673	C580				
TM707	C707				
TM711	C719				
TM719	C719				
TM720	C720				

This interchange should be used for inventory level references only. Use our Replacement Camshaft Catalog to select the proper camshaft for a particular application.

FIGURE 5–16 Interchange lists show what parts can be interchanged with another in actuality or only in reference.

Corrections should be added to the new catalog. If, for whatever reason, the factory neglected to make the correction, the counterperson should mail the old page to the customer service department of the supplier with a short note explaining the situation.

"SELL" PAGES

The last thing the counterperson should do is to review the "sell" pages that might have been added to the front. These are often printed on higher-quality paper than the rest of the catalog, and they might have color pictures of new products.

NEW VENDOR'S CATALOGS

All of the points listed apply to a new version of a catalog that the store has been using in the past. If management should choose to change vendors for a product, the counter personnel must become familiar with any changes in format in the new vendor's catalog.

A counterperson should scan a numerical listing section in the back, or possibly an illustrated buyer's guide, to determine the level of coverage available from the new supplier.

If the new catalog includes a cross-reference to the line the store formerly carried, counter personnel might find it easier to memorize those popular numbers.

A counterperson should also transfer any notes written in the old vendor's catalog to the new catalog (Figure 5–17). For example, by using the dimensional interchange in the back of the catalog, he or she might have found that the wheel bearing for a Mazda RX7, which was not in the catalog, was the same bearing for a Renault R18, which was in the catalog.

USING BULLETINS FOR CATALOG MAINTENANCE

Bulletins are lists of updated information that the manufacturer sends to jobbers between times of publishing their catalogs. They are the primary tool for conducting catalog maintenance, which involves continually updating and revising the catalog racks and making sure counter personnel are using the most up-to-date information from those catalogs.

Bulletins are usually one of the five following types:

1. New item availability bulletins list items now in stock.

2. Supersession bulletins note part numbers that now supersede previously noted numbers.
3. Product information bulletins provide specific information about a product such as a manufacturing defect or a unique method of installation.
4. Technical bulletins alert counter personnel to any unusual installation or fit problems or unique maintenance tips.
5. Correction bulletins refer to catalog errors due to typing errors or due to inaccurately assigned parts numbers.

Bulletins present a lot of information that would be difficult to retain, so counter personnel should write the information into the catalog where applicable. Problems that could arise from a particular customer's situation should also be written into the catalog to avoid making the same mistakes. Almost all catalogs have note pages for counterpeople to personalize the catalog with pertinent notes.

ARRANGING THE CATALOG RACK

Like any skilled technician, a counterperson relies on "tools" to help do the best job in serving customers. Those "tools" must be kept handy and in the right place so information can be obtained quickly and efficiently (Figure 5–18). Most shops and stores have a system for organizing catalogs that a new counterperson must learn. The following are some of the most common methods.

FIGURE 5-17 A counterperson should transfer any notes written in old catalogs to new ones.

FIGURE 5-18 The catalog rack must be handy and kept organized so counter personnel can find information quickly and efficiently.

ALPHABETICAL APPROACH

Probably the most commonly practiced catalog rack arrangement is an alphabetical approach— either by manufacturer/supplier order or by product type orders. Variations on this approach are virtually limitless. A typical set-up can be arranged in general alphabetical order—with a few exceptions.

ONE-RACK APPROACH

In many stores, one set of catalogs is used to a much greater degree than any other in the shop. Whether due to location, convenience, or various other reasons, one rack is preferred. Consequently, that one rack is the most updated and reliable rack to utilize.

WEATHERLY INDEX

One possibility for arranging the catalog rack is to use the Weatherly Index numbers in the upper right-hand corner of some product catalogs. However, because many counter personnel have found that the Weatherly Index is not always practical or workable, some manufacturers no longer print the number on the cover.

SUPERSEDED CATALOGS

Any catalog that has been superseded by another might not have any value. All information should be picked up by the new one. Superseded catalogs can be discarded as long as they are a direct replacement for the old one and not merely a partial one. If a superseded catalog does not completely replace the old one, then the old catalog might have to be retained for limited utilization in a back rack, under the counter, or another out-of-the-way location known to personnel.

SPECIALTY CATALOGS

Specialty catalogs—whether for high-performance speed applications or for custom body panels—probably should be directed to the back rack for at least two reasons: infrequent usage and limited space. Antique-type catalogs or even those that cover pre-1960 vehicles can be relocated to an out-of-the-way section. Frequency of usage is the primary determining factor.

MISCELLANEOUS LISTS

Price lists, supersession lists, and new items each have their niche in the catalog rack. Price lists can be placed in front or back of their corresponding catalog, placed in their own price list section, or maybe even not be used in the rack at all if a computer system handles pricing.

Supersession lists, on the other hand, have a home with the alphabetical manufacturer and new items must be earmarked for the front of the catalogs they pertain to. Once again, the arrangement really depends on what the store is accustomed to using on an ongoing basis. In some cases, manufacturers will include all supersession information on each individual price list, which, on occasion, will save you some needless page flipping to a supersession list.

Other types of information that must be routinely accessed might include core charges, product information sheets, classification and popularity information, interchange listings, return policies, special order instructions, etc., all quite pertinent information that must be filed in the proper slot.

REVIEW QUESTIONS

1. Which of the following is not included on the cover of a catalog?
   a. form number
   b. table of contents
   c. date of issue
   d. product line

2. What is the name given to a list of three-digit codes organizing catalog categories?
   a. Weatherly Index
   b. supersession list

c. bulletin
d. parts numbers

3. Which of the following is not encoded in the vehicle identification number?
   a. model
   b. year
   c. Weatherly Index
   d. plant of manufacture

4. Which of the following can be acquired through use of illustrations?
   a. dimensions
   b. correct identification
   c. specifications
   d. all of the above

5. What information should the counterperson obtain from the customer before attempting to research a part?
   a. make and model of vehicle
   b. parts number from worn part, if possible
   c. none of the above
   d. all of the above

6. Which of the following is not true concerning footnotes?
   a. Footnotes can be noted with letters or numbers.
   b. Footnotes can be noted with symbols or shapes.
   c. Footnotes are usually located in a separate section of the catalog.
   d. Footnotes contain additional information, exceptions, or alternate parts.

7. Which of the following is not true concerning supplements?
   a. Supplements provide missing information.
   b. Supplements arrive at regular intervals.
   c. Supplements provide information that supersedes previous information.
   d. Supplements usually match the format of the "mother" catalog.

8. Which of the following can cause catalog confusion?
   a. exceptions
   b. vehicles too new to be covered in the catalog
   c. unique vehicle characteristics
   d. all of the above

9. Which of the following is not a common exception listed in a catalog?

a. with original equipment, without OE
b. with gas engine, with diesel engine
c. 2-wheel drive, 4-wheel drive
d. automatic transmission, standard transmission

10. Which of the following should be checked in a new catalog?
    a. index
    b. new numbers
    c. corrections
    d. all of the above

11. Which of the following is not a common type of bulletin?
    a. new item availability bulletin
    b. new vehicle bulletin
    c. product information bulletin
    d. supersession bulletin

12. Which of the following can be discarded after reading?
    a. bulletins
    b. supplements
    c. interchange lists
    d. none of the above

13. Which of the following is not a common method for arranging the catalog rack?
    a. alphabetical approach
    b. Weatherly Index
    c. coverage years
    d. one-rack approach

14. Which of the following cannot be placed in an out-of-the-way place?
    a. price lists
    b. superseded catalogs
    c. specialty catalogs
    d. antique catalogs

15. Which of the following must be located for easy accessibility?
    a. price lists
    b. current catalogs
    c. interchange listings
    d. all of the above

16. When researching parts in a catalog, Counterperson A always looks up the main or largest item first. Counterperson B looks up parts in the order requiring the least amount of time. Who is right?
    a. Counterperson A

b. Counterperson B
c. Both A and B
d. Neither A nor B

17. If a customer has the old part to show, Counterperson A first examines it for a part number. Counterperson B first compares it to diagrams in a catalog. Who is right?
    a. Counterperson A
    b. Counterperson B
    c. Both A and B
    d. Neither A nor B

18. Which of the following is an example of a related item that can be added on to most sales?

a. Selling wheel covers along with tires
b. Selling hangers and gaskets along with a muffler and pipe
c. Selling washers and springs along with drum brakes
d. All of the above

19. When a new catalog arrives, Counterperson A determines whether or not it replaces the previous catalog by looking at the cover. Counterperson B determines whether or not it replaces the previous catalog by looking in the applications section. Who is right?
    a. Counterperson A
    b. Counterperson B
    c. Both A and B
    d. Neither A nor B

CHAPTER SIX

THE COMPUTER IN PARTS MANAGEMENT

Objectives

After reading this chapter, you should be able to:
- Explain how the computer can save time and increase sales for the counterperson.
- Explain the function of basic computer hardware, including CRT, CPU, and printer.
- Define basic computer jargon, including command, cursor, downtime, drive, disk, hardware, main storage, menu, peripherals, software, and terminal.
- Explain the process of information storage and retrieval.
- Explain how the computer can assist the counterperson in the following tasks: inventory control, bookkeeping, filing lost sales reports, sales analysis, and cataloging.
- Define the direct terminal ordering system and explain how it can assist the counterperson.
- Explain how the computer can be used for catalog distribution.
- Explain CRT shortcuts, including programmable keys.
- Describe symptoms of CRT fatigue and methods to alleviate it.

Computers today provide immeasurable benefits to almost every facet of daily activities. More and more the computer is moving into the work environment, including the automotive industry. Regardless of size, jobbers are facing concerns that demand some sort of computerized assistance for their businesses. The average age of automobiles on the road continues to increase, necessitating more repairs and parts replacements. The complexity of these autos also has increased, often requiring broader inventory per car. In addition, the average price of each inventory item has increased in proportion to inflation.

A computer can simplify and quicken almost all responsibilities of the counterperson. It increases counter productivity in the following ways:

- Eliminates time-consuming task of finding a specific application in the catalog rack
- Eliminates lookup errors
- Insures that all of the correct information is gathered so the correct part can be found the first time.
- Displays the most common footnotes
- Automatically cross references parts for different manufacturers.

A computer improves customer service by

- Displaying the correct price for a particular sale
- Providing fast service
- Asking the customer the right questions
- Not leaving the customer waiting at the counter or on the phone
- Selling the customer the right part for the application.
- Providing helpful suggestions on related parts the customer may require
- Not suggeseting a substitute item when it is temporarily out of stock

A computer can also increase sales by identifying related items and showing substitute parts, prices, and quantity on hand.

This chapter explains the abilities of the computer and how this technology makes the counterperson's job easier.

COMPUTER COMPONENTS

The computer is simply a machine. Just as a driver does not have to understand the various components of an automobile to drive it, the counterperson need not thoroughly understand the workings of a computer to use it. However, the driver who does

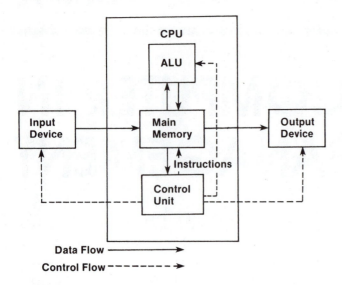

FIGURE 6-1 Computer systems and components

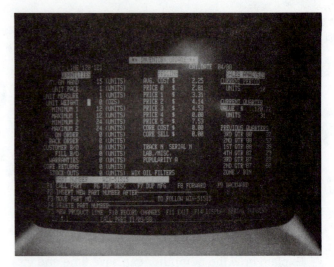

FIGURE 6-2 The movable, blinking marker on the screen is the cursor. It defines the next point of character entry or change.

have some knowledge of the mechanics of a car might notice early warning signals of a mechanical problem. And the person who understands the computer might detect problems sooner as well.

Like the auto, the computer is made up of various systems and components (Figure 6-1). The input/output devices are items such as cathode ray tubes (CRTs) and printers. The central processing unit (CPU) is where the work—processing data—actually takes place.

Most people have probably seen a CRT in the form of a screen. This device serves as both an input device where the data is keyed in and as an output device where the information is displayed on the screen. Information that is keyed in is sent to the CPU to be processed, then the information is returned to the screen for viewing.

This process can assist the counterperson in many ways. For example, if a customer requests a particular part, the counterperson simply keys in the parts number to determine if the part is in stock. If the person is a regular customer, the counterperson can also key in the name and account number along with other pertinent information. Depending on the particular system in use, the counterperson will then be able to rapidly determine the status of the particular inventory item, automatically charge the amount to the customer, and also produce an invoice of sales.

COMPUTER JARGON

An awareness of computer jargon is essential to understanding the computer. The following is a list of definitions to acquaint the counterperson with computer terminology, increase an understanding of the computer, and serve as a reference guide for future questions.

Command   An instruction to the computer to perform a predefined operation.

Control Unit (of CPU)   The section of the CPU that directs the sequence of operations by electrical signals and governs the actions of the units that make up the computer.

Cursor   A movable, blinking marker on the terminal video screen that defines the next point of character entry or change (Figure 6-2).

FIGURE 6-3 A floppy disk is erasable and reusable.

178

Downtime   The period of time when a device is not working.

Drive   A device that holds a disk or diskette so that the computer can read data from and write data onto them.

Floppy Disk   A thin magnetic disk (in protective paper jacket) that stores data or programs. It is erasable and reusable and comes in several diameters (Figure 6–3).

Hard Disk   A device to store data on magnetic disks that, unlike floppies, is rigid and not readily interchangeable but can store and retrieve data faster than a floppy (Figure 6–4).

Hardware   The physical parts of a computer, such as keyboard, screen, disk drives, printer, and so on.

Main Storage   Also known as internal storage, memory, or primary storage unit; the section of the

A

B

FIGURE 6–5   Peripherals are components added to the basic computer, such as (A) printer and (B) disk drive used to accept floppy disks.

CPU that holds instructions, data, and intermediate and final results during processing.

Menu   A displayed list of options from which the user selects an action to be performed.

Peripherals   Hardware that can be added to a basic computer, such as a printer, disk drive, or modem, for transferring information by telephone (Figure 6–5).

Software (also programs)   Instructions written in a computer language that tell the computer what the user wants to do. Most programs come on floppy disks for small computers and on magnetic tape for larger ones.

Terminal   An input/output device used to enter data into a computer and record the output. Terminals are divided into two categories: hard copy (e.g., printers) and soft copy (e.g., video terminals).

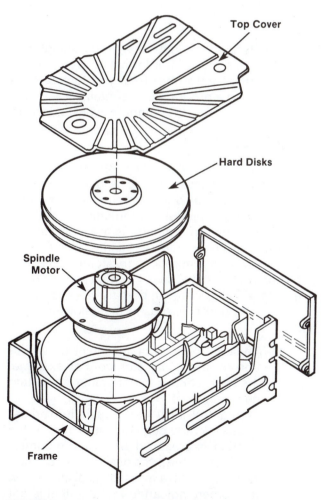

FIGURE 6–4   A hard disk mounts onto the spindle motor of a system. They are not readily interchangeable but can store and retrieve data faster than a floppy disk.

INFORMATION STORAGE AND RETRIEVAL

An understanding of how information is stored and retrieved within a computer will enable the counterperson to determine the cause of certain problems encountered within the system.

Figure 6-6 shows the general configuration of the workings of most computer systems. The major difference between this diagram and Figure 6-1 is the addition of the secondary storage block and the narrative below many blocks. When not in use, information is stored in secondary storage. It is permanent storage as compared to primary storage, which is temporary. Examples of devices used for secondary storage are magnetic tape and magnetic disk.

Information is retrieved from secondary storage and brought into the CPU for processing when requested by the user. For example, when a customer requests an item, the counterperson keys in the request on the CRT or, in some stores, via a point-of-sales (POS) terminal, commonly called a cash register.

These devices are the user's link with the CPU. They do no processing themselves, but serve as

FIGURE 6-7 A computer or electronic cash register can determine if an item is available and calculate the cost including sales tax as fast as the user can push the keys.

display units for the dialogue that the user holds with the CPU. The request for information is sent to the CPU where the control unit takes over. Someone has already written a program that contains a series of instructions directing the computer what to do next.

THE PROCESS

The control unit issues an electronic command that causes the requested data to be found in secondary storage and copied into primary storage. After this transfer is complete, the information is available for use. Note that information is available for use only after it has been transferred into primary storage, never while it is in secondary storage. The program (instructions) that specifies what must be done to this data is also stored in primary storage.

The computer proceeds in the following manner:

1. The control unit obtains the next instruction from primary storage. For example, the instruction is to determine if a certain item is in stock.
2. The control unit finds the quantity in primary storage and compares it to zero. If the number is greater than zero, the control unit locates the price of an item in primary storage. If not, a message is displayed on the screen indicating that the item is not available.
3. The computer multiplies the quantity of that particular item requested times the price to give the subtotal.

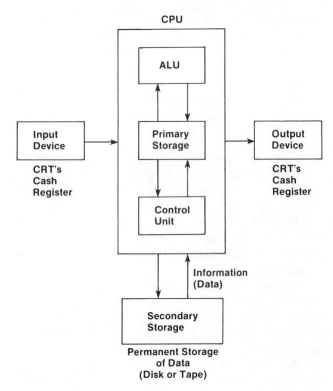

FIGURE 6-6 General configuration of most computer systems

4. The control unit decodes the next instruction, which is designed to calculate sales tax.
5. The control unit multiplies the percentage of sales tax by the subtotal to give the amount of sales tax. This figure is stored in primary storage.
6. The control unit decodes the next instruction, which involves adding the sales tax to the subtotal to give the total.
7. The control unit obtains the next instruction, which is to display the computed information on the CRT or possibly on the cash register (Figure 6–7).

The computer can perform millions of these steps in just 1 second. The above example can be duplicated for as many items as necessary and provide the customer with a sales receipt for one hundred items as easily as for one.

Problems

Sometimes the system might appear to be dysfunctional. For example, the computer might show that an item is available when in fact it is not. Many times, the problem is that inventory is not being updated (changed) instantly; a record is kept by the machine of each transaction and items withdrawn or added and not subtracted or added back until the end of the business day. Remember, the information about a particular part is stored permanently in secondary storage. All the information calculated in the previous example that was in primary storage is destroyed when the next request is made. So even though the system might report the quantity of a particular item remaining at the end of the transaction, that does not mean that the number on hand is being changed in secondary storage.

An example of a system that might not be precisely up-to-date often involves two counterpeople working at the counter, each with a CRT and/or cash register. Customers coming to two different counterpersons might deplete the supply of a particular item before the inventory in the computer is updated.

Many computer systems do not instantly change information in the secondary storage for two reasons. The first is cost. The type of system that will automatically update secondary storage is much more costly than one that will not. Also, security must be considered. For manangement control purposes, it is unwise to allow many users to be able to update information in secondary storage. Management is able to retain much more control, and in the long run, inventory records are usually more complete and accurate.

However, this situation is uncommon. Speed and accuracy—two of the main advantages of computerization—would not serve the user as well if inventory records were not instantly accurate and if other paper records had to be double-checked in order to best serve the customer. Most computer systems in the automotive aftermarket update inventory levels without allowing uncertified users to gain access to confidential information.

COMPUTER USES

According to a recent study by the Motor and Equipment Manufacturers Association (MEMA) and Lang Marketing Resources, Inc., jobbers today are much more sophisticated operationally than they were a few years ago. Today, more than 60 percent of jobber stores use computers. This figure shows more than three times as many jobbers use computers today than used computers just seven years ago (Figure 6–8).

The intensely competitive market has forced jobbers to take advantage of any tool that can assist in controlling expenses or increasing profitability. Many aspects of a jobber operation, if not managed properly, can adversely affect the business. Inventory control, accounts receivable/payable, customer service, purchasing, timely product deliveries, sales analysis, core banking, and counter sales are among the areas that can be managed with greater accuracy and efficiency with the aid of a computer.

INVENTORY CONTROL

Maintaining and controlling inventory is a vital part of a counterperson's everyday activities. Almost every aspect of the counterperson's job affects inventory.

It is the counterperson's responsibility to accurately record every sale against inventory. For example, if an automobile dealer or large garage account orders six of a particular part and the store normally stocks eight of the item, only one sale can deplete the stock to the reorder level.

In most jobber stores, inventory of sold items is taken every night. In other stores, an inventory of sales is computed twice a day.

A computer can help a jobber to better manage inventory and so help to reduce inventory carrying costs. It can help to better balance on-hand inventories, which results in fewer lost sales and backorders. When a sale is keyed into the computer, the items will be subtracted from the existing inventory. The CRT screen will also display the number of items remaining in stock.

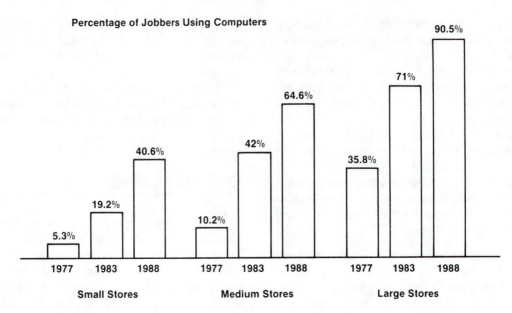

Percentage of Jobbers Using Computers

FIGURE 6-8 More than three times as many jobbers use computers today than used computers seven years ago. *(Statistics courtesy of Jobber Topics Magazine)*

FIGURE 6-9 With a computer system, the counterperson can quickly scan inventory levels.

With a computer system, the counterperson can quickly scan inventory levels to determine items that have reached their reorder levels (Figure 6-9). And if the system is linked to the store's warehouse distributor's computer, it can carry inventory control one step further. In this case, the jobber's computer "talks" to the distributor's computer and automatically orders the low inventory parts so they can be delivered the next day.

Computerized inventory can also help to increase working capital by reducing slow-moving inventories. With the computer, the jobber can maintain inventory at optimum levels, minimizing investment and maximizing profits.

A computer can only react to its programming and record the sales and data entered into it. Incorrect entries and poor programming will result in the computer displaying inaccurate inventory records. A computer is a tool that can make the counterperson's job easier, but it merely complements a counterperson's talents at inventory control.

BOOKKEEPING

A computer system can also automate tedious bookkeeping and accounting tasks. The payroll and accounts receivable and payable can be handled more quickly by computer. Cash flow increases with automatic generation of customer statements, aged balanced reports, and automatic addition of finance charges on delinquent accounts. It can also inform a counterperson when it is necessary to limit a customer's purchases.

On the other hand, a computer can help a jobber organize payables in a planned, orderly manner. As a result a good credit rating is maintained, and the jobber is also able to take advantage of discount incentives and special payment opportunities.

FILING LOST SALES REPORTS

For many years, lost sales reports were normally handled by filling in a short report sheet such as the one illustrated in Figure 6-10. Management periodically reviewed the file of lost sales hoping to spot weak areas in the store's inventory.

Today, part number proliferation has made the job almost impossible to do by hand. Fortunately, the speed of modern computers has refined the value of the lost sales report.

To understand the value of computers in handling the volume of possible lost sales, consider the following. The average warehouse distributor now carries more than 90,000 separate parts and still is not able to fill all orders from the average jobber. The average jobber store is in much the same situation, only on a smaller scale.

Lost sales can be entered into the computer data base in one of two ways: manually or automatically. Systems that create computer-generated invoices generally have the ability to transfer lost sales information from the sales ticket to a separate printout. For example, the system scans the information entered in the sales ticket, and any part number that is entered, but not filled (i.e., the computer lists the item as out of stock or an item the store does not carry) is automatically transferred into the lost sales file.

Automatic systems are good because they eliminate the need to punch in the lost sales item in a separate operation. However, an automatic system is less flexible in that it does not allow for the times the counterperson can special order items. In other words, this system will print out all out-of-stock sales as lost when a number of items will probably be special ordered.

Manual computer systems, on the other hand, require that all lost sales be keyed into the files in a separate operation. The computer then generates a list from this data. Systems that key in lost sales items one at a time are more flexible and accurate. For example, special order items need not be keyed in. However, it requires more work and effort on the part of the operator.

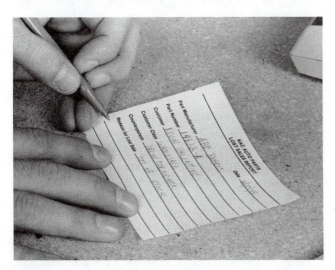

FIGURE 6–10 The report sheet used to file lost sales reports without computerization

Most jobbers who use a manual entry system have a special place for recording lost sales, and then enter them into the computer files when time allows. Both manual and automatic systems work well if the user has a good understanding of the basic lost sales report. The important factors in any system are consistency and accuracy.

SALES ANALYSIS

A computer can also be used for sales analysis. It can produce financial reports suitable for an audit. The counterperson can build an accurate historical file on valuable customers and their vehicles. The computer can take this file and generate customer reminder letters of scheduled service and preventive maintenance. It can also generate letters for advising customers of seasonal sales and other discounts, increasing repeat business and profits.

ELECTRONIC CATALOGING

Computers are automating the counterperson's most time-consuming task—researching parts. Electronic cataloging is much faster than leafing through paper catalogs. However, because electronic cataloging does not contain the diversity of information found in paper catalogs, catalog racks are still often used with the computer terminals.

Also, using electronic cataloging can result in fewer lost sales, more related sales, better customer service, increased employee productivity, and reduced liability.

For decades, it had been an axiom in the automotive industry that a counterperson could produce about $10,000 to $12,000 in monthly sales. With the introduction of POS computer products, that figure has doubled.

ADVANTAGES

Although speed is the most obvious benefit of the electronic catalog, other advantages exist as well. Electronic cataloging reduces errors. It is often easy to read the wrong line with the small print in a paper catalog. With electronic cataloging, the only numbers that are displayed are the ones that fit the application.

Electronic cataloging also eliminates the need or desire to memorize common parts numbers, a task that is often difficult with today's rapidly changing automobile systems. In addition, a counterperson with several numbers memorized might have the tendency, especially at times when the shop is crowded, to concentrate on selling only those items which he or she has memorized.

In addition, electronic cataloging enables a counterperson to sell related items to a customer's request without taking a great deal of the customer's or the counterperson's time.

Using computers resolves the problems associated with inconsistently presented information or information that changes from manufacturer to manufacturer. Parts are looked up, not by manufacturer, but by category. From category to category, regardless of who made the part, the information is displayed the same way, and the process of selecting the part and adding it to the invoice is consistent throughout the system. A system that requires consistent actions by the user is known as a user-friendly interface.

LIMITATIONS

Electronic catalog systems are often based on the "80/20 rule." That is, 80 percent of the counter sales come from 20 percent of the parts. These systems are the most useful when utilizing a relatively small portion of the parts or total possible accumulation of data. Therefore, at present, electronic cataloging remains incomplete.

Applications are totally absent for esoteric cars, early models, medium and heavy trucks, marine, agricultural, and industrial equipment because the expense of adding this data does not meet the comparatively small demand for it.

Also, electronic cataloging is limited to the display of text. When it is necessary to view a drawing—as is often the case with exhaust systems, suspension components, carburetor and transmission parts—paper books are still the only alternative.

Other information available in paper catalogs and not in computers includes various procedures such as surface preparation prior to painting or how to attach a piece of 5/16-inch copper tubing to a 1/4-inch male pipe fitting.

ACCURACY

The process of taking data from printed catalogs and keying it into a massive database affords many opportunities for errors to creep in. To prevent this, many companies have built a system of checks and double checks into their procedures.

One of these checks is called a *valid table*. It is a huge table of valid makes, models, years, and options that is stored in the master computer system. Every application that is entered into the electronic catalog system is filtered through this table, and it

prevents a lot of errors that are in the paper catalogs from being duplicated in the electronic version.

The people who key the catalog information into the database are divided into groups that specialize in specific product categories. Over a period of time, a person who works exclusively with ignition catalogs, for example, becomes so familiar with that category that many discrepancies in the paper catalog are caught and corrected before being placed into electronic cataloging. Similarly, errors noted by electronic catalog users can be rapidly addressed.

With paper catalogs, it takes a long time for errors and omissions to be corrected. Likewise, supersessions, consolidations, and application expansions usually have to wait for the next catalog reprint, which can be as much as a year or two away. With the electronic catalog, changes are continually being made to the database, and updates are distributed monthly.

PROCEDURE

While counterpeople must still use paper catalogs along with computers, productivity in those shops utilizing electronic cataloging can more than double.

Various electronic systems differ in some ways, and the counterperson must become familiar with the systems he or she will use. However, most systems require that each item of data be entered as a selection from among only the valid choices that are displayed on the screen to prevent keying errors and to ensure that all pertinent information is gathered.

The following steps are a common procedure for filling a customer's order with an electronic cataloging system. (For a more detailed explanation, utilize the computer disk included in the appendix of this book.)

1. The catalog program might have to be invoked by first entering the customer's account number, salesman's code, reference information, and perhaps a purchase order number.
2. After the catalog program is invoked, select a category from a list that includes the majority of often-replaced parts and type in the year and make of the car. In this example, choose ignition and engine filters for a 1982 Chevrolet (Figure 6–11).
3. Enter information about the model/engine combination from among the choices displayed on the screen and choose specific parts groups. Figure 6–12 shows the screen if the choices were a Monte Carlo V8-305 5.0L with the user requesting engine filters

FIGURE 6-11 With electronic cataloging, the counterperson selects a category from a list that includes the majority of often-replaced parts and types in the year and make of the car. *(Courtesy of Arturo Vera)*

FIGURE 6-12 The counterperson chooses the model/engine combination and the specific parts group from the choices displayed on the screen. *(Courtesy of Arturo Vera)*

and PCV, spark plugs, tune-up ignition, and specifications.

4. Figure 6–13 shows that the screen will then display manufacturer's parts numbes, quantities on hand, per-car quantity, price, and related tools and chemicals. Note that reminders for related sales or customer advice are often included at the bottom of the screen.

5. Make the specific selections to print on the invoice (Figure 6–14).

6. The system generates an invoice listing the car's make, model, year, and engine; the list of parts ordered; and tune-up specifications for the application (Figure 6–15).

This electronic catalog interfaces with the user's inventory files, so in addition to presenting descriptions and part numbers, it shows quantity on hand, quantity per car, along with the list, core, and selling prices.

At a glance, the counterperson can answer the customer's questions about price because the system accesses the user's customer file and displays the correct price for that customer. The counterperson is able to determine whether or not the part is on the shelf and to offer alternatives in case of multiple lines.

RELATED-ITEM SELLING

Electronic cataloging removes much of the drudgery from related-item selling. The counterperson no longer has to look up prices for each of the related items or to key in the part numbers; they are all there automatically. The system also helps prevent suggesting a related item that is not in stock.

To add an item to the customer's invoice, the counterperson selects the item and enters the quantity, again making a menu selection rather than by keying in the part number.

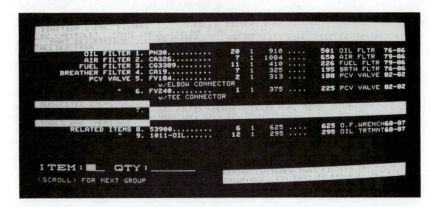

FIGURE 6-13 The screen displays manufacturer's parts numbers, quantities on hand, per-car quantity, price, and related tools and chemicals. *(Courtesy of Arturo Vera)*

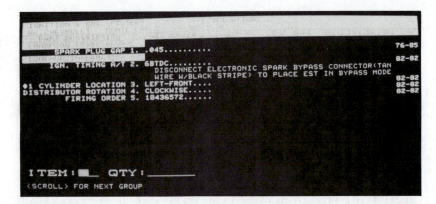

FIGURE 6-14 The counterperson moves the cursor to make the specific selections to print on the invoice. *(Courtesy of Arturo Vera)*

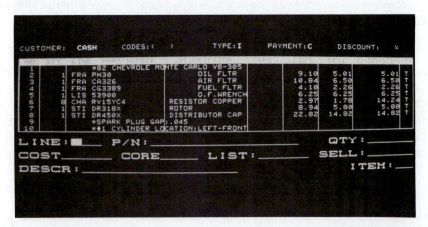

FIGURE 6-15 The screen shows the invoice that can be printed listing vehicle details, parts, and specifications. *(Courtesy of Arturo Vera)*

DIRECT TERMINAL ORDERING SYSTEM

With a direct terminal ordering system, the consumer can be linked into the electronic cataloging system. For example, a computer terminal can be set up right in a mechanic's shop. With a few keystrokes he or she can cause a modem attached to the terminal to dial the computer system at the jobber's store. When the telephone link is established, the mechanic has complete access to the electronic cataloging program. The mechanic can look up parts, check stock, get price quotes, and even create an invoice, all without the assistance of a counterperson. When finished selecting the desired parts, the push of a button prints the invoice out of the jobber's printer.

The remote site terminal also saves the mechanic time when writing up the customer's bill. All of the information is right in front of the mechanic. It also increases accuracy in parts selection and pricing.

The mechanic can do a lot more than just order parts with a direct terminal ordering system. If a customer is scheduled for service, he or she can order the parts before the job comes in. But before ordering the parts, the mechanic can see what they cost and what the markup will be. Just as important, the mechanic knows if the part is in stock or if it will be back ordered. The computer tells the quantity on hand. With this kind of information, the mechanic can shop around for a part not in stock or reschedule the job for a later date, eliminating having cars and the shop tied up waiting for parts.

The mechanic can also use the system to get information that the counterperson does not have time to provide. The mechanic can compare new versus rebuilt parts. He or she can see the price differential and make a choice based on differences in cost and quality. Being able to compare brands, prices, availability, and quality enables the mechanic to order those items that best serve the needs of the shop and customer.

Shops where mechanics have access to electronic cataloging allow counterpersons to be free to serve walk-in customers instead of the mechanics, and the counterperson can concentrate on merchandising the store and cultivating other dealer businesses.

CATALOG DISTRIBUTION BY COMPUTER

Since only a small percentage of jobber stores are currently using electronic cataloging, the counterperson might still have the responsibility of paper catalog distribution.

Knowing when to order catalogs and price sheets, how many to order, and distributing them to the right customers is a demanding job. Most jobber stores currently use a computer to keep track of parts. The system in use was probably not programmed specifically for the task of catalog distribution, but it can give reports of which documents are needed and how many and which customers get them if the data is listed as if the catalogs and price sheets were parts (Figure 6–16).

Even though the documents are treated as parts, some obvious differences arise. For example, since price sheets are not bought or sold, sales history and price information need not be recorded.

The only essential information for document control includes

- A unique part number for each document
- Distribution list for each item
- Quantity to order with each new issue

Some desirable additions might be a description of each document and its effective date.

Document records can be included among all the other parts in each line or the document records can have a separate line(s). Assigning documents a separate line allows them to be conveniently segregated from the other records and makes it easier to assign each document a unique part number.

When the store receives documents, the counterperson simply changes the on-hand quantity in the computer back to the order-point, or shelf level, quantity so that future runs of the reports will not reorder these documents.

THE COMPUTER AND THE PARTS MANAGER

The computer can also rapidly provide various services associated with a manager's position, such as writing reports, improving cash management, and increasing control over cash flow. Computers with management programs provide the following:

- Immediate information and updating of all invoicing

FIGURE 6–16 Documents, catalogs, and price sheets can be listed in the program as parts.

- Inventory and accounts receivable activities
- Detailed sales analysis
- Pricing controls
- Profit planning
- Purchasing and receiving functions
- Customer back orders
- Core accounting
- Sales tax acounting
- Serial number tracking
- Customer histories

Many programs provide some or all of the following lists and reports in only a few simple steps:

- Complete vendor information
- New invoices
- Checks written during current month
- All purchase and payment activity for current month
- Chart of accounts printed with or without balances

In addition, most programs protect all confidential matters with a unique user or password security system. Many systems require different passwords to gain access to different levels of confidentiality.

The benefits of using a computer for management functions can be recognized in speed and accuracy but can also be totaled in dollars and cents. Management systems can reduce inventory carrying costs through better inventory management, reduce lost sales and back orders through better balancing of on-hand inventories, reduce outstanding receivables through better credit management control, increase profits through higher margins and service levels, increase working capital by reduction of slow-moving inventories, and increase overall return-on-investment.

COMPUTER COMPANY BENEFITS

All computers are not the same, and all computer developers, manufacturers and vendors are not the same as well. A computer is not an item that is purchased without follow-up and repeated contact with the vendor or manufacturer. Those systems designed specifically for the automotive parts business offer the highest and fastest return on the investment.

Most computer companies fall into one of these three categories:

1. Software company, offering specialized programs designed to run on hardware not sold by the software company.
2. Hardware company, usually offering some general business software programs as well, but allowing users to tailor their pro-grams specifically to the automotive aftermarket.
3. Single source company, offering an integrated and complete hardware and software package, including installation, training, and support.

A computer should be chosen not only for its merits but also for the merits of the computer company as well. Each computer vendor should be evaluated on the basis of financial strength, equipment, usage, programs, service and maintenance, financing, trial period, customer support, and warranty.

Financial Strength

A computer company's reputation and financial strength are important to a computer buyer because if the computer company does not stay in business or disappears, the computer user will be left with no contacts for a piece of equipment designed to change his or her business forever. Without updates in programming, answers to usage dilemmas, and service and maintenance provisions, a particular computer rapidly falls behind others in the field, becoming outdated.

Equipment

With the rapidly growing computer industry, a computer buyer should consider equipment that is as modern as possible. In addition, the buyer should look for equipment that can be updated or expanded in the future with additional programs, manuals, printers, terminals, and so on to meet projected needs of the automotive aftermarket and the buyer's particular business. Equipment that fits the area where it is to be installed and which can be used comfortably by personnel should also be considered.

Usage

Because data storage capacity depends on the computer language and the programs, a computer buyer should ask if a computer can actually store and recall the number of parts numbers required (with room to grow), store and maintain the inventory, and all other operations that the buyer would have the computer do. Also, a computer with the connections to add hardware to expand capabilities is desirable.

Programs

A computer with seemingly limitless memory capacity will not serve the user well if the programs in use do not fit the user's needs. It is the programs that actually benefit a business. A computer buyer should ask if the software is ready to load up on the system or if someone else has to customize the programs for his or her particular business. The buyer should also ask: Who pays for modifications and updates? Who will actually own the software? What provisions exist for data security?

Service and Maintenance

A computer buyer should find out what provisions are offered for service and maintenance. Does the company service all components or just the main ones? Is service limited to certain times of the day or certain days of the week? How quickly can service be provided?

Financing

Financing the original purchase as well as future updates and add-ons must be considered and weighed against the projected profits.

Trial Period

A computer buyer must examine what type of installation assistance will be available and what the new owner can do if the assistance or sales representatives do not meet expectations after the system has been purchased.

Customer Support

Initial training and future expansion training for the computer customer is as important as the computer itself because the efficiency and profitability of any machine depends on the person using it.

Warranty

Because the computer is becoming the heart of automotive businesses, the buyer must know the equipment will be dependable and what recourses can be taken if it is not.

CRT SHORTCUTS

Keyboards on CRTs are not completely standardized, just as carburetors on automobiles are not standardized, but the concepts among models are the same and usually the same principles that apply to one apply to them all. Information entered by pushing keys is converted directly into codes understood by the computer. There are two standard coding schemes that are used almost universally. The first code is the American Standard Code for Information Exchange (ASCII, pronounced "askey") and

the second widely used code is Extended Binary Coded Decimal Interchange Code (EBCDIC, pronounced "eb-se-dick"). These codes are used for information exchange among data processing systems, communications systems, and associated equipment such as CRTs and printers.

Each time a key is pressed, a message is generated that results in an action by some element in the computer system. The same code is generated each time a certain key or combination of keys is pressed.

PROGRAMMABLE KEYS

The same instructions could be entered with the use of programmable keys. The user programs the keys by storing keystrokes in them. These programmed user keys are useful for storing keystroke sequences that are typed often. These sequences can be phrases, cursor movements, and/or command sequences. The end result is that only one key must be pressed instead of a number of keys to accomplish the same result.

Many other procedures can be used to make work on the CRT easier. A counterperson should consult the computer manual or other employees for more information specific to the terminal in a particular business.

SECURITY BACKUPS

A security backup is simply a record of the data and transactions inputted into the system. Performing security backups is essential to ensure the store can generate a hard copy of all needed data.

Fortunately, many computer programs are designed to automatically perform security backups at the end of each workday as a part of the overall procedure for shutting down the system.

If backups are not performed automatically, every employee responsible for computer operation should know the procedure for generating a security backup copy. Of course the exact procedure varies from system to system.

CRT FATIGUE

About 90 percent of CRT operators reportedly suffer from one or more of the following problems:

- Eye strain, including blurred vision, burning eyes, double images, and headaches
- Muscular or postural problems resulting in fatigue and discomfort

Fortunately, most counter personnel do not constantly work at CRT screens for extended periods of time. But there are some basic rules to follow when setting up a CRT workstation. Problems usually arise for one or more of the following reasons:

- Since the CRT is difficult to move around, the person using the CRT must assume awkward body positions in order to see what is displayed on the CRT screen.
- The CRT is usually set on a counter, which is often either too high or too low, placing excessive strain on the arms and wrists.
- Excessive glare on the screen and the mere process of viewing the data through a curved reflective surface put added strain on the operator's eyes, resulting in headaches, eye strain, and, in some cases, nausea.

ERGONOMICS

The physical and mental ailments that result from using the new technology are diverse and the symptoms vary widely in appearance and degree of intensity, but there is little doubt that they do occur. In fact, a new science has developed called "ergonomics" to deal with the new jobs and tasks workers must perform.

The dictionary definition of ergonomics is "the science that seeks to adapt work or working conditions to suit the worker." The term is a combination of two Greek words—ergon (work) and ockonomia (to manage). The real significance of the concept is that it is finally being realized that the work environment should be adapted to meet the needs of the worker and not the other way around.

Manufacturers of CRTs are confronting a problem faced by the automobile industry in the past. The situation centers on the sacrifice of efficiency for the sake of operator ease and comfort. For example, when Americans were demanding large comfortable cars, sacrifices had to be made in the areas of fuel efficiency and wind resistance. This same situation now confronts manufacturers of CRTs. Users of CRTs increasingly demand operator ease and comfort. Thus the problems associated with the new technologies are a growing concern.

Correcting the problems associated with the use of CRTs can be approached from two directions—changing the environment or changing the CRT itself.

CHANGING THE ENVIRONMENT

The two major concerns associated with environmental problems are viewing problems and postural problems. To reduce viewing problems, the work environment can be modified in several ways.

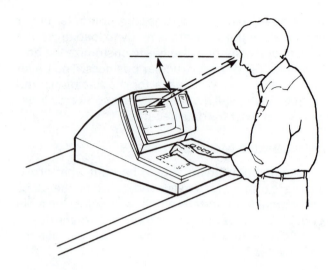

FIGURE 6-17 CRT fatigue can be alleviated by using movable screens and keyboard and minimizing reflection.

First, screen off bright light sources such as windows and bright overhead lights from the field or vision. Never position a CRT so the light reflects directly on the screen.

Another option is to minimize reflectance levels of large reflective surfaces adjacent to the CRT. For example, light walls tend to reflect more light than dark walls. In addition, large metal cabinets should never be placed so the light reflects off them onto the screen.

Also, reduce building illumination levels when possible.

To reduce postural problems, adjustable work surfaces and adequate work space should be provided. Adjustable work surfaces permit a person to assume a comfortable operating position in relation to both the keyboard and the display screen (Figure 6-17). Recent investigations have found increases in productivity of 10 to 24 percent through the use of adjustable workstations.

CHANGING THE CRT

The second method that can be used to overcome the problems associated with using a CRT involves changing the terminal itself. CRTs should be chosen that possess tiltable and rotatable display faces. Tiltable/rotatable display faces can be adjusted for viewing comfort and for reducing reflections from environmental light sources.

Detachable keyboards make keying more comfortable and allow the user to adjust the keyboard in a suitable location.

Brighter Background

Another way the CRT can be changed for the benefit of the user is by buying a brighter display background. Brighter display backgrounds bring the screen to an acceptable luminance. Computer manufacturers and users do not agree on whether the screen or the characters should be dark; however, it has been found that dark characters displayed on a light background might not have the visual clarity of light characters on a dark background.

Antireflective Surfaces

The final way to change the CRT to accommodate the user involves using antireflective display surfaces. Antireflective display surfaces include polymer films and etched glass CRT faces. Extreme care should be exercised in the selection of either one of these methods to reduce reflection. Polymer films, while yielding an image of high quality, are susceptible to smearing from touching. And, etched glass faces, while reducing reflections, also reduce image quality.

REVIEW QUESTIONS

1. Which of the following is not an input/output device on a computer?
   a. CRT
   b. printer
   c. CPU
   d. cathode ray tube

2. What computer component processes the data?
   a. CPU
   b. CRT
   c. disk
   d. cursor

3. Which of the following tasks of a counterperson's job can be made easier by the use of computers?
   a. inventory control
   b. cataloging
   c. related item selling
   d. all of the above

4. What is the name of the movable, blinking marker on the terminal video screen that defines the next point of character entry or change?
   a. command
   b. drive
   c. cursor
   d. CRT

5. What device is used for storing data or programs to be used again?
   a. disk
   b. drive
   c. menu
   d. cursor

6. What is a displayed list of options from which the user selects an action to be performed?
   a. drive
   b. menu
   c. CRT
   d. cursor

7. What is the name given to hardware that can be added to the basic computer, such as a printer, disk drive, or modem, for transferring information by telephone?
   a. peripherals
   b. terminal
   c. main storage
   d. menu

8. What is the common term for a point-of sales (POS) terminal?
   a. direct terminal ordering system
   b. CRT
   c. ergonomics
   d. cash register

9. Information in the computer is available for use only after it has been transferred into what form?
   a. secondary storage
   b. primary storage
   c. menu
   d. CPU

10. What are the advantages of using electronic cataloging?
    a. speed
    b. ease of selling related items
    c. accuracy
    d. all of the above

11. Counterperson A uses only electronic cataloging to ensure accuracy and consistency when researching parts. Counterperson B uses paper catalogs in conjunction with electronic cataloging. Who is right?
    a. Counterperson A
    b. Counterperson B
    c. Both A and B
    d. Neither A nor B

12. What is a computer terminal set up in a mechanic's shop and linked into the computer system at his jobber's store called?
    a. central processing unit
    b. modem
    c. direct terminal ordering system
    d. ergonomics

13. Which of the following is not a symptom of CRT fatigue?

    a. blackouts
    b. eye strain
    c. muscle aches
    d. headaches

14. Counterperson A eliminates glare on the computer screen in order to reduce CRT fatigue. Counterperson B maintains comfortable posture when using a computer in order to reduce CRT fatigue. Who is right?
    a. Counterperson A
    b. Counterperson B
    c. Both A and B
    d. Neither A nor B

15. Which of the following will not reduce CRT fatigue?
    a. reducing glare on the screen
    b. placing the terminal on the counter
    c. using movable display screens
    d. using detachable keyboards

CHAPTER SEVEN

BASIC PARTS MANAGEMENT PROCEDURES

Objectives

After reading this chapter, you should be able to:
- Explain the importance of balanced inventory in terms of parts carried and quantities of parts carried.
- Name aids used on the jobber level for establishing parts to stock in inventory.
- Describe the dangers of over- and understocking inventory.
- Explain inventory turnover and the basic methods of calculating turnover rates.
- Explain acquisition and possession cost as they relate to store inventory and how they affect jobber finances.
- Explain stock ceilings and reorder points and their use in inventory control.
- Explain how sales are recorded against inventory in both manual and computer-controlled systems.
- Explain lost sales and how lost sales reports are used to study and structure inventory.
- Define inventory cleanup and how it is conducted.
- Define inventory changeovers and how they are conducted.
- Define and explain the use of supersedures, renumbers, and obsolete part numbers in inventory control.
- Describe store operations that affect inventory control and parts management including checking in stock orders, handling cores, defects, distress merchandise, broken kits, special order parts, swapped parts, returned items, and independent orders.
- Explain the importance of neat, accurate invoices and record-keeping.

The art of parts management involves the skills needed to establish, control, and document the movement of parts through the aftermarket distribution chain. On the jobber store level, these parts are continually in motion. Parts are ordered from suppliers, delivered by suppliers, and the orders are checked in by the jobber staff, entered into the store's inventory files, and stocked on the store shelves. At the same time sales are made, items are removed from stockroom shelves and inventory records. Parts are then picked up by customers or delivered to customer locations. Sold parts are reordered from suppliers, and the cycle continues.

At the same time, parts might also be returned by customers and re-entered into the jobber's stock and inventory records. Jobbers might also return parts to their suppliers for a variety of reasons, including straight returns on sales that have been canceled by customers, obsolete parts that have outlived their life as movable stock, or for product line changeovers or cleanups (two methods of replacing slow-moving inventory with items that will sell better).

THE TEAM EFFORT

The most popular misconception concerning parts management is that it is the sole concern of jobber store owners and executive managers, and that counter and other store personnel need not be concerned.

Such is not the case. Countermen and women work on the front line of the aftermarket parts industry. Eight-five percent of all jobber sales are handled by counter personnel. Counter personnel handle an average of forty-five transactions per working day. This translates into $162,413 in average headquarter store sales per counter employee in 1987.

Counterpeople have first-hand knowledge of which part numbers and lines are moving, which are faltering, and customer reactions to product performance and quality. This is extremely valuable information in running a competitive parts supply business.

It is also the reason that two out of every three jobbers in business today include counter personnel on their management staffs. And more than 85 percent of all jobbers consult their counterpeople before making product line changes or additions to existing inventory. Nearly all of the duties performed by a store's counter staff directly affect inventory and parts management. For although it is owner/management's job to establish the store's philosophy on inventory control, parts pricing, and parts management procedures, it is the counter staff's responsibility to understand this philosophy and how it relates to store operations.

THE IMPORTANCE OF INVENTORY BALANCE

The main function of a jobber store is to supply parts, supplies, and services to its customers. To meet this goal, jobbers must maintain a balanced and adequate inventory of parts on hand. This inventory is commonly referred to as *stock*. An individual item is often called a *stock-keeping unit* or *SKU*.

Money must be carefully appropriated for inventory so that it returns a sufficient profit to meet operating and overhead expenses, improves customer service, and expands the business. Achieving the balance between the "right" and "wrong" amounts and types of inventory is an extremely important part of the parts management business. Part proliferation has made this task even more difficult. Larger warehouse distributors commonly stock more than 100,000 different part numbers to meet jobber demands. And it is estimated that nearly 30,000 new part numbers will be added to the aftermarket system in the next two years.

Not even the largest jobbers can hope to carry a complete inventory or available parts for today's applications. As shown in Figure 1–3, on the average, jobbers normally stock between 12,000 to 25,000 individual part numbers. If significantly larger numbers of parts were held in stock, capital investment would be too great, and the volume of storage space needed would be tremendous.

MATCHING STOCK TO THE LOCAL MARKET

Inventory must be very selective depending on the store location and customer base. For example, in Liberty, Illinois, a suburb of Chicago, jobbers cater to Chevrolets and Fords. Fifteen miles to the east, a jobber sells to installers servicing Mercedes and BMWs. Several miles north, near the waterfront, marine customers make up a significant segment of a jobber's market. And in the heart of Chicago, industrial and fleet accounts are intermixed with the regular professional and retail trade.

AID IN SELECTING INVENTORY

In time, every successful jobber determines the correct inventory for location and demand. There are a number of selection tools that can be used to assist in this process.

Manufacturer's Inventory Lists

Many auto parts manufacturers publish lists suggesting what inventory jobber stores should stock (Figure 7–1). Ideally, the manufacturer hopes this list will result in the store carrying its line. A manufacturer's list can be helpful in establishing a starting point for inventory selection, but rarely, if ever, are they accurate enough to meet the needs of a jobber store precisely.

Warehouse Distributor Help

Many WDs and programmed distribution groups offer inventory selection and control assistance to their jobber customers. Because WDs are only one step above the jobber in the distribution chain, their recommendations on part numbers and product lines might be more attuned to the local market forces. Many jobbers rely on WD suppliers for inventory assistance and advice. But as with manufacturer's lists, be sure suggestions can be backed with hard sales data based on the store's location and market.

Counter Experience

Counter personnel are an excellent catalyst for good inventory selection. Counter workers fill customer orders, field requests for parts, and are the first to notice new trends and fading parts lines. That is why it is importance for counter personnel to work with the store owner and management in establishing and maintaining a solid inventory. A counterperson will be the first to know if a store is losing sales

Part No	Application	Veh Type & Nat'l Class	Classification NE	SE	MW	NW	SW	Box No	Wgt Ea	Pieces Per Veh	Per Box
CC837	LI, 80–85 (R)	PN	N	N	N	N	N	P130	28.0	1	1
Coil Spring Stabilizers											
J650	BU, OL, PO, 64–80; C, 67–79; F, 58–66; LI, 60–69	PA	W	A	B	D	B	0260	0.29	1	1
J700	BU, CA, C, OL, PO, 60–80; F, LI, ME, 55–80	PA	C	A	A	B	A	0260	0.33	1	1
J720	C, F, 77–80	PD	W	C	W	W	W	0260	0.28	1	1
J750	AM, 59–80; BU, C, CA, OL, PO, 55–83; F, LI, ME, 55–83	PA	B	A	A	B	A	0260	0.41	1	1
J800	AM, 74–80; F, FOR, ME, 63–83; BU, C, CA, OL, PO, 59–83	PA	C	B	A	C	B	0260	0.45	1	1
J850	AM, 59–83; BU, C, CA, OL, PO, 59–72	PA	W	B	B	C	C	0260	0.51	1	1
J900	AM, 60–83; C, PO, 64–77; F, ME, 65–78	PA	D	B	A	C	C	0260	0.59	1	1
J950	AM, 74–79; BU, CA, C, OL, PO, 71–83; CH, DO, P, 81–83; F, LI, ME, 79–83	PA	C	A	A	W	B	0260	0.63	1	1
J960	CHE, GMC, 60–83; DOD, PLY, 71–83; FOR, 65–83	HD	W	C	W	W	W	0260	0.51	1	1
J970	CHE, GMC, 60–83; DOD, PLY, 71–83; FOR, 65–83	HW	W	W	W	W	W	0260	0.57	1	1

FIGURE 7–1 Example of a typical manufacturer's popularity list. Each item is assigned a code letter designating its popularity ranking.

by not carrying certain items. He or she will also be the first to realize that certain parts are nearing obsolescence and might not justify the expense of keeping them in stock and wasting valuable storage space.

Parts Surveys

The lists presented by manufacturers and WDs are based on part demand surveys conducted on a large scale. Jobbers can conduct their own specific parts surveys, particularly for certain types of accounts.

Obviously, it is not economically practical for an individual jobber to survey even a small portion of the professional trade or retail market in his or her area. Over time, a store's lost sales reports and counter personnel input can serve this purpose.

Jobber surveys are most helpful in servicing or when pursuing specific accounts or extremely localized markets. A good example would be surveying farmers in the area to determine the exact part numbers needed to fill their equipment needs. Boat and marine markets fit into this category. Fleet and industrial are two other types of accounts where selective surveying may be a great assistance to jobbers.

Even with careful planning, no jobber can expect to fill all orders. If a jobber can fill 85 percent of orders on the spot, that is a fairly high average, especially among stores catering to the professional trades. For the do-it-yourself market, a 90 percent fill-rate is average.

MAINTAINING PROPER INVENTORY LEVELS

In addition to structuring inventory to meet local demands, it is also very important to determine the optimum number of all parts to have in stock at a given time. Determining this is highly related to inventory turnover rates, a concept covered later in this chapter. For now, the major problems of over- and understocking will be addressed.

UNDERSTOCKING

The major problem with maintaining low stock levels is lost sales. If you do not have the parts, you lose the business. Consider the customer's viewpoint. If a customer needs an idle stop solenoid and a PCV valve for a late-model Buick and the jobber does not have them in stock, it is probably time for the customer to shop elsewhere. Do not expect endless patience from a do-it-yourselfer, and certainly not from a professional installer who does not have time to wait.

Chronically low inventory levels can also affect related sales efforts. The basic ignition parts a customer requests might be in stock, but the wire set, an item that should be suggested as an additional replacement item, might not be available. The chance for an increased sale is lost.

OVERSTOCKING

The problems caused by overstocking an item are not as obvious. In fact, the tendency is to overstock to avoid irate or dissatisfied customers. But if a jobber is continually overstocked, the resulting expense might eventually cripple the business.

Every SKU is costing the jobber money when it sits on the shelf. These parts make up an investment that is not paying off. Ideally, stock should be maintained at a level that meets demand without overstocking.

But even low levels of overstocking can adversely affect a store's financial stability. Consider what happens when a jobber overstocks by 10 percent on a $200,000 base inventory.

This means the jobber has $20,000 worth of items in inventory that are not really needed to meet customer demands. The annual cost for the loan to purchase this inventory (even at 12 percent interest) would be $2,400. This is just the cost for keeping an extra item on hand, such as maintaining a stock of eleven XYZ fuel filters when ten will do. If the ideal stock level is determined to be ten of an item and the store regularly stocks twelve, it is 20 percent over. That is $40,000 worth of inventory at a cost of $4,800 per year.

Paying a large amount of interest on items that are overstocked will have an adverse effect on the store's financial stability. With too much of the store's cash flow being used to pay interest on stock, the jobber might not be able to take advantage of the 2 percent cash discount offered by many warehouse distributors and suppliers. Thus, the store loses another 2 percent from the parts it buys from distributors. And if the store's customers take advantage of their 2 percent discount for paying cash, more profit is lost on each sale (although cash-paying customers will certainly help the store's overall cash flow).

Consistent overstocking, even in small amounts, can be extremely costly. That is why the store owner, management, and counter personnel must work together to establish optimum stock levels and reorder points for all items in inventory.

INVENTORY INVESTMENT AND TURNOVER

As the above examples indicate, most of a jobber store's capital or money is tied up in inventory. This money, called *investment*, makes more money when the turnover of inventory is increased. Investment dollars make more dollars when stock does not stay on the shelf long.

One of the best ways to measure business efficiency is to determine how fast the inventory moves off the shelf and is replaced with new inventory. This is called *inventory turnover*. Regular inventory turnover means increased business, and increased business usually means better customer service. On the average, a solid jobber store has an inventory turnover of approximately four times a year. However, this average can vary widely from item to item.

In addition to determining how quickly inventory moves through a business from purchase to sales, turnover also gives a good indication of a company's activity within the marketplace, the age of its inventories, and the salability of those items.

An inventory turnover that is too low indicates an unbalanced inventory with items remaining on the shelves too long and undoubtedly running up unnecessarily large inventory costs.

Conversely, a turnover that is too high indicates poor inventory management with a lack of sufficient inventory to maintain proper sales volume, thus resulting in lost sales.

COMPUTING TURNOVER RATES

As mentioned, the turnover rate is usually computed on a yearly basis, although any time frame, such as three months or six months, can be used. Turnover rates can be computed for the entire inventory as a whole, certain parts lines, such as filters or gaskets, or for individual items, such as a particular brake rotor or ignition kit.

The formula for determining turnover for an entire store inventory or a wide line of items is as follows:

$$\frac{\text{Cost of goods sold (in given time period)}}{\text{Cost of average inventory}} = \text{Inventory turnover}$$

The cost of goods sold is the actual dollar value of the parts the store sold in the given time period. Determine this value using the jobber cost to purchase the items. Since various customers are offered professional discounts, using the final sales figures will not produce an accurate turnover figure.

Cost of average inventory is the dollar value of the inventory the store normally stocks. If an accurate average inventory level is not available, add the inventory value for the beginning of the year and the inventory value for the end of the year and divide by two. Again the value of this inventory must be stated in jobber cost.

Dividing the cost of goods sold by the cost of average inventory will determine the inventory turnover.

PROBLEM: For the previous year, J & D Auto Supplies sold $325,000 worth of parts, based on jobber cost. The store's average inventory is valued at $75,000, again at jobber cost. What is the store's turnover rate for the year?

SOLUTION:

$$\frac{\$325,000}{\$75,000} = 4.33 \text{ annual turnover}$$

The same formula is used to determine turnover for individual lines of parts within the store's overall inventory. For example, to determine the turnover of all air filters sold during the previous year, the cost of all filters sold during the year is divided by the cost of the average air filter inventory kept in stock.

The procedure is still the same when computing turnover for individual items within a line.

PROBLEM: For the last calendar year, Joe's Auto Supply sold 2,923 XYZ widgets. Each widget cost $2 and Joe carried an average inventory of 825 widgets in stock throughout the year. What is the yearly turnover for XYZ widgets at Joe's Auto Supply?

SOLUTION:

$$\frac{2,923 \text{ widgets} \times \$2/\text{widget}}{825 \text{ widgets} \times \$2/\text{widget}} = \frac{\$5,846}{\$1,650} = 3.54 \text{ turnover}$$

Notice that when dealing with individual items, the cost factor can drop out of the formula. Simply divide the number of items sold during the year by the average number of items kept in stock.

$$\frac{2,923 \text{ widgets}}{825 \text{ widgets}} = 3.54 \text{ turnover}$$

For individual items or small lines of stock, average inventory can be estimated by taking an inventory count and the midpoint between reordering points.

BALANCING POSSESSION VERSUS ACQUISITION COSTS

Balance is very important when dealing with turnover figures. A turnover figure that is more than 3.0 indicates the turns on a line or item are probably adequate to meet customer demands and avoid overstocking. However, getting more turnover on items is not the overall goal. With too few turns, the store loses money on inventory investment. However, with too many turns, the store loses both sales and operating efficiency.

To illustrate this problem, change the figures in the last sample problem. Assume that the sale of widgets at Joe's Auto Supply remained constant at 2,923 units, but the store lowered its average inventory level on widgets to 100 items. Dividing 2,923 units by 100 gives an annual turnover rate of 29.23. Numerically, this looks like quite an improvement. But in reality, the store has substantially decreased its operating efficiency.

There is such a situation as too much turnover. If an inventory is turning too many times, it is likely the store is running out of the item from time to time and losing sales.

A second problem created by ordering the item twenty-nine times in the course of a year is that the counter personnel are probably spending more time on the phone to the warehouse distributor than they are selling the widgets.

While an increased turnover rate decreases the cost of possession, frequent ordering of parts in small quantities increased the cost of acquisition. These acquisition costs include:

- Labor for placing orders
- Delivery and freight charges
- Phone
- Increased paperwork and documentation

If the jobber's expense of obtaining goods to sell is too large a percentage of the operational budget, the store is really worse off than it would be if it were operating on a lower turnover rate.

To determine the best possible rate of turnover, parts managers and jobber owners must compare the costs of acquisition against the costs of possession. The ideal turnover rate is that rate at which the lowest combined cost of acquisition is obtained.

The chart in Figure 7–2 illustrates this situation. A low rate of turnover increases the costs of possession but drives down the costs of acquistion. The reverse is true with high turnover rates. Acquisition costs are extremely high, while possession costs are low. The important fact in both cases is that their combined costs are quite high and little profit remains.

The chart verifies the advantage of maintaining turnover at a rate of approximatley four times per year. At this rate, both acquisition and possession costs are relatively low and maximum profit is realized.

The latest survey of jobbers indicates they operate with an average annual turnover rate of 3.05. This means the average inventory item remains on the stockroom shelf for roughly 120 days before it is sold. It should also be noted that the turnover rates for stores specializing in heavy equipment or agri-

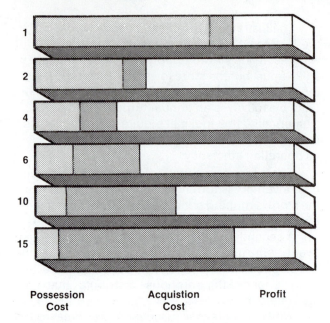

Possession Cost   Acquistion Cost   Profit

FIGURE 7-2 This bar chart indicates the delicate balance that must be struck between possession and acquisition costs. Good balance maximizes profit.

cultural parts are slightly lower than those in standard jobber stores.

STOCK CEILINGS

The ideal stock level of a part, often called its *best stock level* or *BSL*, is very closely tied to turnover rates. As mentioned, the theoretical ideal turnover rate is four times per year or one turn every three months or ninety days.

So ideally, the number of units of a particular part number to keep in stock should equal the store's unit sales of that part for a ninety-day period. For most parts, the order ceiling will be approximately equal to total sales in the past three months. For example:

Sales of part XYZ

July	12
August	9
September	14
Total	35 = Order ceiling for part XYZ

To determine the reorder quantity for part XYZ, subtract the number of parts in stock, plus any back orders from the order ceiling quantity.

Order ceiling	35
Parts in stock	15
Parts back ordered	5
	15 Order quantity

SEASONAL PARTS

For parts and products that sell only during certain times of the year, the order ceiling should equal total sales in the previous year for the three-month period, adjusted by the normal sales growth of the store. For example:

Sales	
November	20 items
December	25 items
January	15 items
Total	60 items

If it is determined that the average growth rate of the store is 8 percent annually, then the quantity should be increased by $60 \times .08 = 4.8$, or roughly five items. So when placing the seasonal order in late October compute quantities as follows:

Sales for previous year	60 items
Adjustment for growth (+)	5 items
Ordering ceiling	64 items
Stock on hand (−)	12 items
Order quantity	52 items

REORDER LEVELS

While stock ceilings are the ideal maximum levels of parts kept in stock, they are usually not the store's reorder level. Placing an order every time a part inventory drops below its stock ceiling would up the cost of acquisition above acceptable levels.

To prevent this, most jobber owners and parts managers establish best reordering points (BRP) for individual part numbers. For example, a part might be assigned a stock ceiling level of 20 and a reorder level of 10. So while the number of units should never exceed 20, stock levels will be allowed to drop to 10 units before the part number is reordered from suppliers.

PROBLEM: Part #123 has been assigned a stock ceiling of 10 units and a reorder level of 4 units. The store's inventory records indicate 8 units of Part #123 in stock. The counterperson fills an order for 2 units. Should a restocking order be placed with the jobber's supplier?

ANSWER: In this case the answer is no. Subtracting 2 units from the 8 units in stock leaves an inventory level of 6 units, 2 units above the level set for reorder.

PROBLEM: Part #567 has been assigned a stock ceiling of 12 units and a reorder level of 6 units. The store's inventory records show 10 units of the part in stock. One of the store's fleet accounts purchases 6 units. Should a reorder be issued? For how many units?

ANSWER: In this case the part needs to be reordered. With 10 units in stock, filling an order for 6 units reduces the stock level to 4 units, 2 units below the set reorder level. At this time, an order should be placed with the supplier to bring the part back up to the latest stock ceiling.

Store's present inventory	10 units
Latest order	6 units
Store's adjusted inventory	4 units
Assigned stock ceiling	12 units
Reorder number	12 units – 4 units = 8 units

FACTORS AFFECTING STOCK CEILINGS AND REORDER LEVELS

Establishing stock ceilings and reorder levels for all parts in inventory is one of the most difficult duties of jobber store owners and parts managers. Continuous monitoring of the store's sales and demands are needed. Every part has an effective life span that matches the present age of the vehicle(s) it fits. Parts for newer vehicles will have limited demand except those used in scheduled maintenance. As vehicles enter into the prime repair and maintenance years (three to eight years for most models), demand for service parts increases. And as vehicles reach retirement age and are removed from use, parts demand will again be reduced.

Stock ceiling and reorder levels must be adjusted to this demand cycle. For example, when a part is nearing obsolescence, the reorder number will be reduced to zero so that existing stock can be depleted and eliminated from inventory.

Not all parts fit into the 90- or 120-day inventory rule. Some parts might sell only once or twice per year. But the jobber might think these parts must be kept in stock for the store to be considered a full-service jobber. Parts managers and owners must carefully evaluate these exceptions on an individual basis.

The store's financial standing is a major factor in determining stock and reorder levels. The store might not have the money needed to stock to theoretically ideal levels. Again, careful assessment of the individual situation is needed and hard choices will have to be made. No jobber with limited inventory capital can afford to make poor inventory selections.

RECORDING SALES AGAINST INVENTORY

Accurately recording sales against inventory is the first important step to inventory control and effective parts management. Every sale that is made affects inventory. An item is no longer in stock, it is sold. It is the counterperson's responsibility to accurately record this sale against inventory. If a large industrial or garage account orders six of a particular part for stock and the jobber normally carries eight of the item, it is easy to see how only one sale can deplete stock to the reorder level.

In most jobber stores, inventory of sold items is taken at the close of the workday. In some stores, inventory of sales is computed twice each day. Both manual and computerized systems are in use today.

MANUAL CARD OR TICKET SYSTEMS

In this type of inventory recording system, when the customer receives his or her sales receipt, the store's copy of the sales ticket is placed in a box, usually near the cash register. At the end of the sales day, these tickets are counted up against existing inventory to determine parts to be reordered for the next day. Accurate tickets must be written up for every item sold. If mistakes are made, inventory will run short or be overstocked.

COMPUTERIZED SYSTEMS

The speed with which inventory control can be performed has been greatly improved through the use of computerized systems. When the sale is keyed into the computer files, the items will be subtracted from the existing inventory. The CRT screen will also display the number of items remaining in stock and perhaps the item's preset reorder number.

With a computer system, it is possible to quickly scan inventory levels to determine items that have reached their reorder levels. And if the system is linked to the store's warehouse distributor's computer, it can carry inventory control one step further. In this case, the jobber's computer "talks" to the distributor's computer and automatically orders the low inventory parts so they can be delivered the next day.

But computerized systems are not the total answer to inventory control. A computer cannot determine parts that a store does not stock but should. It cannot unpack new stock and check the packing slips. It cannot note any errors in the amount of merchandise received and check invoices from the warehouse distributor against the merchandise. And it cannot stock the merchandise on the proper shelf.

A computer can only react to its programming and record the sales and data entered into it. Incorrect entries and poor programming will result in the computer displaying inaccurate inventory records. Never rely totally on a computer system for the essentials of jobber store operation. It is a tool that can make the counterperson's and parts manager's jobs easier, but it merely complements the staff's talents at inventory control.

LOST SALES REPORTS

One of the finest methods of determining some of the gaps in a jobber store's inventory is by maintaining an accurate and consistent record of lost sales. Overall, a lost sales report, diligently worked, can be a lifesaver in the fast-moving competitive aftermarket. With the present proliferation of parts it has become harder to stock all items needed to meet customer demand. And when a store does decide to stock a part, it might carry only one or two of the item, rather than five or six. No jobber store can triple its inventory without a sharp increase in sales volume. Pinpointing areas where the store is consistently losing the opportunity for sales has become extremely important.

DEFINING LOST SALES

The first problem in dealing with a lost sale is defining it. Is it a lost sale when the customer requests a part the jobber does not carry, but allows the counterperson to order from the jobber's WD or manufacturer source? Is it a lost sale when the jobber is temporarily out of a stock number and the customer goes to another source to purchase the part? How about the lost sale that occurs when the customer refuses to buy a stocked part due to the jobber's asking price and purchases the item at a lower cost from a different source? Is it a lost sale if the customer wants twelve of an item, but buys six because that is all the jobber has in stock? What is the proper way to handle situations when the customer's preference is for Brand B when the store only carries Brand A?

First of all, there is no "right" and "wrong" method of tracking lost sales. The individual philosophy of the jobber owner and key management personnel will be reflected in the store's lost sales records. The examples presented below are merely a composite of what many different successful jobbers do. While there are variations within the industry, remember that it is extremely important to consistently track lost sales using the guidelines established by management.

Now that the basics have been discussed, a definition of a lost sale can be proposed. Far and away, most stores describe a lost sale as one in which they do not get the order. Most jobbers record only lost sales that are gone forever, not just out-of-stock

situations where the customer asks the counterperson to order the part. A good example is rebuilt rack-and-pinion steering units. Most remanufacturers want the warehouse to stock only four or five units and have all others drop-shipped to the jobber.

Special orders such as this are usually entered into the store's inventory files, so the store's sales records will indicate some sales movement of the item. So if a counterperson records a lost sale on a rack-and-pinion unit, then special orders the item and sells it, false information will be recorded in the store records. The item will be recorded as both a lost sale and a sale. Special-order items should be noted somewhere in the store's records, but not as a lost sale.

Loss Due to Price or Brand

Is it a lost sale if your store carries a part, but the customer claims he or she can buy it at a lower price elsewhere? Most jobbers will not record this as a lost sale because it was a result of sales policy, not inventory policy, and the lost sales report is generally a function of inventory control. If counter personnel find price or brand objections are consistent problems, the parts manager or owner must be informed, but remember, a store's pricing policy is usually the result of hard work by many employees. Counter staff and parts managers should not panic at the occasional bargain buyer.

However, if a customer requests twelve of an item and the store stocks only six, most jobbers agree that this is a lost sale and should be recorded as such. What they might not agree on is how many sales were lost if the customer did not buy any. Should it be recorded as six or twelve? Jobbers will probably be divided on that score, so management must decide on a policy and consistently follow it.

The next question concerns lost sales due to a customer's brand preference or line preference. Is a lost sale recorded if the jobber carries only new carburetors and the customer wants to buy a rebuilt one? What if the jobber carries ABC brand and the customer wants XYZ? Many stores handle this question by having a separate form for the counterperson to fill out. In this way the store owner or manager will have a written record of sales lost due to brand preference. This data is usually treated separately from true lost sales records. Many factors, such as cost, coverage, manufacturer support, and product reliability, affect the choice of the product lines carried by the store. The relative value of these factors usually offsets the occasional demands for brands the store does not carry. But constant requests for a product line the store does not carry can result in the store management considering a product line change-over. More will be said concerning product change-overs later in this chapter.

FILING LOST SALES REPORTS

For many years, lost sales reports were normally handled by filling in a short report sheet such as the one illustrated in Figure 7–3. Management periodically reviewed the file of lost sales hoping to spot weak areas in the store's inventory. Today, part number proliferation has made the job almost impossible to do by hand. Fortunately, the speed of computers has refined the value of the lost sales report.

How Computerized Systems Work

Lost sales can be entered into the computer data base manually or automatically. Systems that create computer-generated invoices generally have the ability to transfer lost sale information from the sales ticket to a separate printout. For example, the system scans the information entered in the sales ticket, and any part number that is entered but not filled (i.e., the computer lists the item as out of stock or an item the store does not carry) is automatically transferred into the lost sale file. Automatic systems are good because they eliminate the need to punch in the lost sales item in a separate operation. However, an automatic system is less flexible in that it does not allow for the times the counterperson can

special order items. In other words, this system will print out all out-of-stock sales as lost when a number of items will probably be special ordered.

Manual computer systems, on the other hand, require that all lost sales be keyed into the files in a separate operation. The computer then generates a list from this data. Systems that key in lost sales items one at a time are more flexible and accurate. For example, special-order items need not be keyed in. However, it requires more work and effort on the part of the operator. Most jobbers who use a manual entry system have a special place for recording lost sales and then enter them into the computer files when time allows. Both manual and automatic systems work well if the user has a good understanding of the basic lost sales report. The important factors in any system are consistency and accuracy.

ANALYZING LOST SALES REPORTS

The main benefit of using lost sales reports is the ability to spot trends in the aftermarket parts business. They will show new parts numbers the store does not stock or older parts numbers that are increasing in demand. Some part numbers are assured of almost instant demand, such as those found on some of the newer downsized models. Others will grow in demand, but somewhat more slowly. The store owner or buyer must analyze not only the part numbers that are selling, but also those that are not moving and those that are in demand but that the store does not stock.

KAC AUTO PARTS
LOST SALES REPORT

Date _____

Part Manufacturer _____

Part Number _____

Customer _____

Customer Class _____

Counterperson _____

Reason for Lost Sale _____

_____

_____

FIGURE 7-3 Typical lost sales report

For example, a lost sales report might indicate the following part history of a particular item: The store normally stocks two of the item. Four sales were made on the item, two by special order. However, four sales were missed because the customer would not wait for the order and bought elsewhere. This history would indicate that stock levels for that item should be raised to meet demand.

Another example is a part number that the store does not stock but missed a sale on three times in the past six months. This part definitely shows potential for good turnover and should be stocked as standard inventory.

But not all items on the lost sales report are candidates for stocking. A single request for an oddball item the store does not carry is not a signal the item should be added to inventory. As discussed throughout this chapter, there is simply no way for the average jobber to stock all the parts needed to give total coverge in all sales areas. This means that lost sales are a fact of life in most jobber stores. And while the lost sales report helps spot trends and gaps in a store's inventory, it also points out in hard facts exactly how many sales a store really loses each month.

INVENTORY CLEANUPS

An inventory cleanup is a simple concept. A jobber store returns parts that are not selling to the store's WD supplier(s) and replaces them with parts that are expected to sell better. However, the execution of a good cleanup is anything but simple. Masterful inventory cleanups are the result of good inventory control, careful evaluation of sales records, accurate planning, and sometimes tough negotiations with the store's suppliers.

Even with a sophisticated computer system, good cleanups are quite time-consuming and cannot be performed very often. Accomplishing an annual cleanup of the entire stock is a realistic goal for most stores. And a poor cleanup can be just as time-consuming and do little to revitalize a store's inventory.

CONDUCTING CLEANUPS

Cleanups begin by reviewing the detailed sales history and turnover of each part in inventory. This is done by studying the part histories recorded on the store's inventory cards or in the store's computer records. Look for slow-moving items and items that are nearing obsolescence.

Once slow-moving items are identified, the reorder number for these items should be either drastically reduced or eliminated. As the remaining items are sold, stock is not replenished, and eventually the part number is eliminated from inventory.

To decide the exact point at which a part becomes a slow mover is difficult. Store management might decide that a part has to be sold at a turnover of four times per year. Of course, most stores accept higher turnover on some items and tolerate lower rates on others. There are simply many items that the store must stock to be considered a full service jobber, regardless of turnover rates.

With accurate card systems, it is not too difficult to add up the sales for a given period and perform the calculations that indicate yearly returns. It is just very time-consuming.

Computers are thousands of times faster. But before management blindly trusts its computer system to make order-point decisions, it must be aware that the system programmers might have made some dangerous assumptions.

For example, the order point calculation program in one system might assign an order point of at least one to any item that has sold just once in the computer's memory (it maintains sales history for two years). Normally, if a part has sold only once in the past two years (a 0.5 turnover), store management is eager to eliminate it from in-house inventory. Only vigilant editing of the computer's order-point reports will ensure that slow-moving items are spotted and removed from regular inventory.

RETURNS TO SUPPLIERS

The practice of not reordering low turnover items is not enough for most cleanups. For every slow mover the store eventually sells, there are dozens of others in inventory with virtually no chance of being sold in the future.

During cleanups items that are greatly overstocked will also be determined. The best way to bring down these overstocked items to a more accurate reorder point, as well as totally eliminate many of the "dead" items in inventory, is to arrange a return with the store's suppliers.

Offsetting Orders

A return is not as simple as shipping back slow-moving items for a refund. Most suppliers require an offsetting order that is at least equal in value to the parts returned. The offsetting rate can be as high as two-for-one, meaning for every dollar in goods returned, 2 dollars of new merchandise must be ordered.

Preparing the offsetting order is very important. Ideally, it will include a large number of parts already in stock and identified as increasing in demand. With these, the store can be highly confident of improved movement for the affected line. Designing an offsetting order to include a high percentage of "bumped-up" order points on good movers is a way to reduce the line's overall size and bring its turns up to par.

Additionally, an offsetting order should contain some new numbers. These will be determined by records of lost sales from new number announcements, from counter and sales personnel recommendations, and from studying the latest WD and manufacturer inventory stocking lists.

LIMITATIONS OF CLEANUPS AND RETURNS

Cleanups and returns are subject to limitations. For example, very few suppliers will accept a broken package as a returned item. Consider that #44 hose clamps are sold in boxes of ten. The store's stock ceiling is determined to be eighteen, and the store has twenty-three on hand. By returning a box, the store risks running short. So the store maintains a level of twenty-three, or five over ideal maximum stock levels.

Also, some parts might have become obsolete and no longer qualify as returnable. Most manufacturers' price sheets include popularity classifications, and many indicate pending obsolescence. Whenever prices are changed, it is a good practice to update records to indicate manufacturer classification and to reduce to zero the order point of any part number going obsolete. With a computerized system it might be possible to automatically zero the order point for any part number assigned an obsolete classification code.

Another limitation placed on returns is that manufacturers usually limit the size of returns to some percentage of the previous year's sales, minus any returns that have already been made. Although this policy is primarily aimed at the warehouse distributor, many WDs impose the same limitations on their jobber customers. There are several ways to solve the problems, at least partially.

The first step is to eliminate careless ordering that results in day-to-day returns. By eliminating unnecessary returns throughout the year, the store will have a larger return allowance to work with when a truly big cleanup is conducted.

The second step is dealing and negotiating with the WD. If the supplier wants to reduce the size of the cleanup based on the store's previous returns, be prepared to question the numbers. The WD representative might be relying on a computer printout that indicates total yearly credits, without any consideration for the reasons the returns were made.

What about the times the WD shipped 50 rolls of heater hose when the store ordered 50 feet? Or the times parts were boxed incorrectly? What about defectives?

Try to deal with just the WD. The topic of return limits is much more likely to surface if the manufacturer's rep is involved. Also point out that even though the request is to return more than the manufacturer's policy allows, the effect on the WD will probably be negligible. This is for two reasons. First, the returns will, in all probability, include numbers that the WD will not need to return to the manufacturer. Quite a few of the store's low turnover items will still move well at the warehouse level. They are not just right for your store location or customer base. And if store cleanup can be scheduled just prior to the next order the WD places with the manufacturer, many of the returns will simply eliminate items from the WD to manufacturer order.

Use the same argument with manufacturer's high-popularity items. Your store probably stocked them based on manufacturer's popularity guides or on a factory rep's assurance that they would be fast-moving items. With a little pressure, some WDs will not count these items against a jobber's cleanup limits. The fact that the manufacturer considers them good movers should remove any objections the WD has in taking them back.

Another method of negotiating a better return allowance is to increase the offsetting order. Of course, this might not be beneficial if the line being cleaned up is already too big. But there are other ways to solve the problem.

One highly effective technique of cleaning a line of slow-movers and reducing its overall size is to offer to place a generous offsetting order, but ask to be allowed to spend it on other lines. The WD will be particularly receptive if the offer includes taking on new lines. Manufacturers might put a cap on returns, but cleanups are still highly negotiable. So ask—and do not be bashful. When WDs and manufacturer's reps are trying to talk you into taking on their line, everyone promises a "guaranteed sell." Collect it.

INVENTORY CHANGEOVERS

Cleanups eliminate weak parts from the product lines a jobber carries. Changeovers switch product lines within the entire store inventory. When a changeover is made, the store eliminates its line of XYZ filters, brakes, gaskets, etc., and replaces it with a line of ABC filters, brakes, or gaskets.

Changeovers can involve hundreds of individual items. They are major undertakings and should not be taken lightly. A changeover is much more than simply labeling new file cards or entering new computer files and later discarding or erasing the old records. The store must preserve its hard-earned sales history on each item, and this means taking each card or computer file, interchanging the old and new part number, and filing the card or computer file in a new sequence.

Changeovers mean learning a new catalog system for the new part line. Service can suffer as the transition from old to new lines is made. And some customers loyal to the old line of parts might be lost.

Despite these drawbacks, an occasional changeover might be needed for the jobber store's financial security. Poor fill rates, deteriorating product quality, lack of manufacturer's commitment to advertising, reluctance to service the line, such as harsh return policies on cleanups, and loss of local market acceptance are all conditions that can make a changeover necessary.

Choose the new line carefully. Make sure it is backed by a healthy manufacturer who possesses the ability, commitment, and resources to make the new line a winner.

Manufacturers sometimes offer significant incentives to jobbers willing to change to their lines. A changeover offer might include free merchandise or even cash bonuses. The manufacturer might even offer to buy out the jobber's existing product line. But always base decisions on long-term profitability and support, not a one-time bonus.

When discussing potential changeovers with manufacturer's representatives, keep a written record of everything that is promised. Always record the terms of the sale and keep this information with the inventory files.

With any changeover, there is the danger of customer rejection, especially if the main reason for the changeover is not related to product demand or quality. Quietly check potential new lines with the best customers to assure a high level of acceptance. When the changeover takes place, promote it vigorously so the new line takes root.

Immediately prior to the changeover, run an inventory report on the old line to determine its value. Compare this figure with the credit issued by the WD when the old line is returned. When the new line is received, take the time to add up its dollar value to compare against the billing from the WD. With so many items in the line, mistakes can happen.

 SALES TALK ──────────

These value checks should also be conducted for larger cleanup operations.

DETERMINING WEAK LINES

While any changeover should be approached cautiously, no store can afford to carry consistently weak lines. Locating weak lines in a store's total inventory is a time-consuming ordeal, but it is a task successful jobber staffs must perform to stay in business. A single weak line can cost a jobber thousands of dollars in lost sales due to incomplete coverage and low turnover.

When looking for weak lines, begin by studying lines that generate annual sales below a certain level. This cutoff figure will differ from store to store. Obviously the store that does $150,000 of business per year will have a different set of criteria than one averaging $1,000,000 in sales. For example, for the store with $150,000 in sales, $1,000 or less might be a good starting point for the investigation. A store with $1,000,000 in sales might have a $10,000 or less starting point.

Regardless of the cutoff point, the next step is to compute turnover rates for the entire line. The concept of turnover and its calculation were discussed earlier in this chapter. Now we will see how turnover is used to spot weak lines, and how some lines with good turnover rates are actually weak performers.

PROBLEM: A line is located with sales of less than $1,000 per year. Its inventory turnover rate is 2.5. What are the possible recommendations for this line?

SOLUTION: A first impression might be that the line should be dropped or changed. It does not offer the advantages of high sales volume or high turnover. But remember, there are some lines a store must carry to be considered a full-line jobber store. Is this one of them? If it is a must-have line, one solution might be to slightly decrease inventory levels or generate increased sales through promotions, etc.

PROBLEM: Joe's Auto carries two filter lines. Line #1 is a fast seller with an annual turnover of nine. On this line Joe stocks only the fast-moving part numbers, and he discounts the line heavily. His annual sales on the line are $25,000. The second line in the store has annual sales of $12,000. Turnover on filter line #2 is four.

SOLUTION: An initial appraisal indicates that line #1 with annual sales of $25,000 would appear to be a very strong line. But a closer examination of the sales and profit figures reveals that Joe is discounting this line so heavily that he is only making a gross profit of 15 percent, or $3,700. A high turnover rate of nine also indicates that Joe's cost of acquisition for the line might also be costing the store money. When these two factors are considered, the line's "strength" is weakened.

On line #2, Joe does not discount the price below what the price sheet allows. On its sales of $12,000, a 37 percent profit or $4,400 is made. Four turns a year are certainly adequate and keep the cost of acquisition in line.

When profits and acquisition costs are compared, line #2 is actually a stronger money maker. In regard to line #1, Joe has a number of options. He might be able to increase profit on the line without hurting his sales volume. But line #1 might have to be discounted for competitive reasons.

Joe might feel that if he does not really discount the filter line, he will never get the people in his front door. As a result, he misses not only the sale of the filter, but also the sale of many other, higher-profit items.

Again, these considerations have to be acknowledged. It might also be that by buying this line in volume, his purchases get high enough for him to receive a volume rebate from the warehouse.

There are other subtle considerations to evaluate when searching out weak lines. For example, Joe documents another line of brake parts with a sales volume of $10,000 per year, a 35 percent gross profit, and 3.5 turns per year.

Joe feels very satisfied with this line. All vital signs are excellent. But while talking with someone, either a competitor, warehouse salesman, or good customer, Joe learns that his competitor of equal size up the street is doing three times his volume carrying a competitor's line.

Upon further investigation, Joe learns that the competitive line has a very active field sales force, and they really help their jobbers by selling for them in the field and doing stock updates (cleanup).

Now the line that Joe thought was a strong one does not look so good. The point being made is that when evaluating lines for weaknesses, there are many factors to consider.

Not only does turnover have to be considered, but also gross profit, field help, freight requirements, volume discounts, sales, and potential. Only by taking all these things into consideration can parts management teams start replacing those weak lines with some really strong ones.

SUPERSEDURES, RENUMBERS, AND OBSOLETE PART NUMBERS

Supersedures, renumbers, and obsolete part numbers are three factors that often cause confusion in inventory control. Yet suppliers make these changes mainly for the benefit of the jobber store. They are used to help the jobber consolidate stock and eliminate obsolete parts from inventory.

Supersedures are situations where if part B supersedes part A, then part B will work in all applications for A, but A cannot be used in place of part B. In other words, supersedures are a one-way street.

Renumbers are situations where if part B is renumbered to part A, the two parts are actually interchangeable. The supplier or manufacturer has found that one part can fill two applications and decided part A should be the surviving number. Often this decision is based on the supplier's inventory of the two parts and the pricing relationship between the two numbers. An example and explanation of a typical superceded number sheet are shown in Figure 7–4.

Obsolete numbers are issued to announce that the effective life span of a part is over. It is not in the suppliers' or jobbers' interest to carry the inventory burdens of parts with no realistic chance of turnover. Another factor that might lead to a part's obsolescence is that it is no longer available at a price that allows the supplier to make a profit without charging the customer an extremely high price. A high rate of return or warranty claims can also lead to a part becoming obsolete. But the overwhelming majority of parts are made obsolete because there just is not enough demand to justify keeping them in the line.

LINE CONSOLIDATIONS

Supersedures and renumbers are two strong tools the jobber can use to consolidate and streamline the store's product lines. Skillful use of these tools will eliminate overlapping inventory and provide more coverage with fewer parts. While jobbers capable of handling the paperwork and time needed to implement supersedures and renumbers into their records will benefit from consolidation, manufacturers and suppliers take a slightly different view.

Supersedures and renumbers represent a definite lose-lose situation for manufacturers and WDs. While it might be in the supplier's best interest not to

SUPERSEDED NUMBERS

SS — An existing part number being superseded to a new part number not previously listed. The parts are completely interchangeable. Relabel stock to new number immediately.

ST — Both part numbers are completely interchangeable. Supersede immediately, combine and relabel stock.

SB — These part numbers are not interchangeable. The new number will replace the old number, but the old number will **not** replace the new part number.
Sell out old stock. Do **not** combine stocks.

012-4851 ST 012-4853	**037-4282 ST 039-6158**	**037-4419 ST 039-6203**	052-3286 ST 052-3214
012-4852 ST 012-4854	**037-4283 ST 039-6159**	**037-4420 ST 039-6230**	052-3291 ST 052-1260
012-5052 ST 012-4867	**037-4285 ST 039-6160**	**037-4433 ST 039-6204**	052-3313 ST 052-3290
014-6145 ST 014-0566	**037-4286 ST 039-6161**	**037-4467 ST 039-6205**	061-2929 ST 061-2077
014-6146 ST 014-4907	**037-4287 ST 039-6162**	**037-4476 ST 039-6206**	062-1134 SB 062-1123
014-6147 ST 014-0574	**037-4290 ST 039-6163**	**037-4484 ST 039-6207**	**071-0806 ST 071-0657**
014-6148 ST 014-6084	**037-4291 ST 039-6164**	**037-4485 ST 039-6208**	071-6589 SB 071-5284
016-0086 ST 016-0061	**037-4292 ST 039-6165**	**037-4486 ST 039-6209**	**071-6852 ST 071-7291**
022-1468 ST 022-1467	**037-4294 ST 039-6166**	**037-4487 ST 039-6210**	**071-7418 ST 071-5912**
022-1500 ST 022-1501	**037-4299 ST 039-6167**	**037-4488 ST 039-6211**	**072-7404 ST 072-3247**
024-0958 ST 024-0942	**037-4300 ST 039-6168**	**037-4493 ST 039-6212**	072-8296 ST 072-7768
031-1332 ST 031-1985	**037-4302 ST 039-6169**	**037-4495 ST 039-6213**	077-0044 ST 077-0069
031-1922 SB 031-1989	**037-4303 ST 039-6170**	**037-4496 ST 039-6214**	**081-2488 ST 081-2495**
031-1983 SB 031-0318	**037-4304 ST 039-6171**	**037-4509 ST 039-6215**	**081-2489 ST 081-2503**
032-1984 ST 032-2677	**037-4305 ST 039-6172**	**037-4517 ST 039-6216**	**081-2490 ST 081-2511**
032-2222 SB 032-2593	**037-4306 ST 039-6173**	**037-4518 ST 039-6217**	**081-2491 ST 081-2529**
032-2675 SB 032-0812	**037-4307 ST 039-6174**	**037-4523 ST 039-6218**	082-1211 ST 082-1188
035-1353 ST 035-1786	**037-4308 ST 039-6175**	**037-4524 ST 039-6219**	083-1412 ST 083-1081
035-1411 ST 035-1788	**037-4310 ST 039-6176**	**037-4532 ST 039-6220**	083-1735 ST 083-1958
035-1753 ST 035-1544	**037-4317 ST 039-6177**	**037-4533 ST 039-6221**	083-2111 ST 083-2095
035-1785 SB 035-0868	**037-4318 ST 039-6178**	**037-4534 ST 039-6222**	084-0935 ST 084-1060
036-1293 ST 036-0545	**037-4319 ST 039-6179**	**037-4540 ST 039-6223**	086-1211 ST 086-1188
037-0015 ST 039-6000	**037-4320 ST 039-6180**	**037-4541 ST 039-6224**	087-1211 ST 087-1188
037-0742 ST 039-6136	**037-4321 ST 039-6181**	**037-4542 ST 039-6225**	**101-2145 ST 101-2137**

FIGURE 7–4 Typical supercedure listing. Always follow the exact supercedure procedure outlined by the parts manufacturer. Methods can differ between manufacturers.

have the same parts under different numbers, line consolidations are arguably more critical for the jobber than for the supplier. Chances are a supplier is in a much better position to control the accompanying inventory than the jobber.

A case can certainly be made that a supplier will sell more of product A and product B, each with individual, seemingly unrelated applications, than he or she will sell of product "A" if the applications of "B" are merged with it, as they would be in the case of supersedure or renumber.

This phenomenon can be attributed to several factors, most of which have to do with the peculiar nature of a part's life cycle once it gains access to a jobber's shelf.

Once a part is added to the jobber (or warehouse) inventory, it becomes a "citizen" of that inventory control hierarchy, with all the rights and privileges thereof. It is assigned a minimum reorder quantity, below which point it is automatically replenished through the distribution chain. This min-

imum quantity becomes the "base stock," a concept very critical to a supplier. In theory, this base stock is never depleted, especially if it is a much-requested number. The more the number is asked for the higher the minimum reorder quantity is.

To the jobber, base stock parts can no longer be considered inventory items available for resale (and thus, profits). Rather, they are an indirect cost of doing business in this line. They are merely support fixtures, indistinguishable for all practical purposes from the shelves themselves.

This example shows how consolidations affect base stock in the jobber's favor. Consider two parts, A and B. The reorder quantity for A is four. The reorder point for B is three. So there are seven of these units in base stock.

If these two parts are consolidated through the use of a supersedure or renumber, the jobber's base stock will be lowered, but customer demand will remain the same as for the two separate parts. If the jobber increases the base stock level of the consolidated item by only one unit, a net inventory savings of two units is accomplished. Automatically, through no real effort on the part of the jobber, inventory turns have increased while the money committed to inventory has decreased.

If the effects of this one small example are multiplied by the number of jobber stores in which a supplier has products, the magnitude of the supplier's potential lost sales becomes evident. Every consolidation results in a direct reduction of a supplier's potential sales volume to jobbers.

To some extent, the supplier's lost sales are offset by the efficiencies gained in inventory control. Since supplier prices will reflect their inventory commitment, jobbers are actually subsidizing any massive inventory duplication or inefficiencies at the supplier level.

Initially, any supplier will sustain a loss on most consolidations. But to stay in business, suppliers must look out for the best interests of the jobber. Most suppliers try to control the timing of the supersedure or renumber announcement so that their return liability is as small as possible. By waiting until the supplier and his or her customers' inventories of the undesired numbers are exhausted, the supplier avoids these problems. Most companies have a rather elaborate system for accomplishing this. Quite often it takes as long as three years to successfully eliminate a part number.

One much-used system is to begin packaging the new part number in a box labeled with the old part number as soon as supplies of the actual old part are depleted. These upgraded old numbers are controlled either by date codes, red labels, or dual numbering, so that they can be renumbered at a later date. The strategy employed by the supplier is for the jobber is to sell all the old numbers in old boxes before selling the new numbers in old boxes.

OBSOLETE NUMBERS

The process of making numbers obsolete is more straightforward than supersedures or renumbering. By the time the manufacturer or supplier announces the obsolescence of a part, demand for the product has usually been nonexistent for some time. Jobbers who diligently track inventory popularity with price sheets and guides will have returned most fading numbers before they reach obsolescence. Most suppliers allow two years for the return of obsolete parts, but many jobbers still miss this deadline. With demand gone, recouping any investment on obsolete parts is difficult. Donations to vocational schools can generate future customers and community goodwill.

Supersedures, renumbers, and obsolescence are three very strong tools at the jobber's disposal. When used diligently and accurately they can make the job of inventory control much easier. Transferring the numbers from supersede, renumber, and obsolete bulletins into the store's inventory, sales, and cataloging systems takes time, effort, and skill. But the positive effect this work will have on the store's inventory is well worth the effort.

STORE OPERATIONS THAT AFFECT INVENTORY

As mentioned at the beginning of this chapter, parts and merchandise are constantly moving on the jobber level. Incoming shipments from distributors must be checked in, recorded, and stocked on shelves. Customer orders must be filled and recorded. Core exchanges, defective merchandise, and special orders must be handled accurately. All of these transactions affect inventory. A careless or unorganized approach to the daily activities of store operation is the fastest way to destroy any hope of maintaining an accurate accounting of inventory.

CHECKING IN STOCK

Most jobbing stores receive merchandise from their suppliers on a daily basis. Each shipment from the WD or manufacturer must be checked and double-checked. Individual parts in the shipment cannot be mixed with other parts in stock until the entire freight bill has been reconciled with the order. Mistakes in shipping happen, so nothing should be stocked on shelves and entered as inventory until every item is accounted for. If a mistake is found, such as an overage or shortage, notify the supplier immediately.

Warehouse distributors and manufacturers might have back order problems with some parts. For every shipment, note what parts arrived and those the supplier has placed on back order.

Damaged Cartons

Before signing the freight bill of incoming merchandise, inspect all cartons for damage. Note any damage on the freight bill and discuss the problem with the delivery person before signing the freight bill. Collecting a damage claim will be difficult if the bill is not accurately marked with a damage description.

Hidden damage to enclosed parts will not be visible until the freight bill is signed and the package opened. Immediately report any damaged items to the proper store personnel so a damage claim can be filed with the freight company. Unfortunately, damage claims are usually filed for more expensive items, since freight claims for small dollar amounts generally cost more in time and paperwork than they are worth.

Save the package cartons and packing papers. They will be inspected by the freight representative investigating the claim. If the box or papers are discarded, it is unlikely the problem will be resolved in the jobber's favor.

If a satisfactory settlement is not reached with the freight company, contact the parts supplier. The supplier is usually willing to help in these cases because the WD has an ongoing interest in selling to your store. The supplier can exert more pressure on the freight company to reach a settlement in your favor.

Checking in Merchandise

One of the most common and least effective systems is the one that calls for the stockroom worker or counterperson to check off items one at a time as he or she pulls them out of their boxes. It might be fast, but it is also extremely inaccurate. To point out how confusing this system can be, look at a money analogy. If you were trying to check in a pile of money that consisted of a mix of several denominations, you would not pull one bill out of the pile and check it off your list. You would sort through the bills and put like denominations together.

The proper way to check in freight is to sort the different part numbers into stacks and then check in one stack at a time (Figure 7–5). Unpack the entire

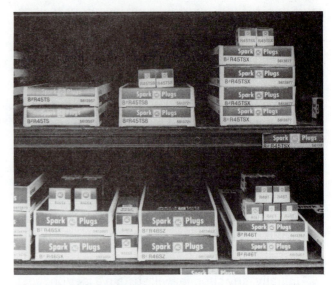

FIGURE 7-6 Most parts lines are arranged in numerical order by their part number designation.

shipment and place the various part numbers in numerical order before check in begins. Following this procedure avoids having a discrepancy between the amount of parts shipped and the amount received. If there is a discrepancy, it is easy to go back to one stack and count those parts again, instead of having to go back through the entire order.

Once all the parts are checked in, they must be shelved. Carefully place the stock in the stocking cart in numerical order. This allows starting at the first shelf and working sequentially to the last. Not only will this method help save time when putting stock on the shelves, it will ensure the stock is put in the proper places on the shelves (Figure 7–6).

Start with the first number and shuffle the empty spots on the shelf toward it so there is room on the shelf for the part numbers that come later. Attempting to fit a part in any available spot on the shelf, rather than the spot where it is supposed to go (because that spot was already filled), runs the risk of losing a sale because the part might not be able to be found.

HANDLING CORES

Many items such as water pumps, fuel pumps, and master cylinders are handled as core items. In exchange for the new part, the customer returns and receives partial credit for the old or damaged part. The jobber then returns the core to the manufacturer where it is rebuilt and resold. Core items must be handled properly. They are part of the store's capital investment and must be returned to the manufacturer so credit and a replacement item can be received.

FIGURE 7-5 Care must be taken to check in and stock inventory carefully.

Obviously, having three cores in stock is not the same as having three replacement parts in stock. Another problem that occurs when cores are carelessly handled is that they might find their way back to the stockroom shelves. Always keep cores in a designated storage area and return them promptly for credit. For more information on cores, refer to Chapter 8.

DAMAGED AND DEFECTIVE MERCHANDISE

Damaged or defective parts are usually spotted at one of two points in the sales transaction. Ideally, the counterperson will determine that the part is defective when he or she examines the part before presenting it to the customer. However, many parts cannot be determined as defective until they are being installed or have been installed.

Handling customer claims of defective parts is a separate skill covered in Chapter 11. When a part is found defective, clearly tag it with the problem, the WD or supplier's name, and the original sales invoice number. Notify the parts manager or owner immediately so the part can be returned to the supplier and credit and a new part received. Store defective items in a separate area. Like cores, defective parts can be mistakenly returned to the stockroom shelves, especially if the part is clean and the packaging undamaged. Handling defective parts claims promptly is the best way to avoid this problem.

If damaged, defective core parts are returned to the regular inventory shelves, they will distort the store's true inventory levels. Compounding the problem is the chance that the defective parts will be resold to another customer at a later date.

DISTRESSED PACKAGING

Distressed packaging is another problem when maintaining clean, organized inventories. In this case, the store still has a good part in stock, but due to damaged packaging it is less salable. Given the choice of presenting a customer with a nice, new package, or a dog-eared, grease-smeared box, most counter personnel will follow the path of least resistance and select the former. The battered package remains on the stockroom shelf as the item of last resort.

The best way to avoid distressed packaging is with careful handling and a little package maintenance. Opening packaging so prospective customers can examine merchandise is part of every day operations. Counter personnel should keep their hands clean to avoid smudging the box. Carefully

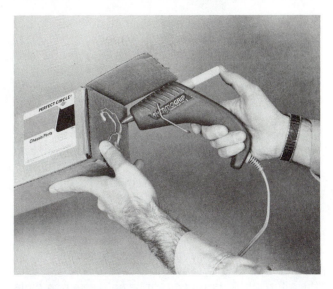

FIGURE 7-7 A hot glue gun is a great method for resealing boxes like new.

open the package for the customer. Handing a sealed box over to a grease-smeared mechanic eager to return to work is the fastest way to create a torn, dirty package. If the sale is not made and the part is reboxed and returned to the shelf, carefully clean and reseal the box. Immediately wipe off any smudges or dirt. If left on a package, grease penetrates deeper into the cardboard surface. It also attracts dust.

Boxes that were glued shut, such as those used to package shock absorbers, wheel cylinders, and gaskets, should be resealed. For best results, use a hot glue gun (Figure 7-7). With hot glue, the package flaps can be resealed like new. Cordless glue guns are handy and easy to use.

Plastic box sealing tape is a good second choice. It is clear, strong, and, if carefully applied, will not hurt the appearance of many packages. However, if the package must be opened a second time, damage is likely to occur.

Cellophane or masking tape should not be used. Cellophane tape is weak and will not stick to oily or dirty surfaces. It also turns yellow quickly. Masking tape is stronger, but it can cover information found on labels and never gives a good appearance.

Of course, sometimes there is nothing that can be done to save a damaged package, particularly on returned items. In this case, get a new box. Most factory representatives are well stocked with empty boxes for changeover purposes, and they are often stored at a WD location for convenience. Ask the store's suppliers about acquiring new packaging for the most distressed items. Since most suppliers view this as a necessary favor to their customers, do not overwork this approach.

Another method of obtaining a good box is to ask for it when a similar item is sold. This is best done with regular professional trade customers with whom the store and the counterperson have a solid working relationship. Many of the best customers will not mind occasionally switching boxes on an item once the actual parts have been examined and accepted. If exact box replacement cannot be acquired, use a similar box and renumber it with a neat sticker.

BROKEN KITS

Broken kits are similar to distressed merchandise and packaging. Inventory records still indicate the kit in stock, but it cannot be sold as a complete kit.

As with distressed packaging and cores, good counter procedures provide the answer. Here is the best system for handling broken kits: the package is opened carefully, the needed items are taken from the kit and sold as the same numbers under which they have to be reordered. Those numbers are immediately put on a purchase order, along with appropriate comments about to whom they were sold and in which kit they belong. A note is attached to the opened package, listing the needed contents, the P.O. number that applies, and the initials of the selling counterperson (Figure 7–8). When the needed parts arrive, they are returned to the set, the note is removed, and the package resealed. Only then it is returned to the shelf.

A similar problem occurs when only several items of a set are sold, such as selling two spark plugs out of a set of six. Obviously two spark plugs will not be reordered to refill the set, but the box must be clearly marked as broken, and the number

of remaining parts indicated. All counter personnel will then know that the set is broken and draw incomplete set orders from that box. Carelessness in this area will result in several broken sets of the same part. Again, there is distressed packaging concerns and the embarrassing problem of selling a broken set as complete.

SPECIAL-ORDER PARTS

This problem usually begins when a special-order item does not remain sold. It might be returned by the customer as the incorrect part for the job. Or the customer might order the part and never return to pick up and pay for the item.

The incorrect way of handling the situation is to place the part on the stockroom shelves. Even if it is correctly entered into inventory and recorded into the store records, most jobbers do not want or need special-order items taking up storage space and inventory dollars.

The correct way of handling special-order items is to require a nonrefundable deposit before the special order is placed. This deposit should equal the cost of phone, freight, and handling charges. Credit customers should be advised that a debit will be charged to their account for these amounts. Promptly return special orders to the supplier for credit. If the special order is nonrefundable to the WD, advise the customer that the sale is absolutely final.

SWAPPED PARTS

Swapping parts with a customer is another way inventories can become confused. For example, suppose a customer purchases a 42-inch fan belt. Later in the day, he or she returns, asking to trade the belt for one that is an inch longer. Unless the first belt is damaged, most stores will make the trade. To keep the inventory accurate in this case, the counterperson must either credit the customer for the first belt and bill for the second, or make an adjustment to the inventory records.

RETURNED ITEMS

Returned merchandise is a fact of business. But if these returns are not accurately recorded or if communication between workers breaks down, the result can be inaccurate inventory levels.

For example, on Monday, two distributor caps are sold to a good customer. On Wednesday, the customer returns one cap for a valid reason. The counterperson promptly puts it back on the shelf, thinking no other action is really needed. However,

FIGURE 7–8 Always clearly mark broken kits. List the individual parts requiring replacement.

on Tuesday, the store manager or buyer noted the sale and reordered two caps to replenish stock. The reordered caps arrive with the Thursday shipment. Now the store is overstocked by one cap.

Multiply that one distributor cap by two or three returns a day, for each of three of four counterpeople, six times a week. The overstocking problem caused by an indifferent approach to returned items quickly becomes a large problem.

So whenever a part is returned, check the computer or card file on the item to determine if it has been reordered. If it has, it might be possible to cancel the order. Otherwise, when the part arrives, it should be returned to the supplier.

If the part has not been reordered, make the adjustment to the inventory records and return the part to stock.

Counter personnel should also know store policy on which items are not acceptable as returns. This list often includes cut-to-order materials, electrical components, machine removals, and items damaged after the sale.

INDEPENDENT ORDERING

The problems caused by independent ordering are similar to those that result from poor handling of returns.

Once again, the problem arises when the counterperson believes he or she is doing the job but is not aware of the total picture. For example, it is early in the day and a big sale of a popular item has been made. The counterperson notices the store's stock is nearly depleted or gone. If he or she calls immediately, the delivery truck can be caught before it leaves with the store's order for the day. The counterperson calls and places an order for the items, and they are slipped into the day's deliveries.

Up to this point, the actions have been those of a quick-thinking, conscientious counterperson. But a problem has been created.

The problem occurs when the person in charge of stock orders is unaware of the independent order. Obviously, he or she is going to consult the computer or card file for reorder levels. Unless the right people in the system are informed that an independent order has been placed, the result will be overstocking.

So any time counter personnel place an independent order, two things must be done. First, check to see if the item is already on a pending stock order. If it is, order only what is needed to fill present orders. Second, whether it is on order or not, the store's inventory control system—computer, card file, or manager's memory—must be informed.

THE IMPORTANCE OF NEAT, ACCURATE INVOICES

Every sales ticket or invoice contains basic information vital to maintaining accurate sales and inventory records. Recording the information in the appropriate columns and spaces is vital in generating the data needed to assess and maintain store operations. Mistakes in recording part numbers, quantities, prices, and other information will result in errors in inventory, billing, and other systems within the store.

Poor penmanship and math errors in adding up prices and figuring discounts are also very costly. Fortunately, computerized systems that automatically print out invoices and compute billing solve some problems. But the data still must be accurately entered into the system. Mistakes made in keying information will not be caught by the computer. The counterperson is still responsible for accuracy.

Figure 7-9 illustrates some of the most common errors made when manually filling out invoices. It is obvious the counterperson hurriedly filled in the ticket. Chances are the store was very busy at the time, but in attempting to save time, the counterperson created a chain reaction of problems.

When it is time to post inventory, invoices are reviewed and recorded. But with this invoice, it is unclear whether four or nine of the item were sold. When questioned about the sale, the counterperson cannot recall the number.

This one mistake can cause several problems. When inventory is updated, a decision must be made on the number of items sold on this ticket. If nine were sold and four are removed from inventory records, inventory will show five more of these items than are actually in stock. If four were sold and nine are removed from the records, the item might be reordered with overstocking as the end result.

The same confusion can also occur with part numbers. Look at the part number on line 3. Does it read 1234 or 7234? Once again, problems result when records are updated. If part number 1234 was sold, but the inventory manager updates part number 7234, a double error results. The inventory records for both parts will be incorrect. Overages and shortages will occur.

Poor penmanship can also cause problems in a number of other ways. Looking back at Figure 7-9, line 4 shows that one of part number 1307 was sold. The counterperson wrote down the part number and priced the merchandise one line at a time. Since there were six lines filled on this particular invoice, a little time passed between when the counterperson wrote the price on line 3 and added up the ticket.

ORD.	SHIP'D	UNIT	PART #	DESCRIPTION	MFR.	LIST PRICE	UNIT PRICE	NET PRICE	
	4		1370					10 60	
	2		2116					9 20	
	1		~~1234~~ 7234					6 14	
	1		7307					2 04	

(Top of form: SHIPPED VIA / FILLED BY Joe / ORDER NO.)

FIGURE 7–9 Example of poorly recorded quantity and parts numbers on handwritten invoices. In line item #1 have 4 or 9 of the item been shipped? In line item #3 is the part number 1234 or 7234?

Hard as it is to believe, when the worker added up the ticket the counterperson could not read his/her own handwriting and misadded the ticket. The item sold for $7.04, but the counterperson mistook the 7 for 2, and lost $5.00.

Mistakes that have been made and corrected by writing over them are another problem that occurs on invoices. In Figure 7–10A it is evident that the counterperson either made a mistake on the quantity entered, or the customer revised the order at the last minute. The problem again is determining how many of the items were actually sold. In Figure 7–10B a problem exists in determining the correct part number.

The proper way to handle the correction is to draw a line through the entire transaction and start over on a different line. In some stores, the policy is to start a fresh ticket with no crossouts. Either method eliminates unnecessary mistakes.

Tips on Filling Out Invoices

The first step a good counterperson will take when filling out an invoice is to write his or her name or initials somewhere on that ticket. Most tickets have a spot designated for just that purpose and, by writing the person's name there, it serves two purposes: First, if a question arises about the ticket at a later date, the store manager knows whom to ask about it. Second, the customer will have a record of who waited on him, and if a good job was done, he or she might come back to that person again. This is especially important if working on commission.

The second thing a counterperson should do is write down the customer's name legibly to avoid billing the wrong person for the merchandise.

When recording the sale, write down the part numbers and quantities neatly and concisely. As part numbers are recorded, consider the following:

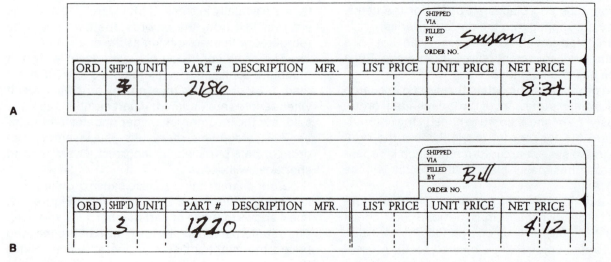

A

B

FIGURE 7–10 (A) The danger of writing over existing numbers. What was the number shipped, 3 or 4? (B) Is the part number on this sales ticket 1720, 1220, or 1770?

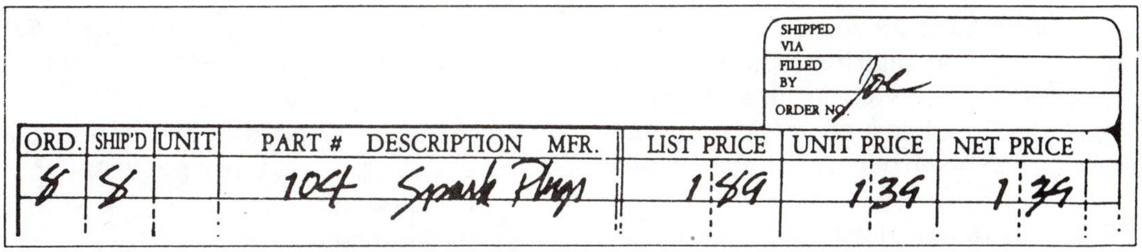

ORD.	SHIP'D	UNIT	PART # DESCRIPTION MFR.	LIST PRICE	UNIT PRICE	NET PRICE
8	8		104 Spark Plugs	1 89	1 39	1 39

SHIPPED VIA / FILLED BY *Joe* / ORDER NO

FIGURE 7-11 Example of a typical extension error on a hand-generated sales ticket. This counterperson sold eight spark plugs for the price of one.

1. Are sizes involved? An example would be 0.010 undersize set of main bearings. The size is just as much a part of that part number as the prefix is.
2. Does the color of the item change the part number? In many instances, seat covers, for example, the color changes the part number completely.
3. Are there any prefixes and suffixes involved? A part number consists of all the information, not just part of it.
4. Is the particular part number sold by the inch, foot, roll, or box? Know products and how they are sold.

Inventory can be affected by improper units of measure just as easily as it can be by illegible numbers. An example would be selling 250 sandpaper discs instead of one roll.

Now that the quantities and part numbers are written down, price the ticket. It is important when doing this that only the unit column of the ticket is filled first. This helps avoid extension errors. An example of a typical extension error is seen in Figure 7-11. The counterperson responsible for this invoice gave a customer eight spark plugs for the price of one. This problem happens often, and it cheats the company out of profits every time. If the counterper-

son had priced only the unit portion of the ticket, then multiplied the units times the unit cost, he or she would not have made the mistake.

It is also an excellent idea to include a short description of the item along with the manufacturer's name and part number. A professional installer might know what a R44TS is, but does a do-it-yourself customer? Many industrial customers also require identification for all items purchased. Rather than try to do it for just those who require it, why not make it a practice to write the description of all the articles on a ticket all the time? Not only will customers appreciate it, it also looks better and has the added advantage of helping detect other errors. For example, if a counterperson wrote down part #1307 = 5/8 wrench but meant to put down part #1370, the mistake would be easier to find.

A good counterperson will also make sure that the customer's order number and method of delivery is written down. Again, industrial customers appreciate it. With just a little practice, it becomes second nature to do it all the time.

If there are transportation charges and/or phone calls involved in the transaction, be sure to include them in the proper columns in the ticket. The store cannot afford to provide these services for free and the customer is not likely to point out this type of error.

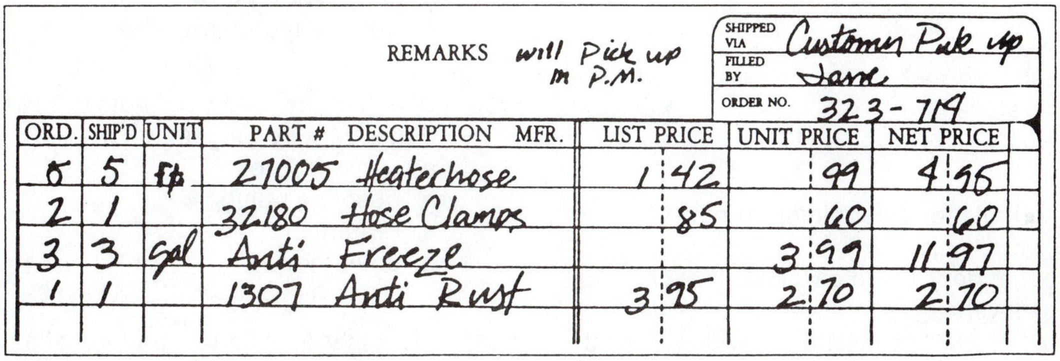

REMARKS *will Pick up in P.M.*
SHIPPED VIA *Customer Puk up* / FILLED BY *Jane* / ORDER NO. *323-719*

ORD.	SHIP'D	UNIT	PART # DESCRIPTION MFR.	LIST PRICE	UNIT PRICE	NET PRICE
5	5	ft	27005 Heater hose	1 42	99	4 95
2	1		32180 Hose Clamps	85	60	60
3	3	gal	Anti Freeze		3 99	11 97
1	1		1307 Anti Rust	3 95	2 70	2 70

FIGURE 7-12 Example of a properly filled out handwritten sales ticket. Penmanship is excellent and all vital information is completed.

On the sample invoice shown, the remarks column is used to note any special instructions, billing terms, etc. Of course, this invoice is for demonstration purposes only. But almost all invoices contain some of the elements shown here and the same rules apply.

Figure 7–12 illustrates an ideal invoice. It is neat, concise, and legible. It contains all the information needed to satisfy the customer, store manager, and inventory clerk, and anyone in the store picking up the ticket can see how it is to be shipped and to whom.

The small amount of extra time it takes to fill out a sales ticket neatly saves a lot of time and frustration for the inventory clerk, the parts manager, and all store personnel who depend on accurate records to perform their jobs.

A neat sales ticket is also more professional in appearance and assures customers that they are doing business with a reliable company.

REVIEW QUESTIONS

1. Eighty-five percent of all jobber sales are handled by _____ .
   a. managers
   b. counter personnel
   c. delivery persons
   d. suppliers

2. What is SKU?
   a. stock-keeping unit
   b. an individual item in an inventory
   c. both a and b
   d. neither a nor b

3. What is a way to measure business efficiency by determining how fast the inventory moves off the shelf and is replaced with new inventory?
   a. SKU
   b. cash flow
   c. overstocking
   d. inventory turnover

4. The formula for determining turnover is _____ .
   a. cost of goods sold divided by cost of average inventory
   b. cost of average inventory divided by cost of goods sold
   c. number of items sold times the number of items in average inventory
   d. number of items in average inventory times the number of items sold

5. Optimum turnover rate is approximately _____ times per year.
   a. three
   b. thirteen
   c. four
   d. fourteen

6. Which of the following is an example of a lost sale?
   a. A requested part is not in stock but is ordered.
   b. A customer wants ten of an item but buys only seven because that is all that is in stock.
   c. A customer refuses a stocked item because of price.
   d. It depends on the jobber's philosophy.

7. Most stores describe a lost sale as one in which _____ .
   a. they do not get the order
   b. they must special order an item
   c. a customer buys only one of an item
   d. no customers request a particular item

8. What is an offsetting order?
   a. an item that must be special ordered
   b. an order made to a supplier when returning slow-moving items
   c. an order of slow-moving items
   d. all of the above

9. What is a changeover?
   a. a cleanup
   b. switching product lines within a store's inventory
   c. an offsetting order
   d. none of the above

10. Which of the following indicates two numbers that are interchangeable?
    a. supersedures
    b. obsolete numbers
    c. renumbers
    d. changeovers

11. What is the minimum quantity of an item allowed in stock before replenishing?
    a. base stock
    b. supersedures

c. line consolidation
d. turnover

12. The proper way to check in freight is to
_____ .
a. check off items as they are pulled from the box
b. sort part numbers into stacks and check in one stack at a time
c. examine items with the delivery person present
d. any of the above

13. What is an old or damaged part that is returned in exchange for partial credit for a new part?
a. remanufactured item
b. defect
c. core item
d. none of the above

14. What must be done whenever an independent order is placed?
a. check to see if the item is already on a pending order.
b. inform the store's inventory control system.
c. neither a nor b
d. both a and b

15. What information should be included on an invoice?
a. names of customer and counterperson
b. quantity and parts numbers
c. method of delivery
d. all of the above

16. Problems caused by understocking include
_____ .
a. tying up store capital (money)
b. lost sales
c. decreased store efficiency
d. both B and C

17. Problems caused by overstocking include
_____ .
a. tying up store capital (money)
b. lost sales
c. decreased store efficiency
d. both B and C

18. Which of the following is not a problem commonly associated with low turnover rates?
a. poor cash flow
b. unbalanced inventory
c. increased cost of acquisition
d. increased cost of possession

19. Counterperson A conducts cleanups to remove slow-moving inventory items. Counterperson B changes product lines to remove slow-moving items. Who is right?
a. Counterperson A
b. Counterperson B
c. Both A and B
d. Neither A nor B

20. Which of the following can be a helpful aid in selecting inventory?
a. manufacturers' popularity lists
b. warehouse distributors
c. counterpersonnel
d. all of the above

21. Which of the following is not a problem commonly associated with high turnover rates?
a. increased cost of possession
b. increased cost of acquisition
c. increased paperwork and documentation
d. all of the above

22. Counterperson A uses the stock ceiling level to determine when a part should be reordered. Counterperson B uses the BRP for individual parts numbers to determine when the part should be reordered. Who is right?
a. Counterperson A
b. Counterperson B
c. Both A and B
d. Neither A nor B

23. What is the first step in conducting a cleanup?
a. preparing an offsetting order
b. changing the reorder number(s)
c. reviewing detailed sales history and turnover
d. calculating BRP

24. Which of the following must be taken into account when considering replacing one product line for another?
a. competition
b. gross profit
c. turnover
d. all of the above

25. How much time do most suppliers allow for the return of obsolete parts?
    a. two months
    b. two years
    c. one year
    d. ten years

26. The first thing a counterperson should write on an invoice is _____ .
    a. name or initials
    b. time of sale
    c. amount of sale
    d. amount of discount

CHAPTER EIGHT

PARTS PRICING AND SERVICE CHARGES

Objectives

After reading this chapter, you should be able to:
- Explain the different types of prices, including list price, net price, and variable pricing.
- Describe price sheets or lists and explain their use.
- Describe bar codes and explain their use.
- Explain different types of accounts, including accounts receivable, accounts payable, active account, dual account, and credit account.
- Explain how to process invoices, shop tickets, cash sales, and charge account sales.
- Explain the function and application of discounts, including cash discounts, trade discounts, chain discounts, extra dating, rebates, and coupons.
- Explain the function and processing of warranties.
- Explain the procedures for handling core charges, returns, and identification.
- Process a return involving return charges, credit memorandums, and/or a defective parts claim.
- Calculate return on investment (ROI) and explain how to improve it and apply it.
- Define common terms for handling paperwork in the stockroom, including bill of lading, break point, freight bill, intermediate order, purchase order, stock order, and turnover.

Pricing is one of the most influential marketing factors at any level of the automotive parts business. Surveys indicate that most auto parts customers look first at price when choosing a seller. Marketing schemes such as sales, advertisements describing low prices, and discounts attract customers. Oddly enough, once a customer is in a particular business, his or her actual buying decision is not necessarily based on price. In effect, although many customers do not choose an item based on price alone, they do often choose what store to shop based on price.

While many automotive jobbers charge the recommended prices as listed on the manufacturer supplied price sheets, a growing number now use their own pricing methods to create a competitive advantage and a positive impression on customers. The use of computerized systems has made it much easier for the jobber to create and adjust a price structure that is competitive in the local market. Once a base price is keyed into the computer system, it is a simple matter to create any number of discount and price levels for the various types of customer classifications.

In an effort to remain competitive, some jobbers buy private label merchandise from direct mail merchandisers or even from local warehouse distributors. In most cases, these products, if they represent reasonable quality, can be sold for less than national brand price and still represent a margin improvement. Using private label merchandise offers customers a price alternative and enables small businesses to be competitive with local chain stores.

Other methods of creative pricing involve marking some prices lower than manufacturers' suggested prices to attract customers and marking some prices higher than suggested to make up the profit difference. Other factors to consider when determining pricing policy include services such as delivery, outside sales assistance, and broad inventories stocked for customer convenience.

There are about fifty products for which consumers have some idea of the value and price. These items are the ones most often advertised or put on special sales. Pricing too high on any of these products can lose sales on other products as well. Pricing on other items might be more discretionary. To be competitive in today's marketplace, a business has to know how to use pricing properly.

Pricing philosophy or policy must be understood and enforced by all employees. However, factors other than pricing affect the total cost of a sale. For example, coupons, discounts, and taxes can either raise or lower the total cost. Counter personnel must be familiar with these and other pricing factors, such as types of prices, accounts, invoices, warranties, core exchanges, credit memos, and stocking figures.

TYPES OF PRICES

Prices vary from product to product and from business to business. Manufacturers often recommend certain prices, but actual pricing is usually up to the business owner or manager with input from counter personnel. Therefore, it is important that counterpersons understand the different types of prices.

List price is the suggested sale price of an item to its final user. List price is also referred to as *retail price* on some price sheets. In the average jobber store, very few items are actually sold at the full list price. Most classifications of customers are given some type of discount below list price. And while list price is the cost theoretically charged to retail or walk-in customers, many stores must sell to walk-ins below list price to stay competitive.

A *net price* is the cost of an item usually determined for a specific class of customer, such as a dealer net price, a stocking dealer net price, a wholesale net price, and a jobber net price.

Velocity pricing or *variable pricing* is a price attached to a particular item to make up for a low profit from another item. This method is also known as the "loss/lost leader" approach to pricing. For example, to compensate for the lack of profit from fast-moving staples (such as oil or filters), some businesses mark up the price on slow-moving items, especially those items that are not easily accessible to customers. Most customers are willing to pay a little more for the convenience the business affords by carrying a certain item instead of having to special order it.

PRICE SHEETS/LISTS

Price sheets or lists contain all the necessary pricing information for accurate invoicing or other records of a sale or return of a particular manufacturer's products. Price sheets are one of the coun-

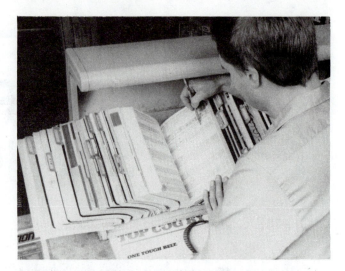

FIGURE 8–1 Price sheets correspond to manufacturer's catalogs and are often filed in the appropriate place in the catalog rack.

terperson's most important tools, and careless use of them can result in lost profits and damaged customer relations.

Price sheets usually correspond to the manufacturer's catalog, listing prices for all part numbers found in the catalog. For this reason, sheets are normally filed in the catalog rack for easy reference (Figure 8–1).

The price sheets indicate how much each classification of customer is to be charged. Like catalogs, there can be major differences in the way various manufacturers organize and print their price sheets. Be sure to always have the manufacturer's representative explain the price sheets in detail so there is no misunderstanding.

Most manufacturers supply different price sheets for different customer classifications. The sheets are usually color coded for easy reference. The following are several examples of what each color can represent:

White. In the past white price sheets normally listed retail or list prices. Today, almost all price sheets indicate list or retail pricing and there is no need for a separate sheet.

Dark Yellow. Often used to give the first discount below list. This price is often offered to wholesale accounts and even walk-in customers in highly competitive markets.

Light Yellow. Price offered to dealers and installers such as garages, service stations, and the like.

Green. Price offered to stocking dealers and installers who purchase parts in volume from the jobber.

Brown. Sometimes used to give prices offered to vehicle dealerships.

Pink. The highest discount available to accounts that buy from the jobber in high volume.

Blue. Prices paid by the jobber to his or her supplier (WD or manufacturer).

Remember, these examples are only generalizations. Many manufacturers combine several of the above classes onto a single sheet. They might also omit a classification from their pricing structure or refer to customer categories by different terms. For example, one manufacturer might refer to garage and installer accounts as "dealer" while another might call them "trade" accounts. Wholesale prices also can be referred to as "user" or "resale" prices, so it is extremely important to understand each manufacturer's setup and system.

A counterperson must know his or her store's discount structure and how to use the proper price sheet for each classification. He or she must also know the proper classification of each of the store's customers or accounts. For example, which garages qualify for stocking dealer discounts? How are the store's industrial and fleet accounts classified? Do they receive wholesale, trade, or volume dealer discounts? What discount is offered to other jobbers? Is there any discount on the everyday walk-in trade, and how are farm accounts classified? Special care must be taken if the store sells to government agencies, which often include a special bid and tax charge. Fortunately, computerized point-of-sales terminals and records can quickly indicate an accounts price classification. The largest problem facing jobbers is keeping prices up-to-date.

UPDATING

Price sheets come in as many different forms as catalogs. They, too, must be kept up-to-date. All price sheets show the effective date of the information and any replaced or superseded lists (Figure 8–2). In the past, a manufacturer's prices might have remained the same for several years, but today it is not uncommon for a single manufacturer to change prices several times a year.

Many WDs compile and distribute indexes of price sheets and dates for the lines they carry. By regularly comparing these indexes with the sheets in the catalog racks, a counterperson can be sure updated price information is available. Some jobbers accumulate the indexes from all their suppliers and update their racks on a regular schedule.

However, even WDs are behind sometimes. If so, the jobber's profits will not be hurt immediately because the warehouse is still selling to the jobber at the old price. A counterperson might consider checking sources from time to time to avoid missing an opportunity for bonus profits if both the WD and the jobber are behind. Whenever a manufacturer's rep visits the business, the counterperson should ask about price changes, especially if a price sheet is almost one year old. Very few price sheets are valid for more than a year.

WEATHERLY INDEX 002, 050

SUGGESTED PRICE LIST

No. PL-3004-88

Effective June 27, 1988

Supersedes No. PL-3004-87A

Dated December 14, 1987

SUGGESTED VOLUME DEALER PRICES

All Part Numbers for Import Applications are Included

Prices are subject to change without notice. No taxes are included.
See separate list for Oil Seal and "O" Ring prices.

For your convenience, this price list has been arranged in numeric sequence; non-prefixed numbers first, followed by single letter prefix numbers and then double letter prefix numbers.

FIGURE 8–2 Price sheets contain important information such as effective dates and supersedure information.

CONFIDENTIALITY

Price lists contain confidential information. Counter personnel should be very careful when billing or quoting prices, especially over the telephone. Allowing a customer to see or hear any price level below what he has been charged or will be charged invites complaints.

Distribution of price sheets should follow the manufacturer's policy and the policy of the particular automotive business.

Many jobbers supply selected garage and other discounted accounts with price sheets and catalogs. These customers can then save time by estimating and pricing their own jobs, and the counterperson in the jobber store can save time giving prices. In this case, the counterperson should make sure the customer has up-to-date price lists. Garage and installer customers could be lost if they discover that they are selling parts for less than they are paying for them. If the jobber store produces its own prices, the counterperson must make sure that the discounted account customers understand how the prices differ from published sheets.

To keep discounted accounts up-to-date, counter personnel will have to use a system to keep track of which accounts receive price sheets and the exact classification of price sheets they receive. For smaller jobbers that supply only a few accounts with pricing, a simple typed checklist will work. But a jobber that supplies numerous accounts with price sheets and other support materials should handle each document as if it were a part in the jobber's inventory.

Each document or price sheet could be given its own inventory card or file listing stating the name of each account it is supplied to. When there is a price change, the appropriate inventory card or file is pulled, the account names are noted, and a mailing or delivery of the materials is made to the appropriate contacts. Computers can greatly decrease the time needed to do this work (see Chapter 6).

With discounted accounts, confidentiality is also very important. All accounts are not the same and, for many reasons, some will be given better prices that others. For example, the garage that buys more than $1,000 worth of products each month will usually receive a better break (stocking dealer or volume dealer) than the garage that occasionally makes a purchase (wholesale or dealer). Therefore, it is best to keep outside knowledge of costs and price levels as low as possible.

With wholesale accounts, confidentiality is also important. All wholesale accounts are not the same and, as with discounted accounts, some are given

better prices than others. For example, the garage that buys more than $1,000 worth of products each month will usually receive a better break than the garage that occasionally makes a purchase. Therefore, it is best to keep outside knowledge of costs and price levels as low as possible.

USAGE

Counter personnel must be able to use these lists quickly and accurately. A counterperson should always refer to a price list for cost; relying on memory can cheat the customer or the business if the amount remembered is from an old price list or for a different product or customer classification.

Various price sheets contain similar pertinent information, but formats vary. For example, a counterperson may encounter price sheets organized for dealers as well as some for wholesalers.

As mentioned earlier, various price sheets contain similar pertinent information, but formats often vary. For example, Figure 8–3A illustrates a typical manufacturer's price sheet. It gives the part number,

No.	List Each	Whsle.	Trade
39695	6.53	5.25 ↓	4.74
39695E	4.58	3.68 ↓	3.32
39745	5.82	4.68 ↑	4.23
39745E	4.07	3.27 ↑	2.95
39751	5.67	4.55 ↑	4.11
39763	5.69	4.57 ↑	4.13
95000	15.09	12.13 ↑	10.95
95005	13.51	10.86 ↑	9.81
95007	12.95	10.41 ↑	9.40
95008	12.95	10.41 ↑	9.40
95009	11.64	9.36 ↑	8.45
95010	10.53	8.47 ↑	7.65
95011	7.33	5.89 ↑	5.32
95013M	16.69	13.41 ↑	12.11
95014	18.60	14.95 ↑	13.50

A

No.	List Each	Volume Dealer	Std Pkg
39695	6.53	4.20 ↓	10
39695E	4.58	2.94 ↓	30
39745	5.82	3.74 ↑	10
39745E	4.07	2.62 ↑	30
39751	5.67	3.64 ↑	10
39763	5.69	3.66 ↑	10
95000	15.09	9.70 ↑	1
95005	13.51	8.69 ↑	1
95007	12.95	8.33 ↑	1
95008	12.95	8.33 ↑	1
95009	11.64	7.49 ↑	1
95010	10.53	6.77 ↑	1
95011	7.33	4.72 ↑	1
95013M	16.69	10.73 ↑	1
95014	18.60	11.96 ↑	1

B

FIGURE 8–3 Typical price sheets for manufacturer's part line: (A) list, wholesale, and trade prices; (B) volume dealer prices for the same part numbers

list price, wholesale price, and trade (also known as dealer) price. The arrows indicate a price increase or decrease over the last price sheet. This is only one of several price sheets this manufacturer offers. Figure 8-3B shows the volume dealer price sheet for the same items. Note that the list price and part numbers are the same, and the price offered to volume dealers is below the wholesale and trade prices in Figure 8-3A. On some price sheets the volume dealer price might correspond to a stocking dealer price.

Figure 8-4A illustrates another example of a dealer price sheet. It lists the part number, suggest-

No.	List	User	Dealer	No.	List	User	Dealer
5188	3.08	2.62	2.05	5282	20.94	17.80	13.94
5189	6.55	5.57	4.36	5283	22.40	19.04	14.92
5193B	13.48	11.46	8.98	5284	13.69	11.63	9.12
5195C	14.81	12.59	9.87	5287A	14.02	11.92	9.34
5196C	13.60	11.56	9.06	5288C	17.66	15.01	11.78
5197D	17.99	15.29	11.98	5289B	32.96	28.01	21.95
5198D	16.32	13.87	10.87	5290A	13.80	11.73	9.19
5200C	12.67	10.77	8.45	5291A	13.28	11.29	8.85
5201B	16.57	14.08	11.03	5292B	28.81	24.50	19.19
5202B	16.55	14.06	11.02	5293B	16.41	13.95	10.93
5204	18.73	15.92	12.47	5294	12.36	10.51	8.23
5208A	13.66	11.62	9.10	5296A	15.93	13.54	10.62
5210C	15.11	12.84	10.06	5297	10.94	9.30	7.29
5211E	20.19	17.16	13.45	5298	2.70	2.30	1.80

FIGURE 8-5 A slightly different price sheet setup using list, user, and dealer price breakdowns

ed list price, and the net price offered to dealers, such as installer and garages. This price sheet also gives a letter popularity rating to each part. The dagger after the part number indicates a price change from the last published price sheet, although it does not specify an increase or decrease. Figure 8-4B illustrates the wholesale discount price sheet for these same parts. Note the slight increase in price over dealer pricing.

Figure 8-5 shows a price sheet giving list price, user price, and dealer price breakdowns. In this case the user price may be offered to wholesale accounts or to the walk-in trade in competitive markets. As you can see, there can be quite a difference in the terms used by various manufacturers. It is the responsibility of store personnel to note and understand these differences.

PREVENTING ERRORS

To prevent inaccurate reading of price lists and, therefore, inaccurate billing, a counterperson must always follow these precautions:

1. Be sure to use the proper price sheet and discount price for each customer.
2. Use a straightedge to read across columns (Figure 8-6).
3. Determine whether the item is for resale or immediate use. This might be difficult to ascertain with a new customer on the telephone.
4. Maintain up-to-date card files on customers' classifications and check this file when in doubt.
5. Know state and local tax laws and what exemption forms apply in exceptions.

No.	List Price	Net Price	Rating	No.	List Price	Net Price	Rating
3153†	74.82	50.32	A	3222†	81.14	54.58	A
3154	72.60	48.83	A	3223†	99.15	66.69	A
3155†	114.65	77.11	D	3224†	51.46	34.61	A
3156	109.73	73.81	S	3225	245.67	163.37	TT
	Relabel to 3157			3226†	63.56	42.75	A
3157†	120.71	81.19	E	3227†	52.56	35.35	A
3158	80.53	54.17	A	3228†	65.97	44.37	A
3159†	27.25	18.33	A	3229†	88.16	59.30	A
3160†	68.99	46.40	B	3231†	62.93	42.32	FA
3161†	41.64	28.01	D	3232†	48.67	32.74	FA
3162†	59.84	40.25	D	3233†	84.92	57.11	FA
3163†	81.64	57.47	FC	3234†	75.37	53.06	FA
3164†	50.70	35.69	FA	3235†	98.29	68.80	FA
3165†	41.30	29.08	FB	3236†	66.38	44.65	A
3166†	57.15	40.23	FC	3237†	49.54	33.32	A
3167†	60.42	42.54	FB	3238†	46.26	31.11	A
3168†	60.23	42.40	FA	3239†	54.65	36.76	A
3169†	61.95	43.61	FA	3240†	53.54	36.01	C
3171†	90.58	63.77	FD	3241†	53.24	35.81	NA
A 3172†	72.88	51.31	FB	3242†	79.96	53.78	NC

No.	List Price	Net Price	Rating	No.	List Price	Net Price	Rating
3153†	74.82	52.97	A	3222†	81.14	57.45	A
3154	72.60	51.40	A	3223†	99.15	70.20	A
3155†	114.65	81.17	D	3224†	51.46	36.43	A
3156	109.73	77.69	S	3225	245.67	171.97	TT
	Relabel to 3157			3226†	63.56	45.00	A
3157†	120.71	85.46	E	3227†	52.56	37.21	A
3158	80.53	57.02	A	3228†	65.97	46.71	A
3159†	27.25	19.29	A	3229†	88.16	62.42	A
3160†	68.99	48.84	B	3231†	62.93	44.55	FA
3161†	41.64	29.48	D	3232†	48.67	34.46	FA
3162†	59.84	42.37	D	3233†	84.92	60.12	FA
3163†	81.64	57.47	FC	3234†	75.37	53.06	FA
3164†	50.70	35.69	FA	3235†	98.29	68.80	FA
3165†	41.30	29.08	FB	3236†	66.38	47.00	A
3166†	57.15	40.23	FC	3237†	49.54	35.07	A
3167†	60.42	42.54	FB	3238†	46.26	32.75	A
3168†	60.23	42.40	FA	3239†	54.65	38.69	A
3169†	61.95	43.61	FA	3240†	53.54	37.91	C
B 3171†	90.58	63.77	FD	3241†	53.24	37.69	NA

FIGURE 8-4 Price sheets often contain popularity and other important information: (A) list and dealer net prices; (B) list and wholesale prices

FIGURE 8-6 Using a straightedge or even an index card will help to accurately read across price sheet columns.

6. Recheck the units used on the price sheet. Units are often different from shipping or packaged quantities. For example, if the price list shows $5 for 100 items and the items are packaged in boxes of 50, make sure the price per 100 is not charged per box.
7. Print to ensure legible sales slips.
8. Double-check sales slips for accuracy and mathematical errors.
9. When unsure about what price to charge, check with the manager.

PRICE REDUCTIONS

Occasionally, a salesperson might consider giving a customer a special price to make a sale. He or she might think that a small profit is better than no sale. However, Table 8-1 shows that reducing a price even a minimum amount can cut into profits greatly. For example, if a salesperson cut a price by 20 percent on a $100 sale, he or she will have to sell

TABLE 8-1:	EFFECTS OF REDUCING PRICE ON A PRODUCT MARKED FOR 30% PROFIT	
Price Reduced	**Dollar Volume Increase Necessary**	**Increase in Amount of Merchandise Handled**
5%	14%	20%
10%	25%	50%
15%	70%	100%
20%	140%	200%

$240 in volume and handle three times as much merchandise to make the $30 profit the original sale should have made.

All salespersons and counterpersons should follow the price sheets and the policies of the business when pricing items. When valid discounts are given to a customer, the counterperson should mention the list price as well as the discounted price. It does not hurt customer relations to point out how much money a customer is saving.

REFERENCE VALUE

Price sheets contain not only important price information, but also information about availability, popularity, obsolescence, shipping source, supersessions, descriptions, catalog page numbers, and policies.

If a customer calls with questions about an item on an invoice that is listed by a part number but not a name or description, the counterperson need only look at a price sheet and not in a catalog to identify the part. Many price sheets also contain descriptions, and some also indicate the catalog page number of each item.

In addition, if a counterperson has the number for a part that does not appear on the most recent price list, then it is safe to assume that the part is no longer supplied by the manufacturer. When using old issues of catalogs or sections pertaining to older cars, always consult the price sheet to determine the availability of a part. Also, superseded and obsolete numbers are usually indicated on price sheets long before new catalogs are printed.

Price sheets also often include popularity classifications for each number. These figures are supplied by the manufacturer and typically indicate how well each item sells nationally. While the movement of parts in a region might differ significantly, popularity classifications can indicate rising or declining trends and can help with decisions about whether or not a part should be stocked.

PRICE QUOTES

The method or procedure used for quoting a price to a customer can make or break a sale. Most people do not like to part with money, so they tend to believe that most prices are too high right from the start. If the counterperson has not fully explained the features and benefits of a product to a customer and cannot support a price that a customer believes to be too high, then the sale will probably be lost. In most situations, to make a sale, the customer must be sold on the benefits before hearing the price. If the customer hears the price first and believes it is too high, then he or she will probably not listen to the explanation of benefits.

A customer buys only when he or she decides that a specific purchase is the best value available and that its value is worth the cost. Therefore, it is extremely important that any prices quoted to prospects are accurate. Quoting a price that is higher than the real price could cause a prospect to decide not to buy at all. Quoting a price lower than the real price could also lose a customer when he or she finds out that they have to pay more than expected.

PROBLEM: Using the price sheet in Figure 8–5, give a price quote to a dealer customer who is interested in purchasing the following:
1 unit: part # 5290A
2 units: part # 5189
1 unit: part # 5210C

ANSWER: Using the price sheet shown in Figure 8–5, you can see that dealer prices are listed last after list and user prices. Compute the price for the part numbers and quantities as follows:

1 unit: part # 5290A = 1 × $ 9.19 = $ 9.19
2 units: part # 5189 = 2 × $ 4.36 = $ 8.72
1 unit: part # 5210C = 1 × $10.06 = $10.06
$27.97

Also, remember to check stock for availability, make certain the price sheet is up-to-date, and double-check quantities and computed price.

FIGURE 8–7 Bar codes are used to accurately and quickly identify an item, read its price, and adjust inventory.

CALCULATING MARKUP

Markup is the amount the jobber increases the price of an item over the price the jobber paid for the item. Various customer classifications receive various markup rates, and standard markup rates are built into the price sheets issued by manufacturers. As mentioned earlier, many jobbers adjust manufacturer's markup to better suit their needs. To determine a manufacturer's suggested markup rate, use the following formula:

markup rate = markup ($) ÷ jobber cost

For example, what is the markup rate of a part that cost the jobber $37.50 and lists for $50.00?

markup rate = markup ($) ÷ jobber cost
= ($50.00 – $37.50) ÷ $37.50
= $12.50   ÷ $37.50
= 0.333 or 33-1/3%

If a jobber store owner does not feel this rate is competitive, he or she may adjust it downward. For example, to calculate the list price for the same item using a 25 percent markup, use this formula:

markup price = (jobber cost × markup rate) + jobber cost
= $37.50 × 0.25   + $37.50
= $9.28   + $37.50
= $46.78

It is faster and easier to simply multiply jobber cost by (1 + markup in percent) or 1.25 × $37.50 = $46.78.

BAR CODES

A bar code, or UPC symbol, is a set of distinctly spaced lines of varying widths located on packages and other items (Figure 8–7). A machine reads the symbol to identify the product and its predetermined price and also usually adjusts inventory records.

Bar codes are used because they are read by a machine more accurately and with less expense than manual techniques. Manually recording price with key strokes on a cash register or keypad generally results in one error in fifty strokes, while bar code entry has less than one error in 3 million. Hence, this technique has been referred to as "key entry bypass."

A bar code is basically a binary code where images are converted to sequences of zeros and ones. Usually, as a scanner passes over the code, the narrow bars and spaces are converted to zeros and the wide bars and spaces are converted to ones. Many grocery stores have successfully used this system for years. The cashier runs the bar code over the scanner, which reads it, adds the price to the bill, and deducts the item from inventory.

Approximately 150 bar code symbologies or languages exist today, each with its own design of bars and spaces representing individual characters. Some symbologies are limited by the number of characters represented. For example, UPC is limited to numeric characters, which means it can only record information in numbers. Code 39 can contain both alpha and numeric characters as well as some special symbols.

Different codes are better suited to different types of industry. The Automotive Industry Action Group (AIAG) supports Code 39 while the retail industry generally supports UPC.

USAGE

Bar codes can be used in the following areas:

- Inventory control
- Work-in process (tracking and cost recording)
- Time and attendance record-keeping
- Document control
- Circulation systems (libraries and tool cribs)
- Point-of-sale automation
- Remote order entry (bar coded sales order books)
- Rental facilities
- I.D. cards and systems
- Shipping control
- Packaging

By using the manufacturers' bar codes together with a retail computer, a counterperson can eliminate mistakes, reduce shipping delays, and streamline inventory management and control. The necessary tools include a portable bar code reader, a hand-held laser scanner together with a portable Tele-Transaction Computer. These tools can be used for fast, accurate data entry anywhere and at anytime. With scanner utility software, the bar code information can be transferred back into some retail computers to be displayed or printed.

Most readers and scanners have a piston-shaped case. Some models do not require direct contact with the bar code surface and can be held up to 6 inches away at a wide range of horizontal and vertical angles. The laser operates automatically, requiring no scanning motion from the operator. Always check with the scanner manufacturer to make sure that the labels in the business can be read by a particular scanner.

Laser scanners used in supermarkets have laboratory first-read rates well above 99 percent. This figure means that for every 100 symbols scanned, only one symbol will not read on the first scan. In actual business applications, first-read rates tend to range between 80 to 90 percent.

First-read rates in the 80 to 90 percent range give the highest productivity. Employees who achieved a higher read rate had moved items over the scanner more slowly, and although they had fewer items to pass over the scanner more than once, they still had fewer packages successfully scanned per hour.

Bar codes can also be used to facilitate taking inventory of items in stock. Instead of manually recording the part number and counting each item, simply scan the bar codes, and the part number and quantity will be recorded quickly and accurately.

TYPES OF ACCOUNTS

Many types of accounts exist in the automotive business, as in any business, and the counterperson will usually have some contact with most of them.

Accounts receivable include those customers' accounts that have not been paid. When totaled they include the balances of money due from purchases. Accounts payable are somewhat the opposite. They include those accounts which the business must pay.

An active account is the account of a particular customer who makes frequent purchases on a regular basis.

A dual account involves a customer who buys for more than one purpose. For example, a customer who buys both for use and for reselling and a customer who buys for both industrial and automotive purposes can be classified as dual accounts.

A credit account is extended to customers who have been approved by the manager or owner to be able to purchase items on credit. Sometimes the credit is limited to certain items or a certain dollar amount. Counter personnel should be very familiar with a business's credit policy.

INVOICES

An invoice or sales slip records a sales transaction and includes all pertinent information and pricing (Figure 8–8). Invoices must be legible, accurate, complete, and mathematically correct. For

KAC AUTO PARTS
SALES INVOICE

Date _11/28/88_

Name _JOE HENKEL – HENKEL'S SPEED SHOP_

Address _506 EAST UNION ST._

City _GRIER CITY_ State _PA_ Zip _18351_

Phone _621-5789_

Order No. _17921_ Filled By _B. DEELEY_

Shipped Via _CUSTOMER PICK-UP_ Remarks _TAX NO. ON FILE_

ORDERED	SHIPPED	UNIT	PART NO.	DESCRIPTION	MANUFACTURER	LIST PRICE	UNIT PRICE	NET PRICE
1	1	KIT	2-397	CARBURETOR KIT	ACE	16.88	11.19	11.19
2	2	GAL	4513	ANTI-FREEZE	PX		4.99	9.98
1	1		393	MUFFLER	AZ	51.43	36.41	36.41

Subtotal	57.58
Tax	—
Trans.	—
Total Invoice	57.58

Customer Signature _Joe Henkel_

FIGURE 8-8 An invoice records all important information in a sales transaction.

KAC AUTO PARTS
MACHINE SHOP ORDER

Date _12/3/88_

Name _CARUSO'S GARAGE_

Address _MAIN ST. AND VINE_

City _PORT CARBON_ State _PA_ Zip _17931_

Phone _366-0146_

Taken By _H. Carl_

Work Order No. _1413-2_

Job Description and Work to be Performed _____

RESURFACE BRAKE ROTORS SUPPLIED

FIGURE 8-9 A shop ticket must be accurately filled out whenever a machine shop job is completed.

more information on invoicing, see Chapter 7. The following steps outline the basic procedure for writing an invoice:

1. Fill in customer information such as name, address, and phone number.
2. Indicate whether the sale is cash or charge.
3. Fill in purchase information: date, quantity of items purchased, part number, name of manufacturer and item, list price, net price, amount of purchase per item.
4. Total the charges.
5. Add a core deposit and/or sales tax if applicable.
6. Sign or initial the invoice.
7. Stamp all copies of the invoice cash or charge.
8. Have the customer sign the invoice if the transaction is a charge sale.

SHOP TICKETS

When a machine shop job is complete, a ticket similar to an invoice must be written to record the transaction and receive payment (Figure 8-9). Follow these steps to write up a ticket for a complete machine shop job:

1. Fill in the following information on the ticket: order number; customer's name and address; whether payment is cash or charge; date; description of labor performed; charge for labor and discount charge for labor; quantity, number, description, and price of parts used in the job.
2. Check to be sure all labor and parts used have been billed.
3. Total the charges.
4. Sign or initial the ticket.

To simplify machine shop billing, a work order is often assigned to each job. As parts are needed to complete numbethe work, they are pulled from inventory and charged against this work order number.

When the job is complete, the work order number is called up and all parts used on the job are listed. This prevents errors and omissions in billing and records.

FILING INVOICES

Methods of filing invoices vary with businesses. Counter personnel must become adept with the method used in his or her workplace. Computerization greatly simplifies the handling, recording, and filing of invoices (see Chapter 6).

The most common methods of manually filing invoices are numerically and alphabetically. To file numerically, arrange the invoices in order with the small number on the bottom and the largest number on top. Place the invoice faceup and file on a metal post or place in a binder. Date the binders to show what business days are included.

To file alphabetically, arrange the invoices in alphabetical order with the beginning of the alphabet first and the end of the alphabet last. Place the invoices faceup and file them in customer account files.

These examples might not match every filing system, so counterpersons must learn the procedure where they work.

CASH SALES

Cash sales are often among the most profitable in a business. Little paperwork is necessary, usually little time is required, and cash flow is not impeded. Offering discounts for cash purchases and making sure cash customers are handled courteously and efficiently encourage customers to pay with cash.

To conduct a cash sale effectively, follow these guidelines:

1. Greet the customer.
2. Determine the customer's needs: what part or parts are involved and get pertinent vehicle information. Write this information down to ensure accuracy.
3. Refer to catalogs and other materials and write down the necessary information. Check inventory.

4. Locate the merchandise in stock and confirm the merchandise requested with the customer.
5. Suggest related items and/or services that can benefit the customer.
6. Fill out the sales invoice, making sure to include any applicable discounts. Accept payment for the purchase and make change correctly if necessary.
7. Wrap or package the purchase.
8. Conclude the transaction by thanking the customer and inviting him or her to return.

CHARGE ACCOUNT SALES

Charge account sales encourage large purchases or purchases from customers who do not want to carry cash or do not have the cash. Counter personnel must be familiar with the charge policy in his or her workplace to ensure accuracy and customer satisfaction.

To conduct a charge account sale, follow steps 1 through 5 for making a cash sale, then proceed with the following steps:

- Fill out the charge slip. Be sure to check to see if the customer's account is past due and obtain the customer's signature on the invoice.
- Wrap or package the merchandise.
- Conclude the transaction by thanking the customer and inviting him or her to return.

TAX

Different taxes apply to different items and customers, and counterpersons should be sure to charge or withhold tax when applicable.

An excise tax is a tax paid by the manufacturer or seller. However, the amount is usually indirectly passed on to the consumer. Sales tax is tax applied directly to an item at the time of purchase.

To compute the amount of sales tax for an order, follow thesse steps:

1. Record each taxable item on the invoice.
2. After all taxable items in the order have been recorded, total their cost.
3. Using a tax chart or tax table, calculate the amount of tax to be charged (Figure 8–10). Be sure to follow existing tax laws for a given area and a given item.

FIGURE 8-10 Counterpersons should use a tax chart or table to be sure to calculate tax accurately.

4. Record the amount of tax.
5. Total the charges on the invoice.

If an item is purchased and is not to be resold, tax must be charged on the item. If a customer is not charged tax on a certain item, he or she must have a tax number on file.

DISCOUNTS

A discount is an allowance or reduction from a given list price or an allowance for prompt payment of an account. Handling discounts is very important to accurately charge customers and accurately receive payment.

Different businesses often have different discount and payment structures. For example, a professional mechanic could purchase an oil filter and a set of brake pads at five different stores and receive five different prices and five different ways to pay for the purchase. Depending on each store's policy, he or she could pay list price, nonstocking dealer price, stocking dealer price, 10 percent off stocking dealer price, or jobber price. In addition, the mechanic could pay for the merchandise with cash, a credit card such as Visa or MasterCard, a hold ticket, or a revolving charge account, depending on the store.

When figuring discounts, counter personnel should always use a calculator to ensure mathematical accuracy. Carelessly calculating discounts results in damaged customer relations and potential loss to the business. Some businesses have calculators on the counter for employees, but it is often a good idea for each counterperson to have a pocket calculator for figuring discounts away from the counter, such as at a display or in the stockroom.

CASH DISCOUNTS

A cash discount is one given for immediate or prompt payment of a bill. Cash discounts originated decades ago as a means of encouraging prompt payment of the charge customer's monthly bill. The usual practice was to take a 2 percent discount off the bill if it was paid by a certain day of the month. However, the cash discount was expanded to mean different terms at different businesses. For example, if a charge customer began getting behind on his or her bills, then the 2 percent discount was lost. But many times the late-paying customer would be placed on cash until the bill payments were current again, then the customer would receive a 2 percent discount for paying cash.

Cash discounts began when almost every business was completely wholesale, so retail sales did not confuse the discount system. Over the years, the cash discount system has been greatly modified, and counter personnel must be very familiar with the procedures and policies in his or her own workplace.

The first real modification of the cash discount was used on equipment sales. Some businesses began to offer a 5 percent discount on certain equipment sales if the purchase was paid in cash. Today this technique is not as common due in part to the increased discounts necessary just to sell a piece of equipment. With today's highly discounted prices, offering even a 5 percent discount might cut into profits too much. In fact, many places will not sell equipment at all anymore unless it is paid for in cash.

Another common cash discount, widely used by service stations for gasoline sales, is to offer a discount for cash purchases and not for credit card sales. Many banks charge anywhere from 1 percent to 5 percent for handling transactions made with cards such as Visa or MasterCard, so businesses who accept these cards often prefer to offer a discount to cash customers. The business still comes out ahead if a customer receives a 2 percent discount, but the business avoids a 4 percent handling fee.

However, when deciding whether or not to offer a cash discount, managers require input from counter personnel to assess customer needs as well as the needs of the business. For example, many customers carry only minimal amounts of cash with them, so if the competition allows credit card purchases, customers might prefer to shop there. In addition, most businesses prefer to work with either credit cards or cash, so if customers refuse to carry cash, the business managers might prefer to continue with credit cards as opposed to checks or open charge accounts.

When figuring cash discounts, be sure to figure the discount on only the items bought, not on other features on a bill such as sales tax, transportation, or phone calls.

Who Gets Cash Discounts

Counter personnel must remember that cash discounts are not always given to everyone who pays with cash. For example, in stores where all equipment purchases must be made in cash, the price quoted for a piece of equipment might already include a cash discount. In some stores, some items always receive a discount while others never do. Sometimes if an item is being sold at net price, no further discounts are being allowed. In addition, if a store is running a promotion on a certain product or line, such as oil, the price might already be listed as lower than cost simply to attract customers into the store. An additional discount would probably not be attached to the oil, even if the customer paid cash.

Another factor that counterpersons must consider when applying discounts is who the customer is. For example, some businesses offer discounts only to customers who have no outstanding debt. A charge customer who is behind in payment might be required to use cash instead of charging more. However, in some businesses, this customer would not receive a cash discount because of the bill pending.

DEALER OR TRADE DISCOUNT

A dealer or trade discount is offered to customers who will use a particular purchase for resale or to make professional repairs. Customers offered trade discounts usually purchase regularly. This discount is for others in the automotive business who would not be able to make a profit if their cost were the same as for retail consumers. Trade discounts vary from 20 to 30 percent for small dealers to as much as 40 percent for large purchasers.

Counter personnel must use extreme caution when applying trade discounts, especially when quoting prices to new customers or to customers on the telephone. A counterperson who quotes a trade discounted price to a nonprofessional customer might take business away from a professional mechanic who buys parts from that counterperson. For example, a prospective customer might call, requesting a rebuilt carburetor for a 1974 Dodge. The customer has all the necessary information pertaining to the engine size and design, so the counterperson suspects that the customer is familiar with ordering automotive parts. The counterperson finds the part in stock and then asks the customer if he or she

is installing it personally. When the customer replies, "Yes," the counterperson quotes a price with a trade discount.

However, after making the sale, the counterperson receives a call from Dave, a professional mechanic who is a regular customer, complaining because that same customer had come to him earlier for a rebuilt carburetor. The professional mechanic had called the counterperson for a price (which included a trade discount) and then quoted list price to the customer. The customer accused the mechanic of trying to cheat him when he called the same counterperson and received a price quote less than list price.

When situations such as this arise, the counterperson should attempt to explain the reasons for trade discounts. Auto repair facilities must get $50 to $60 an hour to make a profit, and most mechanics need thousands of dollars worth of tools, not including the large equipment usually owned by the repair facility or garage. Because the public will not accept paying $50 to $60 an hour to have vehicles repaired, the business must make the rest of the profit on the parts that are sold.

In addition, customers should understand that most repair businesses will not guarantee the whole job, just the labor, if the parts are bought elsewhere and brought to the garage to be installed. In the previous example, if something wrong was discovered with the carburetor, the garage would have to charge the customer for the labor again because it was not bought there and, therefore, could not be guaranteed. Customers who understand this policy will usually pay a little more to buy the part from the garage in order to receive the guarantee, and the problem of offering trade discounts to customers not in the automotive business will be avoided.

CHAIN DISCOUNT

A chain discount is two successive discounts extended on a single purchase. Counter personnel should be careful not to calculate a chain discount as a straight discount. A chain discount of 10 percent and 5 percent is not the same as a straight 15 percent discount. Chain discounts are calculated by finding the first discounted price and then discounting this figure by applying the second discount.

EXTRA DATING

Extra dating is a discount made available for purchased items that are delivered at a given date and marked payable in a given period of time, such as 30, 60, or 90 days. Extra dating, also known as deferred billing, is used to spread payment without losing the cash discount. It is popular for large quantity purchases or seasonal items.

CALCULATING DISCOUNTS

To figure a discount, follow these steps:

1. Determine the type and amount of discount to be given (for example, cash, trade, or chain).
2. Subtract the amount of discount from 100 percent. (For example, a 20% discount would be calculated 100% – 20% = 80%.)
3. Multiply the result by the list price. (For example, an item costing $10 with a discount of 20% would be calcualted $10 × 80% = $8.)
4. If the final two figures are less than 1/2 cent, drop them. If the final figures are more than 1/2 cent, raise the price to the next penny.

PROBLEM: Calculate a trade discount of 25% on a purchase of $310.65.

ANSWER: 100% – 25% = 75%
$310.65 × 75% = $232.99

PROBLEM: Calculate a cash discount of 2% on a purchase of $89.54.

ANSWER: 100% – 2% = 98%
$89.54 × 98% = $87.75

To calculate a chain discount, complete these steps with the first discount, then repeat the procedure applying the second discount to the end result of the first calculations. For example, if a $500 purchase is marked with a 10% discount and the customer normally receives a 5% discount, figure the cost as follows:

100% – 10% = 90%
$500 × 90% = $450
100% – 5% = 95%
$450 × 95% = $427.50

The customer will pay $427.50. If the counterperson had calculated the discount as a straight 15%, the customer would have paid $425, as follows:

100% – 15% = 85%
$500 × 85% = $425

DISCOUNT STORES

More and more parts stores are including the word *discount* in the name of the store. This name gives many customers the impression that most, if not all, parts are discounted. Some stores specialize in a particular part or automotive system, such as exhaust or brakes, and these components are usually discounted. These special features attract customers who then buy additional items. However, quite often other items are not discounted and might even cost more than in other parts stores.

REBATES

The term *rebate* has begun making an impact on the automotive business from sales to repairs. Rebates are usually promises of money returned if a consumer buys the product and mails in a coupon or form from its packaging or labeling (Figure 8–11). Rebates are generally offered by manufacturers to boost sales, attract new customers, and/or introduce a new product. In the parts business, rebates increase sales not only at the time of the rebate, but also in the future when satisfied customers buy the part again without the rebate. Manufacturers can also use the number of returned rebate tickets to accurately assess the success of a particular type of promotion.

Rebates usually benefit all concerned. Manufacturers and dealers enjoy increased sales and promotion, and consumers receive some type of refund. For example, a shock absorber company running a three-month rebate could receive far more than 150,000 rebate coupons in one month alone. In addition, the dealers might receive a price break,

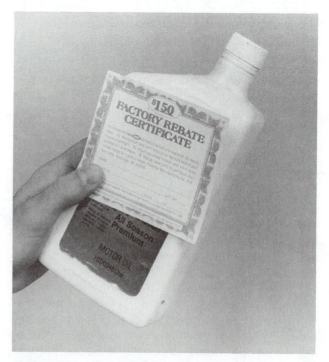

FIGURE 8–11 Rebates usually promise money back if a buyer returns the rebate ticket to the manufacturer.

and if salespersons and counterperson do their jobs, then related sales and additional sales generated by an increase in customers will also increase.

COUPONS

Coupons, a small printed piece of paper or cardboard offering a discount on the price of a specific product, are offered by manufacturers to generate sales in much the same way as rebates. In addition, placing a clip-out coupon in a printed advertisement can tell the manufacturer how many people actually have access to or read that type of advertisement. Long a significant aspect of marketing groceries and household items, coupons are more recently being offered for automotive parts and services as well.

Most businesses accept coupons, and counter personnel must be familiar with both the manufacturer's policy and the store's policy for handling coupons and returning them to the manufacturer for credit.

WARRANTIES

Warranties are promises of refund or replacement if a product or service does not live up to its designated function (Figure 8–12). They are often used as a sales tool in promotions and have become an effective device in the market. One major parts chain offers a seven-year, 70,000-mile warranty on their auto parts, while individual manufacturers and suppliers offer their own warranties.

However, warranties can be more perception than substance in some cases. The average car is kept about five or six years; therefore, the real value of a lengthy warranty to the customer is in the perception of added quality. A warranty lends credibility to a product.

To avoid warranty problems in the future, some counterpersons take a few minutes to explain the basics of a warranty that comes with a particular purchase. For example, some warranties are issued by the manufacturer and some by the jobber, and all warranties have conditions. Even lifetime warranties have limitations, so customers should understand that the part might not necessarily be replaced at any point in the future no matter what. In addition, customers should know that it is not up to the counterperson or the jobber if a manufacturer decides not to honor a warranty because of abuse. In situations where a manufacturer will not offer credit easily, the jobber often offers credit to the customer anyway to ensure his or her return. Many times a jobber will switch lines for this reason.

Some manufacturers have begun to require that the consumer deal directly with the manufacturer when handling a part problem. Because some companies are shifting the responsibility onto the consumer, the counterperson is occasionally spared the hassle. However, the customer will often attempt to involve the person who sold the part anyway and, therefore, many counterpersons believe that the best warranty system allows the counterperson (not the customer or the manufacturer) to decide if credit should be offered.

Sooner or later, counter personnel will encounter a situation requiring that a warranty be fulfilled, and it is the counterperson's responsibility to handle both the customer and the claim. There are many different types of warranties with varying limitations and conditions such as time limitations, use limitations, and abuse conditions. The counterperson must determine whether or not a warranty is in effect and, if so, just how much in effect. To decide, the counterperson must examine the appearance of the item, question the customer, and also often rely on memory for the conditions and date of purchase.

Most customers, both professionals and do-it-yourselfers, assume that every product and part has a warranty. Because of this belief, most customers with a complaint will enter the business believing

FIGURE 8–12 Warranties promise a refund or replacement if a product or service is defective. Counter staff must be familiar with all warranty provisions for the products they carry.

they are entitled to something in return, and the counterperson must be able to handle any attitude.

Today, efficient testing equipment makes it possible to test many parts in front of the customer to determine if a part is faulty or if the customer's vehicle has a different problem. However, even if testing equipment indicates that a part is not functioning properly, the part itself might not necessarily be at fault. A malfunction of another component in the vehicle might cause damage to other parts. Showing a stubborn customer supporting written evidence, such as manuals or bulletins, can help convince him or her of an error.

CARRYING LINES WITH WARRANTIES

Sometimes a business might choose to carry a certain line based on the manufacturer's or supplier's warranty procedures. Some warranties seem to be designed to discourage jobbers from returning items. Given two very similar lines, counter personnel might suggest that the line with the simplest warranty procedure should be carried. Other hassle-reducing features to look for in products are extended free replacement periods on things like batteries, simple procedures for prorating the goods after the free replacement period has expired, and generous warranty periods on other merchandise.

What should a counterperson do when a customer wants retribution not only for a part, but also for other parts he or she claims were damaged by that particular part? The counterperson must weigh each case individually. In many instances, there are enough warning gauges and signals in vehicles to indicate to the user that something is wrong before serious damage occurs.

For example, suppose a customer who purchased a defective thermostat wanted the business to replace not only the thermostat, but also the engine because the thermostat had stuck and the engine overheated, causing extensive damage. What should be the counterperson's course of action? The counterperson should agree to submit the thermostat to the manufacturer for testing and also agree that if the part were indeed defective, then the business would replace the thermostat as well as a gasket, coolant, and the labor for replacing the thermostat.

However, the counterperson should not concede responsibility for the engine damage because the heat gauge should have indicated a problem long before extensive damage occurred. If the gauge malfunctioned, then the gauge manufacturer is responsible. If not, then the operator is to blame. In addition, the counterperson might quote the cus-

tomer an excellent price on the engine work. In this manner, the counterperson can handle the thermostat warranty without costing the business unnecessary expense, perhaps even creating sales of engine components or labor.

TERMS AND CLAIMS

One of the surest ways to avoid unnecessary red tape with warranties is to work with WDs that are concerned with customer relations. Most reputable suppliers are more concerned with customer satisfaction than with the exact terms of a published warranty. For example, 90-day warranties will, in most cases, still be honored after 95 or possibly even 195 days because suppliers do not know how long the part sat on a shelf.

Sometimes it is impossible to determine if the part or if the installer is at fault, or if some other problem caused the part to fail. Most of the time, this ambiguity will not matter. As long as there is no labor claim at issue or a claim for other damages, most suppliers will give credit without asking questions.

A difficult situation for a counterperson to handle might be one in which the customer wants a replacement part before removing the bad part. A customer might need the vehicle in a hurry and might want to condense the time involved in rectifying the situation. In this case, the counterperson should sell the customer a second part and reimburse for one when the suspect part is returned. If a customer cannot afford to pay for the second part without receiving reimbursement on the first, the counterperson might be able to accept a check or credit card transaction and hold it until either the part is returned or a designated amount of time has elapsed.

Labor Claims

One of the more complex issues in dealing with faulty parts is labor claims. Labor is rarely a problem with do-it-yourself customers, but lost labor is an important financial concern with professional mechanics. Unfortunately, with most suppliers, the procedures for filing a labor claim are cumbersome and credits are slow to come.

At times, it might be simpler for the jobber to handle a labor claim without involving any real dollars. For example, a regular professional customer might be offered credits toward machine shop work or electronic component testing. However, consistently striking such deals is not a good business practice and every effort should be made to solve the problem through the proper channels. If the problem cannot be resolved easily, counterpeople should always consult their manager or store owner before negotiating a special settlement.

Safety

Counter personnel should also use caution when handling components that are safety related. For example, a counterperson must obtain the whole story before simply offering to replace a defective ball joint or bad set of brake pads, especially if the defect caused an accident or injury. Replacing safety related parts could possibly later be construed as an admission of responsibility and become an important element in a legal action against the business.

Even if the failure resulted in no harm, the manufacturer has to know if something is wrong with a product. And counterpersons must know whether or not more of the same problems will arise if the line in question continues to be sold.

Necessary Red Tape

No matter how many precautions a counterperson takes against having items returned and no matter how many creative ways a counterperson comes up with to satisfy a customer, there is still a certain amount of paperwork and other procedures that will have to be adhered to for proper credit and satisfaction to be insured. Some red tape cannot be cut out. For example, some warranties require a request for authorization to return goods, and some require that a warranty tag be attached to the part. Steps such as these cannot be eliminated.

If a business replaces a part or gives a customer a refund and then has to wait for credit, the business has, in effect, given the supplier an interest-free loan. It is important to keep these loans to a minimum and to keep their terms as short as possible by doing the necessary paperwork, returning parts to suppliers promptly, getting a receipt, and tracking the returns until credit is given. With the exception of parts that must be inspected by the manufacturer, credits for defective merchandise should be issued in the same billing period as they are returned.

CORE CHARGES

Remanufactured units have a long history in the parts business. Rebuilding and remanufacturing are an important part of daily automotive business. In the last ten years, there has been a resurgence in the rebuilding of some lines, such as air conditioning, power steering, and carburetors, and the beginning

of rebuilding new items such as rack-and-pinion steering.

Most rebuilders go through the same general steps to get the core, or casting, to their place of business so that it can be processed. There are "core collectors" who scavenge the countryside looking for rebuildable pieces that they can purchase. They frequent salvage yards, junk dealers, machine shops, car dealerships, and auto repair shops looking for hard-to-find castings. They, in turn, take the junk units to the remanufacturer to rebuild.

Counterpersons and customers should realize the difference between a rebuilder and a remanufacturer. The rebuilder finds the defect in the unit, cleans it, and repairs it. The remanufacturer disassembles the unit completely, then checks all major components and automatically replaces all minor components.

Both the rebuilder and remanufacturer sell the units to the WD who sells to a jobber who sells to a customer. However, there is one major difference: the customer at each level puts a deposit on the unit called a *core charge*. The core charge separates rebuilt units from all other phases of the automotive business (Figure 8–13).

There is a core charge on rebuilt units because the rebuilder does not have the facilities for making the product. All he or she can do is take a product that someone else has made and rebuild it to his or her specifications. Since the rebuilder does not make the unit, he or she has to buy junkers from a core seller or from another source. If the rebuilder had to re-buy a core every time he or she built one,

FIGURE 8–13 The counterperson must inspect and identify all cores returned for credit.

the selling price would not be far off from a new unit. Core charges keep the price of rebuilt units down.

Most core charge systems are similar. The rebuilder sells a rebuilt unit to a WD, then places a charge on the casting similar to a deposit. The rebuilder keeps track of these core charges through what is known as a core bank. For example, if a WD buys a hundred alternators, he or she is also charged for a hundred alternator cores. If the core charge for each alternator is $20, the WD would have to pay a collective core charge of $2,000 for 100 alternators. These charges are kept in a separate account—the core bank—and as the WD returns the castings, he or she receives credit against that account.

The WD, in turn, sets up a core bank to keep track of the cores the jobber buys. It operates in a similar manner to the rebuilder's core bank.

The jobber usually does not have the means to set up a core bank so he or she charges the customer for the core when the unit is bought and credits him or her for it when it is returned.

CORE RETURN AND IDENTIFICATION

Identifying cores is a much more difficult job than it was twenty years ago when four major automobile manufacturers dominated the market and parts were not as numerous or intricate. Twenty years ago, for example, specialized parts such as internal regulators did not exist, alternators were new, and three numbers covered most applications. Cores were easy to identify and procure and were inexpensive. However, computer control and other involved and expensive components, the increase use of foreign automobiles, and the constant model changing by the manufacturer have made cores expensive.

The counterperson must be careful when handling cores and accepting core returns from customers. A counterperson should never say to a customer, "It does not matter if you lost the starter core, just bring me another one; any starter core will do." They are not all alike. Cores are the same as money; a lost core is lost money.

To help ensure that customers return their cores, a counterperson should always require a core charge. Some counterpersons believe that charging for the core would damage customer relations. While some customers might reliably return a casting of good value, there will always be those who do not return any castings, who return castings of little or no value, or who return some and resell others to a core scavenger. All of these transactions cost the business money.

After a customer returns a casting, how can the counterperson be sure it is the right one? Unfortunately, he or she probably cannot be sure. It is not feasible to train a counterperson to be as adept at identifying a casting by appearance as the core inspector working for the remanufacturer. However, taking the following steps will protect the counterperson in most instances:

- Billing methods for cores must be consistent.
- Try to ensure that the core returned is the same type as the one that went out by marking each box with a stamp that includes a place for the invoice number, the dollar amount of core charge, and the date. The stamp might also include a message stating that the casting is valuable and to receive credit for it, it must be returned assembled in the original box. Having this information on the box informs the customer of store policy, saves time when the item is returned, and makes damages more readily apparent than if the customer returned pieces. In addition, many times a different component or even the same component from a different manufacturer or model will not fit in the same box.
- Check for broken or damaged parts. The penalty for broken or missing parts varies from rebuilder to rebuilder, but almost all of them assess some sort of charge for damage such as cracks in the housing, missing springs, and heat checking.

The counterperson who does not follow these steps can cost the business money when handling cores because many manufacturers require that all cores must be returned in the box to receive full credit, and some remanufacturers even disallow credit for disassembled units.

To accept a core for a rebuilt or exchange item, follow these steps:

1. Identify the core unit by inspection, identification chart, cross-reference to the original part number on the core unit, or by asking the manager.
2. Inspect the core to determine if it is undamaged, complete, and the same unit as the one being purchased.
3. Check the manufacturer's references or guides to determine the amount to charge for any missing or damaged parts.
4. Record the amount allowed for the core on the invoice or core ticket.
5. Store the unit and tag it for identification.

When a customer purchases an item and does not have the core with him, request that he return the ticket with the core.

HANDLING CORE EXCHANGES AND WARRANTIES

Cores and exchanges are an important part of the automotive business and should, therefore, be handled carefully. Usually more than 15 percent of a jobber's business involves exchange merchandise. Since exchange items make up a sizable amount of sales, they should not be referred to as "junk." Customers who believe that employees and managers consider the cores to be junk will not take care of them. And counterpersons themselves should never toss or mishandle cores in the workplace.

In some areas, rebuilt parts tend to generate more complaints than new parts, not because they are defective but because many people assume rebuilt parts are to blame whenever something goes wrong. In many instances, this assumption is incorrect. Another reason some people complain about the rebuilt item is because, like new parts, they carry a warranty; and many people use warranties as an excuse to bring a part back with any dissatisfaction, whether justified or not.

Sometimes customers purchase rebuilt parts when they are not sure of their vehicle's actual problem or when they do not own the proper equipment to pinpoint the problem. They use the trial-and-error method, replacing each suspect part until one of the replacements solves the problem. Then they return all the other rebuilt parts they purchased as defective.

To prevent unnecessary returns of rebuilt parts on warranty, counter personnel should follow these guidelines:

1. Inform customers of the policies in that business. For example, if a customer knows that the business does not accept returns on electrical parts, then he or she is less likely to attempt the trial-and-error method when solving an electrical problem.
2. When a customer wants to purchase a rebuilt item, ask him or her what procedure was followed to ascertain that this particular component is needed. If the customer appears to be guessing, recommend that he or she take the vehicle to a service outlet equipped with the right diagnostic instruments. In the long run, this step will be cheaper for the customer, a point that the counterperson should not omit.
3. When it is necessary to accept a warranty return from a customer, do not use the word "defective." The word implies that the part should work and that it has broken down prematurely, pointing fault at the rebuilder. Fault is difficult to determine at the counter. Use the word "warranty" instead because it indicates the supplier will make good on the part as a matter of policy and customer courtesy without admitting fault.

RETURNS

Every business in any line of industry that supplies products for the public or for other businesses has to deal with returns. Some products will be defective and some will be incorrectly purchased. In addition, some returned items can be simply reshelved and resold, some must be returned to suppliers, and some are no good at all.

To deal with inevitable returns, most businesses have a return policy that should be made apparent to all employees and customers. For example, a common policy among parts stores and departments is to accept no returns on installed parts. For a policy such as this to work, it should be clearly stated on a sign within easy view of customers, perhaps stated on invoices, and it should be repeated and explained to customers at the time of purchase.

However, with a policy such as this one, some customers will still insist on returning an item because they claim they did not install it even if the item looks like it had been installed. The possibility exists that an employee inadvertently reshelved a returned defective item with good stock and it was resold. The matter could become the counterperson's hunch versus the customer's word. In most cases, the counterperson will probably have to let the customer be right.

The best plan is to avoid this type of situation by opening boxes and examining the item in front of the customer before the customer buys it. Pointing out what that item would look like if it had been installed might further help to explain the appearance of the item to the customer if he or she still tries to return it as if it had not been installed.

RETURN CHARGES

Some businesses have policies restricting what item or type of items can be returned for any reason.

And some businesses charge customers for returns. A handling charge is a nominal charge for processing an item for return through the normal channels at a manufacturer or supplier.

Some suppliers might charge a restocking fee for any returned merchandise, and some might charge only for items special ordered. If a jobber store is charged a restocking fee by its supplier, then the store too may charge the customer for at least a portion of that fee. However, in determining how much, if any, of the charge can be passed onto the customer, consider the following factors: store policy, customer relations, frequency of the return, and competitors' policies.

Whether a return charge is established or not, counter personnel should try to limit returnable merchandise in some manner.

Sometimes having a free refund policy can increase sales. For example, if a customer is not quite sure that a certain number of quarts of paint will be enough, a counterperson can recommend that he or she buy the extra quart with the option to return it for a full refund if it is not needed. Usually, the customer will decide to keep it, just in case, even if he or she does not need it for that particular job.

CREDIT AND DEBIT MEMORANDUMS

A credit memorandum, or credit memo, is a record of what is owed to a customer in a transaction, sale, or return. Whenever a customer returns an item, the counterperson must prepare a credit memo (Figure 8–14).

Steps in accepting a return and preparing a credit memo include the following:

1. Verify the authenticity of the original purchase. Most businesses require the invoice as proof of purchase. Verify the completeness of kits and check quantities.
2. Fill in the following information on the credit memo: order number, date, customer's name and address, and salesperson.
3. Clearly indicate "credit memo."
4. Fill in the following return information: quantity, part number, description or manufacturer, list price, net price, and amount of purchase.
5. Subtotal the ticket, add sales tax if applicable, and total the ticket.
6. Make sure the customer signs the ticket.
7. Cross-reference the invoice with the credit memo by writing the original invoice number and date on the credit memo.
8. Tag the claimed defective items.

Credit memos can also be issued when improper charges for sales and taxes have been made, when overshipments to a store's customers are returned, or when shortages of charged merchandise occur.

When issuing credit memos or handling returns, it is extremely important that the store's cash refund policies are understood and followed. All items might not qualify for a straight cash refund. Sale items or special order items are good examples.

A debit memo is the opposite of a credit memo. It indicates a transaction or charge (as for services) to be charged against a customer's account.

DEFECTIVE PARTS CLAIMS

When a defective part is returned to the manufacturer, a claim must be completed accurately and legibly on the proper form. To process a defective parts claim to the manufacturer, follow these steps:

1. Assemble the parts to be returned in the proper order by part number. This step assists in assessing the cost or in identifying parts on the return goods notice at a later time.
2. Complete the return goods notice or the manufacturer's form listing the quantity and part number of each item being returned.
3. Fill in the name of the business and the date and sign it.
4. Attach a letter of explanation, if necessary, and mail the forms.
5. Place a copy of the return goods notice with the shipment to be used as a packing slip.
6. File a copy of the return goods notice as credit follow-up and attach it to the credit memo when it is received from the manufacturer.

RETURN ON INVESTMENT

Return on investment (ROI) is a measurement of business performance, and some managers and business experts agree that is it the ultimate test of business efficiency. While ROI is an important concern of management, it should also be important to counter personnel because it can be greatly affected by the performance of the counter team.

ROI can be defined in many different ways, but it is most generally defined as $\dfrac{\text{Profit}}{\text{Assets}} \times 100$, which

KAC AUTO PARTS
CREDIT MEMO

Issue Date _12/1/88_

Name _JOHN'S AUTO SERVICE_

Address _144 WALNUT ST._

City _DELANO_   State _PA_   Zip _18918_

Phone _729-0811_

Original Invoice No. _68921_   Classification _DEALER_

Reason Issued _DEFECT IN HOSE_

Issued by _B. DEELEY_   Terms _CREDIT TO ACCOUNT_

RETURN DATE	PART NO.	DESCRIPTION	MANUFACTURER	UNIT	UNIT PRICE	NET PRICE
12/1/88	3142	HOSE	GTB	1	5.61	5.16

Subtotal	5.16
Tax	—
Handling	—
Total Return	5.16

Customer Signature _John Smith_

Issuer's Signature _B. Deeley_

FIGURE 8-14 Counterpersons must fill out a credit memo whenever a customer returns an item.

amounts to $\dfrac{\text{Return}}{\text{Investment}} \times 100$. However, before ROI can be measured and applied to a particular business, the counterperson and/or manager must decide which profits and assets will be considered. For example, the formula could be used to relate gross profits to inventory or to relate net profit after taxes to total assets.

Considering gross profit and inventory is often a good place to start because most counterpersons take on inventory management responsibilities before managing other assets and expenses and because basing ROI on inventory and gross profit allows the counterperson to evaluate the performance of each line or each part in terms of its contribution to the overall ROI.

To calculate ROI, use the following formula:

$$\frac{\text{Gross Profit}}{\text{Cost of Inventory}} \times 100$$

which translates to

$$\frac{\text{Sales} - \text{Cost of Sales}}{\text{Cost of Inventory}} \times 100.$$

PROBLEM: A business has an inventory of motor oil worth $1,000. In one year, the business sells $8,000, and the cost on those sales was $6,000. What is the ROI for the motor oil?

ANSWER:

$$\frac{\$8,000 - \$6,000}{\$1,000} \times 100 = 200\%$$

This figure means that in one year every dollar invested in oil returned two dollars in profit.

IMPROVING ROI

A counterperson can change the outcome of the equation by changing either the top (gross profit) or the bottom (cost of inventory) or both. However, each measure taken to raise ROI should be examined carefully to be sure that any additional expense involved in implementing a plan does not eat up the gains achieved.

Increase Sales

One of the most obvious ways to improve ROI is to sell more. An inventory that gives a ROI of 100 percent at three turns a years will give 133 percent if sales can be boosted to the point that it turns four times. Sales can be increased by improving or increasing advertising, sometimes without additional expense. Improved merchandising efforts can also increase sales (see Chapter 9). Better service, improving outside sales efforts, employee training, and internal promotions to motivate employees can also boost sales.

Another way to increase sales is to lower prices. Lowering prices on selected items can generate interest and lead to increased sales on more profitable items. However, caution must be used when lowering prices because greater sales at a lower profit margin might not raise ROI.

Profit Margin

Selling more items raises overall profits, but sometimes the same result can be achieved by increasing the profit margin. Margins can be increased by raising prices and cutting costs.

Raising prices is the simplest way to increase margins, but unless the business has no or few competitors, raising prices will damage a business's position in the market. Some businesses raise prices only on those slow-moving items that require more time and knowledge to research. Raising prices on common items with prices generally known among customers could cost more in lost sales than the increased price would generate. Many businesses (especially those utilizing a computer) do not use manufacturers' price sheets; they create their own prices based on cost, desired profit margin, and degree of competition.

Examining which customers get which discounts is another way to increase margin by getting the highest possible price. Some customers are efficient, always ready with every bit of information necessary for parts that they truly need. Other customers require a lot of research time and return half of what they buy. A sensible pricing policy bases discounts on customer performance rather than on customer demands.

Lowering costs can also serve to increase profit margin, sometimes even more than raising prices. Figure 8–15 illustrates ROI on a part that costs $70 and sells for $100, turning four times a year.

Reducing Investment

Other ways to improve ROI include changing lines, buying from a supplier who sells the desired lines at the best price, and buying distressed stock that can be cleaned up and reboxed. ROI can also be raised by reducing the size of investment. If a particular product or line is not moving well, it makes sense to free up some of the money invested in these parts and use it to do something more productive. If

ROI for a part that costs $70, sells for $100, and turns four times a year:

$$\frac{(4 \text{ turns}) \times (\$100 - \$70)}{\$70} \times 100 = 171\% \text{ ROI}$$

ROI if the selling price is raised 10%:

$$\frac{(4 \text{ turns}) \times (\$110 - \$70)}{\$70} \times 100 = 229\% \text{ ROI}$$

ROI if the cost is lowered 10%:

$$\frac{(4 \text{ turns}) \times (\$100 - \$63)}{\$63} \times 100 = 235\% \text{ ROI}$$

FIGURE 8–15 Cost improvements often increase ROI better than raising prices.

a targeted ROI requires four turns, but these parts are turning only three times, then counter personnel should attempt to reduce shelf quantities by one quarter.

The excess stock can be removed in several ways. Over a period of time, excess inventory will be sold and not reordered. But selling it all at once is a better way. Unwanted merchandise can be returned or sold to the vendor and replaced with a product that is expected to do better.

Disadvantages of reduced inventory include the possibility of lost sales if sales for a particular week or month are unexpectedly high, the need to order more often, and the need to deal with suppliers that can provide the fastest service.

Other ways to reduce inventory investment include taking advantage of lines that are available on consignment and avoiding double lining (the practice of carrying more than one brand of a particular product).

SHIPPER PLEASE NOTE ▶ FREIGHT CHARGES ARE PREPAID ON THIS BILL OF LADING UNLESS MARKED COLLECT

DRIVER PLEASE NOTE ▶ IF SINGLE SHIPMENT CHECK BOX BELOW

STRAIGHT BILL OF LADING
ORIGINAL - NOT NEGOTIABLE
04032 - 07 (3 83)

CF CONSOLIDATED FREIGHTWAYS
CFWY

SINGLE SHIPMENT PICKUP ☒

SHIPPER NO.	16542
CARRIER NO.	47-10
DATE	12/15/88

CONSIGNEE (TO): BAUMANN'S SPEED SHOP		SHIPPER (FROM): KAC AUTO PARTS	
STREET: 304 PARKWAY		STREET: 506 EAST UNION	
CITY, STATE: SCHUYLKILL HAVEN PA	ZIP: 17972	CITY, STATE: PLAINVILLE PA	ZIP: 17852
ROUTE: A-14		VEHICLE NO: 3645189	

NUMBER SHIPPING UNITS	HM	Kind of Packaging, Description of Articles, Special Marks and Exceptions	Weight (SUBJECT TO CORRECTION)	Rate	CHARGES (for Carrier use only)
1		SKID (Engine parts)	150		
2		CARTONS (5 gallons paint each)	50		
1		CARTON (auto parts)	25		
		1			

Note—Where the rate is dependent on value, shippers are required to state specifically in writing the agreed or declared value of the property.
The agreed or declared value of the property is hereby specifically stated by the shipper to be not exceeding

$ ______________ per ______________

Subject to section 7 of the conditions, if this shipment is to be delivered to the consignee without recourse on the consignor, the consignor shall sign the following statement:
The carrier shall not make delivery of this shipment without payment of freight and all other lawful charges.

______________ (Signature of Consignor)

TOTAL CHARGES ▶

FREIGHT CHARGES: FREIGHT PREPAID except when box at right is checked. Check box if charges are to be collect ☐

The agreed value on household goods or personal effects does not exceed 10¢ per lb. per article, unless otherwise specified

"THIS IS TO CERTIFY THAT THE ABOVE-NAMED MATERIALS ARE PROPERLY CLASSIFIED, DESCRIBED, PACKAGED, MARKED AND LABELED AND ARE IN PROPER CONDITION FOR TRANSPORTATION ACCORDING TO THE APPLICABLE REGULATIONS OF THE DEPARTMENT OF TRANSPORTATION."

RECEIVED, subject to the classifications and tariffs in effect on the date of the issue of this Bill of Lading, the property described above in apparent good order, except as noted (contents and condition of contents of packages unknown), marked, consigned, and destined as indicated above which said carrier (the word carrier being understood throughout this contract as meaning any person or corporation in possession of the property under the contract) agrees to carry to its usual place of delivery at said destination, if on its route, otherwise to deliver to another carrier on the route to said destination. It is mutually agreed as to each carrier of all or any of said property, over all or any portion of said route to destination and as to each party at any time interested in all or any of said property, that every service to be performed hereunder shall be subject to all the bill of lading terms and conditions in the governing classification on the date of shipment.

Shipper hereby certifies that he is familiar with all the bill of lading terms and conditions in the governing classification and the said terms and conditions are hereby agreed to by the shipper and accepted for himself and his assigns.

SHIPPER: KAC AUTO PARTS (see above)	CARRIER: CONSOLIDATED FREIGHTWAYS
AUTHORIZED SIGNATURE	AUTHORIZED SIGNATURE / DATE

FIGURE 8-16 Typical bill of lading

APPLYING ROI

For the greatest benefit to ROI, many of the previously discussed methods must be applied together. For example, a parts store with $100,000 inventory, getting 2.88 turns (industry median figure) and a gross profit of 35 percent (also the median) would have an ROI equation as follows:

$$\frac{\$444{,}000 \text{ (sales)} - \$288{,}000 \text{ (cost of sales)}}{\$100{,}000 \text{ (inventory)}} \times 100 = 155\% \text{ ROI}$$

Then sales were increased by 10 percent through a combination of a better selling effort, selected small price increases, and inflation; profit margin increased by 3 percent through cost reductions and price increases; and the business got an additional half turn out of the inventory by increasing sales, lowering costs, and carefully evaluating stock levels. The ROI equation would then read as follows:

$$\frac{\$487{,}300 - \$302{,}126}{\$89{,}383} \times 100 = 207\%$$

When this store reached these easily attainable levels in sales, margin, and cost of inventory, the increase in gross profit was $30,174 and the difference in inventory was $10,614, for a total cash improvement of $40,788 or almost $3,400 per month.

SHIPPING AND RECEIVING

In some businesses, counter personnel are responsible for handling and channeling the paperwork involved with ordering, receiving, and shipping out parts. The following is a list of terms and definitions dealing with the receiving and shipment of stock.

Bill of Lading A written document issued by the carrier acknowledging the receipt of the goods from the shipper (consignor) and setting forth the terms of the contract for delivery. The bill of lading is a written agreement between the sender of the goods and the carrier of the goods (Figure 8-16).

Break Point The point at which the cost of shipping by a certain method or carrier changes significantly because of the size or weight classifications. For example, a type of shipment might not

ROBERT SCHARFF & ASSOCIATES, LTD.
R. D. 1, BOX 276
NEW PHILADELPHIA ROAD
NEW RINGGOLD, PA 17960

THIS NUMBER MUST APPEAR ON ALL INVOICES, SHIPPING PAPERS, PACKAGES, ETC.
PURCHASE ORDER NO.
DATE 12/5/88 DATE REQ. 12/8/88
SHIP VIA Regular Delivery FOB
TERMS 30 day

TO
DOWNTOWN AUTO SUPPLY
406 WEST FRONT ST.
ROCKVIEW PA 18351

SHIP TO
above address

QUANTITY	UNIT	DESCRIPTION	UNIT PRICE	TOTAL PRICE
1		XYX Thermostat #9007	6.06	6.06
2	pair	ACE Wiper refills #103-34	4.46	8.92
5	tube	Silicone sealant (RTV) # SV2	5.13	25.65

AUTHORIZED SIGNATURE Jack Brown

FIGURE 8-17 A purchase order requests items and specifies the conditions of a transaction.

be able to exceed 40 pounds without a substantial increase in shipping cost.

Common Carrier   A carrier such as a trucking firm, railway company, or air freight company that offers its services to all customers for the interstate transport of goods. In contrast, a contract carrier serves only one or a limited number of customers.

Consignee   A person or company to whom goods are delivered. It is easier to understand the terms **consignor** and **consignee** when considering a shipment. For example, when a WD sends a shipment of parts to a jobber store by truck, the WD is the consignor, the jobber is the consignee, and the trucking firm is the common carrier. When a jobber ships an order to an out-of-town customer, the jobber is the consignor and the customer is the consignee.

Consignor   A person or company that delivers goods for resale.

Emergency Order   A purchase order placed for special requirements that cannot be met by the regular stock order or an intermediate order.

Intermediate Order   A purchase order placed between regular stock orders to replenish stock until the regular stock order shipment arrives.

Purchase Order   A formal request from a buyer to a seller that specifies conditions of sale or delivery. To be valid, the seller must accept it and the buyer must acknowledge the terms (Figure 8–17). A restocking dealer may issue a jobber a purchase order for a shipment of parts. A jobber may issue a purchase order to the store's suppliers to replenish the stock sold.

Stock Order   A purchase order regularly issued by a wholesaler to bring stocks to desired levels.

Waybill   A written document prepared by the carrier and rendered to the consignee of the shipment. Sometimes referred to as a freight bill or probill, it contains such information as the nature of the shipment, its weight, the name of its consignor and consignee, its origin, route, destination, and the charges paid. It serves as a means of identification, a guide to routing, and a basis of freight accounting for both the carrier and consignee.

REVIEW QUESTIONS

1. Which of the following is the manufacturer's suggested sale price of an item to the final consumer?
   a. list price
   b. net price
   c. variable price
   d. wholesaler price

2. Which of the following statements is untrue concerning price sheets?
   a. Price sheets contain confidential information.
   b. Price sheets are usually valid for several years.
   c. Price sheets usually correspond to a manufacturer's catalog.
   d. Price sheets contain reference information.

3. What is a set of distinctly spaced lines of varying widths on packages?
   a. price quote
   b. shop ticket
   c. extra dating
   d. bar code

4. Which of the following accounts involves a customer who makes frequent purchases on a regular basis?
   a. active account
   b. dual account
   c. account receivable
   d. credit account

5. Counterperson A fills out a credit memo when a machine shop job is completed. Counterperson B fills out a shop ticket. Who is right?
   a. Counterperson A
   b. Counterperson B
   c. Both A and B
   d. Neither A nor B

6. What is a tax paid by the manufacturer or seller but often passed onto consumers?
   a. excise tax
   b. sales tax
   c. rebate
   d. all of the above

7. Counterperson A offers a trade discount to customers who will use a particular purchase for resale to make professional repairs. Counterperson B offers a rebate to those customers. Who is right?
   a. Counterperson A
   b. Counterperson B
   c. Both A and B
   d. Neither A nor B

8. What is a discount made available for purchased items that are delivered at a given date and marked payable in a given period of time?
   a. cash discount
   b. trade discount
   c. extra dating
   d. rebate

9. Rebates usually benefit whom?
   a. manufacturers
   b. dealers
   c. consumers
   d. all of the above

10. What is a promise of refund or replacement if a product or service does not live up to its designated function?
    a. rebate
    b. discount
    c. warranty
    d. bar code

11. Which of the following statements defines a remanufacturer?
    a. Someone who finds the defect in the unit, cleans it, and repairs it.
    b. Someone who completely disassembles the unit, checks all major components, and automatically replaces minor components.

c. Both a and b
d. Neither a nor b

12. What is another term for *core*?
    a. casting
    b. exchange
    c. defect
    d. component

13. What is a record of what is owed to a customer in a transaction, sale, or return?
    a. defective parts claim
    b. invoice
    c. core
    d. credit memo

14. What is possibly the ultimate test of business efficiency?
    a. profit margin
    b. return on investment
    c. gross profit
    d. assets

15. What is a formal request from buyer to seller that specifies conditions of sale or delivery?
    a. freight bill
    b. turnover
    c. purchase order
    d. intermediate order

CHAPTER NINE

MARKETING AND THE PARTS DEPARTMENT

Objectives

After reading this chapter, you should be able to:
- Explain how to identify a customer base using surveys and popularity guides.
- Explain the role of these five components in any effective marketing scheme: product, price, place, performance, and promotion.
- Explain the significance of forecasting the market in outlining a marketing scheme.
- Explain how the exterior and interior of a business can affect sales.
- Explain how the following attributes of a counterperson's performance can affect customers' attitudes: product knowledge, offering related items, delivery service.
- Describe the advantages and disadvantages of three methods of delivery service, including hot-shot, scheduled, and prorated delivery.
- Explain the advantages and disadvantages of the following advertising media or methods: television, radio, newspapers, direct mail, specialty items, telephone directory ads, co-op funds, and programmed distribution groups.
- Define telemarketing and describe the systems of proactive and reactive telemarketing.
- Explain how promoting a business's machine shop can help increase sales.
- Explain the importance of price indexes, purchasing philosophies, displays, and advertising when marketing and merchandising oil.
- Explain the importance of related sales, cheat sheets, and serving professional customers and DIY customers differently when marketing ignition parts.
- Explain the importance of choosing a line, good rep, and location and display when marketing filters.
- Explain the importance of location and display, related sales, inventory, and servicing fleets and industry when marketing belts and hoses.
- Explain the importance of delivery, technical information, problem-solving, and color matching when marketing paint and body equipment.
- Explain the importance of understanding the advantages of front-wheel drive, encouraging boot inspection, suggesting repair alternatives, diagnosing problems, selling to professionals, and utilizing displays when marketing front-wheel drive parts.
- Explain the importance of service manuals, sales opportunities, and related items when marketing brake parts.
- Explain how selling seasonal items can boost sales and make the surges of both receivables and payables less severe.

Marketing is a total system of business activities designed to facilitate distribution of goods and services from seller to buyer. Any effective marketing scheme must take the following components into consideration:

- Product
- Price
- Place
- Performance
- Promotion

These components can all be controlled by the seller, who, in the automotive aftermarket, is usually the counterperson.

However, another factor involved in marketing has a great impact on the shape of the components mentioned above. It is the target of the marketing scheme—the customer. An entire marketing strategy is focused upon the customer.

IDENTIFYING THE CUSTOMER BASE

Discount parts stores, programmed distribution jobbers, mail-order houses, traditional parts stores, and volume distributors are just some of the forces competing for the aftermarket dollar. Because of this fierce competition, the counterperson must work closely with the management in targeting the needs of the customers. Lack of attention to this matter will result in an inaccurate marketing strategy that can kill profits.

The first step in updating a marketing strategy is to clearly define the store's customer base. Once the customer base is defined, then the inventory and advertising can be targeted toward that base. Making marketing changes before the customer base is defined could end up being more of a hindrance than a help.

Defining the customer base requires the analysis of a store's present operation, customers, and asking questions such as these:

- Who are the customers in our area?
- Are we carrying the parts numbers those customers need most?
- Are the customers predominantly industrial in nature, are they agricultural, or are they all automotive garages and service stations?
- Are they staffed by aggressive, state-of-the-art technical people or are they more content to let other people take on all the headaches of the new cars while working on the older models brought in to them?
- Is the store in an area that is predominantly minority-oriented, and if so, are the people in the store qualified to handle their business, either through language skills or knowledge of the areas of particular interest to the customer?

The list of possible questions is endless. However, the more questions that are asked and answered, the more information that is available for defining the customer base. The clearer definition affords a more accurate marketing strategy.

SURVEYS

Another method for gathering information is the survey. The counterperson should ask the professional customers why they do business with the competitors down the street. Perhaps they have a

FIGURE 9-1 Offering special discounts or services to a specific group of people, such as employees at a particular factory, can widen a store's customer base.

convenient delivery service or offer inferior-quality merchandise at lower prices. Any information like this will help in making sound marketing decisions.

For those stores that are strictly retail-oriented, the same type of information can be gained by examining the type of customers that are going to the competing store. For instance, are there workers at the local factory? Maybe a note could be put on the bulletin board at the factory offering a special discount to employees at the plant (Figure 9-1). If the customers at the competitor's store are primarily farmers, it might help to step up advertising in the local paper or farm trade journal.

POPULARITY GUIDES

Because there are so many cars, jobbers would be buried in parts if they attempted to service all the people all the time. Whether the choice is made to offer popular, fast-moving, quick turnover parts or a more complete line, one valuable tool is required to intelligently shape and mold that inventory into a particular marketplace. This essential tool, referred to as the "popularity guide," helps the jobber tailor inventory to a particular territory. Without certain guidelines, the task of developing a proper inventory would be difficult.

This information is usually—although not exclusively—provided by the manufacturer and presented and developed in a variety of ways. Normally, once a year, around January, popularity guides start appearing at the jobber and distribution center level (Figure 9-2). The content includes a particular sup-

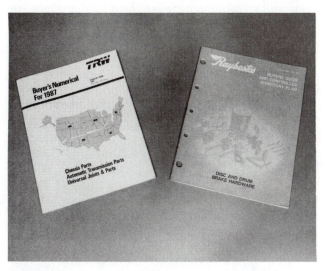

FIGURE 9-2 Typical example of manufacturer's popularity guides

plier's entire product offering with the added enhancement of local, regional, or overall popularity or classification. Each user must adapt the information provided to his or her circumstance.

It would be ideal to have individually tailored classification systems for each level of distribution at every location in existence. Some manufacturers have attempted to achieve this level of service.

Counter personnel must be aware of the various forms in which an update can be presented. Popularity guides might be presented as separate publications or as an ongoing information service such as in a price list or bulletin form. Each form has its advantages. One might have the ability to provide new information frequently, while the other might provide annual comprehensive data. Each might work more to a counterperson's satisfaction in different environments, and the counterperson might voice his or her preference to the manufacturer.

Using Popularity Guides

The mechanics of popularity guides vary from manufacturer to manufacturer, but generally, they consist of some type of classification assignment (Figure 9-3). Significance is applied either to letters of the alphabet or to numbers. An "A" class item usually designates the fast-moving or more popular part, and other letters designate less popular parts.

ADJUSTING SCREW ASSEMBLIES

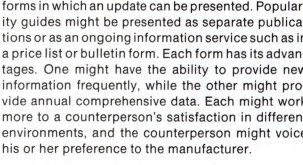

Date Columns

Part No. / Class. Application	Size	Loc.	Std. Pkg. Qty.	Base Stk.													
H 1527-2 B Adjusting Screw Assy. - Blister Packed Chrysler Products w/Spring Loaded Cable		R-F R-RR	1 CD.$(1)^2$														
H 1528-2 C Adjusting Screw Assy. - Blister Packed Pinto, Mustang, Bobcat, Fairmont, Zephyr		R-F R-RR	1 CD.$(1)^2$														
H 1529-2 C Adjusting Screw Assy. - Blister Packed Pinto, Mustang, Bobcat, Fairmont, Zephyr		L-F L-RR	1 CD.$(1)^2$														
H 1530-2 A 11-12"S.A. Buick 84-77 Eldorado, Toronado Chevrolet 84-77 & GM Trks. 84-63 Chevrolet 1 T. 84-76 GM 70-63 Chevrolet ¾ T. 84-71 Jeep 84-74		R-F R-RR	1 CD.$(1)^2$														
H 1531-2 A 11-12" S.A. Buick 84-77 Eldorado, Toronado Chevrolet 84-77 & GM Trks. 84-73 Chevrolet 1 T. 84-76 GM 84-63 Chevrolet ¾ T. 84-71 Jeep 84-74		L-F L-RR	1 CD.$(1)^2$														
H 1532-2 B 9-9½" S.A. General Motors		R-F R-RR	1 CD.$(1)^2$														

FIGURE 9-3 Typical popularity guide listing shows letter classification keyed to the part number's market demand.

Although most manufacturers use a descending type of ranking system, at least one major manufacturer classifies in reverse sequence, so the counterperson must be alert to the sequence used in each guide.

The remainder of the ranking significance is quite particular to the individual suppliers. Such classes as "W"—Warehouse Only; "O"—Obsolete; "T"—Tractor/Agricultural; or "N"—New might mean something entirely different from supplier to supplier. The counterperson must know lines and what classification system is used.

Generally speaking, the top three classifications account for the majority of product movement; usually up to 80 percent of the parts sold from day to day fall into these three classifications. Although accounting for 80 percent of sales, these top three might represent only 25 percent of actual product part numbers in a particular product line. Therefore, it is necessary to have a reliable source selecting the balance of inventory.

The popularity guide provides accurate, statistically backed information that will help to define the needs of the customers. Knowing, understanding, and working with a popularity guide is essential for an automotive aftermarket person at any level of the distribution chain.

PRODUCTS

The products offered must meet the needs of the customers in a particular market area. The counterperson's responsibilities include recording lost sales, reporting demand for new items, seeing that items in low supply are reordered, and helping identify and replacing slow-moving parts. Because the counterperson actually handles the merchandise and sees when stock levels are getting low, much of the responsibility for maintaining full shelves will always be the counterperson's. And as one who most often hears what the customers are saying, the counterperson will also be responsible for seeing to it that the right products are in stock.

FORECASTING THE MARKET

Fortunately, the counterperson does not have to rely on customer response alone as he or she attempts to supply the most suitable product mix for the area. Careful planning can help insure a bright future for a company. In fact, in today's market, careful long-range planning has become a matter of survival.

The world's body of knowledge has been growing so fast and the technological advances have come out so rapidly that even the manufacturers are hard-pressed to keep up with it. Forecasting does not require an expert on marketing and its variables. Long-range planning for the needs of mechanic customers requires anticipating the trend of the market. A counterperson should remember that he or she is not trying to nail down every little detail about the future, only the overall trend. A forecast can be broken down into three steps.

Step 1. The counterperson should consider the environment of the market in the future. For instance, barring any catastrophic events, the market will probably continue on the course it has taken in recent years, and the following trends will continue to shape the market:

- Engine and suspension systems are going to be controlled more and more by computers, and the size of computers will be smaller.
- Weight will continue to be pared off the automobile as more and more plastics and lightweight metals are used.
- Parts are also going to last longer, causing the automobiles to be driven longer before being serviced.
- Pollution controls will continue to be expanded as more and more vehicles clog the turnpikes, and the EPA is going to take an ever increasing role in monitoring hazardous wastes: those in the mechanic's shop as well as those emitted from the automobile.

Step 2. Counter personnel must examine the ways to meet this particular set of circumstances. For example, it would probably help to become more familiar with electronics since the ability to overhaul a Marvel-Schebles carburetor will not be nearly as important as understanding the circuitry of a sequential-port injection system. A familiarity with the tools used to work on the new vehicles will also be helpful.

When planning for the mechanic in the future, the counterperson will have to consider the possibility of having more dealer clinics, since older mechanics might not have been exposed to much of the new machinery.

The counterperson already has to supply mechanic customers with material safety data sheets that detail the chemical composition of a particular product, and in the future the counterperson might have to supply information concerning licensed hazardous waste haulers, disposal sites, and the proper means of handling the wastes that the mechanic is generating.

Step 3. The counterperson must forecast the market reaction to the proposed solutions. Some mechanics may be reluctant to implement changes.

The counterperson must be prepared to convince these mechanics that the proposed changes need to be made in order to remain competitive. On the other hand, the up-to-date customer will ask for more information on a given system, trying to keep up with everything that goes on in the area. This customer will be favorably impressed with an organization that is actively looking forward instead of passively looking back.

One of the most important aspects of a counterperson's job is matching the appropriate products to customer needs. Bringing the customer in line with the product does not imply a practice of coercing customers into buying products that they do not need. Effective merchandising serves only to awaken or remind customers of their present needs. One way in which this can be done is through showing the features and benefits of the products being sold. Another is suggesting related items to the customers so that they can do the complete job. Merchandising can also plant ideas for the customer's future needs.

PRICE

The counter person can in some ways affect another of the elements of successful marketing—price. No one, except perhaps the outside salesperson, is in a better position to know how customers feel about a store's prices. The counterperson gives the quotes and hears the customers' reactions. Both the counter personnel and outside sales people must keep management informed about how the store is doing on price.

The counterperson can also affect price through purchasing. Someone else might have the purchasing responsibility, but when a nonstocking or out-of-stock item is requested it is usually the counterperson's job to get it. Where an item is bought has a lot to do with its eventual selling price and, therefore, the customer's perception of a store's prices.

The counterperson has to cultivate as many sources as possible, understand the pricing differences between competing lines, and monitor ever-changing discounts and specials. This means talking with WD salesmen, reading bulletins from suppliers, and listening to co-workers.

Another source of price advantages lies in the "fleet rebate" programs offered by some manufacturers. Several of the spark plug, filter, rubber, ignition, and lighting companies offer rebate discounts to jobbers who sell to qualified accounts. Originally limited to fleet and farm accounts, many of these programs are being enlarged to cover the "master" or "stocking" installer.

Like the other elements of marketing, price is not an isolated issue. Minor deviations from prevailing prices will seldom damage an otherwise good relationship.

However, price is more important to consumers in recent years than a decade or two ago. A survey taken about twenty years ago showed that among the reasons people bought their goods and services at a particular store, price was number eight on the list behind such considerations as service, availability, location, parking, etc. If the same survey were taken today, price probably would not be number eight anymore. With the changes in the economy, price has become a prime consideration in the buyer's mind. However, even though price has become more important, it still is not the *most* important consideration.

The counterperson's expertise and willingness to help, the availability of parts to do a complete job, and the location of the store are the most vital factors. As long as the prices the store offers are competitive, these factors rank well ahead of the price factor in the buyer's mind (Figure 9–4).

Note that to be competitive does not necessarily mean to be the cheapest. For example, if spark plugs are selling for 89 cents everywhere, the counterperson probably will find it hard to get $1.49 for them without making the customer feel cheated. The counterperson probably can, however, get $1.09 for them without making the customer feel exploited.

All in all, people generally know the prices of only about twenty items in any parts store. That list usually includes plugs, oil filters, shocks, and batter-

FIGURE 9–4 Consistent, knowledgeable service on the part of a counter staff will build and keep both the professional and DIY trade.

ies. Ball joints, wire sets, and wheel bearings are normally not price-sensitive items. Therefore, if the store remains competitive—but not always the cheapest—on those items, it will get its share of the business on the remaining items.

PLACE

Place is the third element of successful marketing that the counterperson can affect. There is much that can be done, both inside and outside of the jobber store site, to encourage traffic flow.

In some businesses, a counterperson might have little to do with the exterior or interior of the building. In such cases, a counterperson can do little but make suggestions or pass ideas to a manager. However, passing on an innovative idea can prove a particular counterperson to be a valuable employee and important parts team member.

EXTERIOR

Exterior appearance is the first thing a customer considers when choosing a store. Store personnel should take advantage of free space on the front of the store and in the parking lot to put up some attractive signs telling potential customers what the name of the store is and what products are sold. Printed metal signs on the side of a building do an effective job. But for maximum attraction, illuminated signs work the best—24 hours a day. The sign must be clear, easy-to-read, and attractive to draw in the potential customer. An unattractive sign, one in poor repair, or one that is hard to read might actually drive the customer away.

Make sure that windows are clear and free from clutter and that sidewalks are in good condition. Keep sidewalks swept and, when it snows, shoveled. The building's paint should be fresh and free from chips, and the parking lot should be free of litter. Even if a store is well kept on the inside, customers will not be drawn to it when the exterior looks dirty and run down.

Parking

Adequate parking is also important to the customer. A poorly laid out parking lot frustrates customers. If the space is available, put all customer parking stalls where they are visible from the street. Rear parking should be avoided whenever possible. Employees should be assigned parking in the most inconvenient stalls in the lot or in another parking lot to allow convenient parking for customers. Also, easy access to adjacent streets and effortless circu-

lation of cars are just as important as the number and size of parking stalls.

Lighting

Finally, the use of effective exterior lighting is another way to draw in the potential customer. Retail experience shows that people are attracted to brightly lighted parking areas because they feel more secure in these lots.

INTERIOR

There are many things the counterperson can do within the store to positively influence the customer's decision to buy. Before trying to sell a product, a counterperson must first sell the store. Housekeeping creates an attractive appearance for the store and, thus, an orderly and safe place to do business. Counter personnel should make sure that aisles and counters are kept clear and that unsold parts are returned to the stockroom. Any way in which a counterperson can make a store or department more inviting will increase sales potential.

Just as every person possesses a particular personality, so does every store—good or bad, intended or not. The mood of the shopper is going to be set by the mood of the store. Since this phenomenon cannot be avoided, the jobber should take advantage of it (Figure 9–5).

Consider the general mood of customers during grand opening festivities. A happy, exciting, carefree atmosphere is created to insure that people are

FIGURE 9–5 Keeping displays stocked, neat, and organized is the best method of maintaining store appearance.

having a good time as well as becoming familiar with the store and the products sold. This type of atmosphere should always be maintained. By using properly planned color schemes and decor packages, the skilled store designer tries to achieve a permanent "grand opening" feeling for the store.

The true secret to a professionally planned store is to have one person responsible for harmonizing everything that will be seen in the retail area. After the initial design, this task could become the counterperson's responsibility. First of all, the wall finish colors should be coordinated with decorative letters and graphics. Then all countertops, cornices, and tile accent strips should be blended into the overall color theme.

Shelving

Shelving is another aspect that must be considered in creating a pleasant atmosphere. A few years ago, almost all jobber stores had 85 to 90 percent of the inventory behind the counter with only 10 to 15 percent of the inventory investment displayed for impulse merchandising. In years past, this technique worked adequately.

Now, even though there has been resistance in the industry to moving toward high-powered merchandising techniques, the automotive merchandising pioneers have brought both competitive and financial pressures into the industry. Modern display fixtures are now tools of the trade. For automotive merchandisers to be able to perform their primary task of presenting products in a favorable manner, they must utilize a variety of store fixtures and display accessories.

Stores once stocked with old grocery gondolas are replacing this shelving design. The problems with the old gondolas are simple: the tool design for the grocery trade is not at all the tool design necessary for jobber stores. As a rule, most grocery gondolas are much too high for automotive parts operations. It takes a lot of inventory to fill up those shelves.

To get the most for the store's money, the counterperson could suggest that the employer purchase a good heavy-duty system rather than a point of purchase or light-duty end display type. A system that adjusts on a one-inch increment has proven, as well, to be much more flexible than the older style fixtures.

An even more important consideration than the shelves themselves is the arrangement of the products on the shelves. Related products should be grouped together within the retail area of the store. Good merchandising takes a little time for the devel-

FIGURE 9-6 "Ribbon merchandising," or arranging products vertically on shelves, usually creates an attractive display.

opment of an overall plan based upon common sense.

One contemporary method of arranging products to make them more attractive to potential purchasers is known as "ribbon merchandising"—placing a vertical "ribbon" of a product on three shelves, one right above the other. The same square footage of the product is provided, but by running the product on the shelves vertically rather than horizontally, the merchandiser has better control of the color coordination of products. It makes the presentation much more dramatic and effective (Figure 9-6).

Most merchandisers agree that the best location for attracting impulse sales is at a point between the chest- and eye-level of the customer (Figure 9-7). By carefully planning the placement of products within the store and by utilizing merchandising tricks such as ribbon placement of products, the counterperson can make every shelf of products work toward the profitability of the store. A concentration of demand items on each wall away from the front of the store induces customer traffic throughout the entire store, rather than to only one or two primary aisles.

Lighting

Many times the most often overlooked ingredient of a well-designed store is proper lighting. Stores that are well lighted sell more products than those that are not. Stores that are not well lit appear to be dark and drab. They have dark spots in certain departments and shadows on everything below the top shelf. Also, insufficient lighting tends to wash

FIGURE 9-7 The best location for products to attract impulse sales is at a point between the chest and eyes of the customer.

out the colors on the packaging, making the presentation of products on display lifeless.

The owner of the store should consider perimeter lighting above all of the wall displays to highlight the merchandise shown on the walls. This can be accomplished by installing standard fluorescent fixtures or track lights. In the same manner, overhead lighting can be arranged to eliminate both shadows and dark spots.

Special attention should also be given to the lighting above the parts service counter because lighting and retail sales are related. Many of the store operators are more concerned about electricity costs than the negative effect of poor lighting upon sales figures.

These design techniques are just some of the tools available to insure that the customer will want to spend time in the store. A recent survey, carried out by the Point of Purchase Advertising Institute, suggests that by just browsing, nearly one-quarter of the customers of auto parts stores bought items they did not intend to buy when they entered the store. A favorable impression can also encourage the customer to return the next time it is necessary to make a purchase. Greater shopping time per customer should translate into increased sales.

PERFORMANCE

The neatness and organization of the store and merchandise will play a significant role in the customer's overall impression. However, equally impor-

tant as design is the personnel who work in the store. Apart from the surroundings, the way that the customer is treated by store personnel can have even a greater impact.

Any favorable impressions formed in a customer's mind by a store's cheerful atmosphere will quickly be dissipated if he or she is not taken care of in a prompt and courteous manner at the counter. Successful marketing requires more in the way of performance than prompt and courteous service. The jobber store must also provide knowledgeable service in order to succeed.

PROVIDING PRODUCT KNOWLEDGE

Counterpeople, with their product knowledge, are among a store's best assets. Without the counterpeople who know the products and how to match them to the customer's problems and needs, an extensive inventory would be suicidal for any store. It is the knowledgeable counterpeople that enable the jobber store to hold its unique place in the market.

OFFERING RELATED SALES ITEMS

The counterperson must also have adequate product knowledge to offer the customer related sales items. The selling of related parts could increase gross sales to dealer accounts by at least 30 percent (Figure 9-8). Obviously, the practice of offering related parts should not be overlooked in any marketing scheme.

Selling related items does not mean pushing unneeded parts sales on the consumer. The selling of the additional parts needed to make a complete, reliable repair not only yields a larger profit, it also increases customer satisfaction.

The most successful counterpeople avoid the limitations of the parts catalog when selling related parts. In the catalog that lists rebuilt water pumps, for instance, the belts, hoses, clamps, antifreeze, and fan clutch are not included in the rebuilt line. Related items should not be missed just because they exist in separate lines and catalogs.

Related parts expertise does not take time to acquire. The secret is, through technical knowledge, to think of related parts systems. If the counterperson lacks direct mechanical experience, he or she can find outside help by taking some general consumer auto repair courses at a local community college.

Barring this, the next best way to acquire knowledge is to work closely with the most technically qualified and prosperous mechanic/technicians in the dealer base. For example, a counterperson might wonder why the best mechanics always have

FIGURE 9–8 Brake service is one area where related hardware and parts are always needed to perform a complete, professional repair.

the flywheel ground and install a new pilot bushing on each clutch job, or why many senior mechanics seem to use an inordinate number of plug wire sets.

In working daily with mechanic customers, the counterperson should examine each mechanic's particular pattern of buying. For example, mechanics who use a large number of lubricant seals have a reason. Discovering that reason will provide the key to extra related sales to other mechanics who might not be building the profit per job they should be through selling related parts.

Selling related parts is a game of percentages. The counterperson who is familiar with the physical relationship of the parts, the possible technical failures attributed to these related parts, the mechanic's in-shop inventory, and the buying patterns of the mechanic customer can indeed increase dealer sales.

STARTING A DELIVERY SERVICE

A delivery service is another valuable service that can be rendered by the jobber store to the customer.

Whether or not to offer a delivery service or whether to make wholesale changes in the existing service is, ultimately, a decision to be made by the employer. However, the counterperson's contact with customers and the feedback received from them is the employer's most valued assets. Thus, the counterperson's input is necessary in order for any delivery service to be profitable. Trying to decide whether having a service "pays" or "costs" is not an easy task.

The most obvious benefit of a delivery system for customers is that it presents an added convenience to valuable business clients (Figure 9–9). Another benefit is that a well-decorated delivery truck provides additional advertising. Also, the person in charge of running the truck is going to have increased contact with professional customers and will be able to watch the situation of competitors.

However, some facts might deter the immediate establishing of a delivery service. Such a service is an extremely expensive operation. The cost of keeping a delivery truck on the road for a year—excluding the salary of the driver—could run anywhere between $8,000 and $15,000. The income for supporting a service must come from either increased volume, higher prices, or a combination of both. Ideally, the former would be preferred, but a combination would most likely be the reality.

Another possible problem is that some customers might find it is no longer necessary to be absolutely positive about an order. In other words, since the customer is no longer making the trip, he or she might figure that if the order is wrong or incomplete, it will be no problem for the delivery truck to make another trip.

Finally, it might be surprisingly difficult to locate a driver who will be responsible with equipment and proficient at public relations. The delivery driver will quickly represent the image of the store. A good impression put forth by such a driver is great; a bad one disastrous.

Once the benefits and detriments of a delivery service are examined, the jobber can more easily make a decision based upon individual competitive circumstances. If a store is located in a small town

FIGURE 9–9 A delivery service makes it convenient for dealers to do business with your store.

and the competition does not deliver, it might not pay to have a delivery service. If a store is in a large city where all of the competitors deliver, the jobber will either have to offer a similar service or redirect business to the walk-in trade and hope some of the dealers will patronize the store.

Once the jobber decides to implement a delivery service, a method of delivery must be chosen.

Hot-Shot Delivery

Hot-shot or demand delivery is probably the easiest of all to implement. The counterperson just picks up the phone, writes down the customer's order, and sends it out with the driver as soon as possible. There are, however, a few problems associated with such a service.

This system, without a doubt, becomes the most expensive form of delivery. It does not matter whether the product on the truck is a 15-cent cotter key or a $50 master cylinder; both products go out the door just as quickly. However, the jobber loses money on any delivery under $10.

Such a system is also expensive because it offers the customer no incentive to place a large order or, for that matter, a complete order. The customer can simply order the parts as he or she gets ready for them, knowing that they will be delivered in just a short time. In addition, unscheduled service almost always requires that the store add an extra employee.

Scheduled Delivery

The scheduled delivery is less costly, gives the customer an idea when to expect the parts, and allows him or her to work around the schedule.

The first decision for such a system is the scheduling. The schedule will be largely decided on two factors: the competition and common economics. If the competitor does not offer delivery service, the jobber might be able to offer delivery, perhaps four times daily. This schedule will meet the needs of a majority of customers, and the delivery person can avoid needless second trips to the same shop. It might also allow the store to add delivery service without having to hire an extra employee, since existing store personnel could handle the scheduled deliveries.

An example of the time savings a scheduled delivery offers is a customer who, when doing a brake job, calls in an order for shoes, then orders the hardware 30 minutes later and waits another 15 minutes to order some brake fluid. With scheduled deliveries, only one trip would probably have to be made to this customer.

A scheduled delivery provides a buffer zone between customer orders and delivery time, and it trains the customer to make the complete order at one time or pay the consequence of having to wait until the next scheduled delivery. The end result is the avoidance of unprofitable repeat trips and a much-simplified operation.

Prorated Delivery

Another often-used delivery technique is to offer free delivery of all orders more than $50 or so and to institute a graduated charge for orders of less than the established amount. For example, a store might charge $2 for making deliveries for orders in the range of $25 to $49 and charge $3 to make deliveries for those in the range of $10 to $24. Some stores will flatly refuse to deliver anything less than $10.

This system has many advantages over the scheduling system because it almost totally eliminates the small orders, and it recognizes that many customers might find the need to wait until a scheduled delivery inconvenient. There are always mistakes and absolute emergencies that require a return trip to certain customers, anyway. But, through this system, the jobber will receive renumeration for the expense of making the trip unless the order is one of more than $50, in which case a profit would be made that is sufficient to support the extra service.

This system is also beneficial for the customer. If that customer is careful to consolidate orders, he or she will receive more free deliveries as the orders become fewer but larger.

A number of variations are possible in this system, one of which would set a certain dollar figure necessary for free delivery by the month for each account. The jobber might decide, for example, to offer free delivery service for all accounts doing an average of $500 of business each month. In essence, the jobber is recognizing and rewarding customers for being good ones. The jobber is also providing the smaller accounts with another reason to direct more business toward this store.

PROMOTION

The last component of marketing is promotion. There are a number of ways to promote products, prices, and performance to customers and prospects.

THE ADVERTISING CAMPAIGN

Perhaps one of the best ways for a store to attract new customers or keep present ones is through some sort of advertising program. An effective advertising campaign considers the specific objectives of this campaign and the type of customer at which the campaign is aimed.

There are two primary types of advertising: ads designed to create an immediate customer response and ads designed to keep the company's name before the buyers and build its image.

Immediate response advertising is used to generate quick results. The usual goal is increased store traffic. Examples are low prices on items that are in demand (for example, brand name oil filters for $1.59), clearances, coupon offers, special prices, and seasonal items.

Image advertising promotes a firm's regular products and services and stresses everyday benefits of doing business with that store. It builds an identity for a particular business so that when customers see immediate response ads, they will already have formed a positive attitude about the store.

Most companies use a combination of both image and immediate response advertising, often blending the two. For example, some companies create an image of low prices by consistently stressing price in their immediate response ads. Their immediate response ads are image advertising.

Television

Television works well for both image and immediate response advertising. For example, on TV, deodorant and soft drink companies do much image advertising, almost never stressing price, while the furniture outlets talk about almost nothing but price.

Of all the available media, television is the most striking and dramatic, and perhaps reaches the largest audience.

If a company has several outlets in an area, TV may well be the best choice. However, production costs are high. If TV is chosen, the message should be kept simple. Brevity is important to both the budget and the audience's patience.

Radio

A well-produced radio spot can deliver the same excitement as a good television ad at less expense.

Radio, like TV, is intrusive—only more so. It reaches listeners wherever they are, whatever they are doing. Whether they are at home, work, driving, or playing tennis, people listen to the radio.

Unlike TV, radio delivers a fairly even-sized audience throughout the day, although the largest and most captive group is the "drive time" crowd—people on their way to and from work. They do not step out of the room during commercials!

Like TV, radio messages should be simple—limited to one basic product or message. Since the

FIGURE 9-10 In newspaper ads, products can be illustrated and important features pointed out.

viewer cannot see the details of a product, radio is no place for complicated messages.

Finally, the stations chosen should reach the targeted customers. For instance, no one should choose a station with a news/talk format when trying to reach high school students.

Newspapers

Newspapers reach large audiences and, unlike radio and television, are read by many for the ads.

In the newspaper and other print media, products can be illustrated and important details pointed out (Figure 9–10). By choosing the section—sports, automotive, agricultural, home, and garden—the advertiser can be fairly selective about the readership.

Similar to a newspaper but more limited in scope are the "pennysavers," small weekly tabloids that are advertising vehicles rather than journalistic efforts. In any given city, there might be a dozen editions of the same publication, each delivered to a different area.

They are read for their ads, offer very reasonable rates to advertisers, and are just the ticket for merchants who want to reach people in their own area.

However, the readers are, by definition, bargain hunters.

Other media that can be used to reach selected audiences include car club and trade publications.

Direct Mail

To really pinpoint a market, nothing but telemarketing and outside sales can beat direct mail.

FIGURE 9-11 Advertising on specialty items can get the message to the customer on a daily basis.

With direct mail, the advertiser has complete control over the market. Audience can be selected based on age, zip code, type of car owned, previous purchases, income, or whatever criteria the advertiser considers important. The jobber can direct ads at homeowners only and avoid apartment complexes where auto repair work is not allowed.

As effective as direct mail is, it is only as good as the mailing list, and nobody has the perfect list for sale. The advertiser can start with a purchased list, but refinements will have to be made over a period of time.

Building and maintaining a good custom mailing list is a time-consuming task. And addressing mail by hand, or even typing the addresses each month, is a tedious chore. Some stores have a computer with enough extra storage capacity for lots of new customer records. However, at other stores, the work must be done by hand or with a small personal computer.

OTHER ADVERTISING

In all the media mentioned so far, the advertisement has a very short life span. Unless the viewer/listener/reader responds immediately, the message is lost.

Two other types of advertising media that get the message to the customer day after day, with only an initial effort, are specialty item advertising and telephone directory ads.

Specialty Items

Specialty item advertising is image advertising, and it involves putting the message on objects the customer will use day after day: pens, lighters, key rings, calendars, screwdrivers, coffee mugs, baseball caps, tee shirts, and many more (Figure 9-11). These items are available through calendar publishing companies, specialty advertising companies, and even from manufacturers.

The key to specialty item advertising is to select items worth keeping.

Telephone Directory Ads

Telephone directory advertising is unique and indispensible. Every other form of advertising is, to a certain extent, random. No matter how narrowly a market has been defined, the advertiser can never be sure the message reaches the prospect when he or she is ready to buy. With a Yellow Pages listing, the message is sought out by customers who want to buy (Figure 9-12).

Directory advertising supports the ads placed in other media. Listeners who are not ready to buy when the ads run can find the advertiser when they are ready. And directory advertising allows the advertiser to reach a broad group of prospects by placing the ad under several headings such as auto parts, paint and body supplies, sound systems, and so forth.

ADVERTISING BUDGET

Developing a budget for advertising efforts is as much a part of a campaign as targeting the market and defining objectives. Businesses use three basic methods to arrive at advertising budgets: unit of sales, objective and task, and percentage of sales.

FIGURE 9-12 A telephone directory ad is the only advertising medium that ensures that customers will see the message when they need it.

Unit of Sales

This is a method that takes into account previous advertising outlays and results and arrives at a figure by multiplying former costs by projected sales. For example, if it took 4 cents of advertising to sell each gallon of antifreeze in the past, the jobber can expect to spend $160 to sell 4,000 gallons.

The unit of sales method is difficult to apply if no previous measurements are available, and it does not really take into account budgeting image advertising, which is difficult to measure.

Objective and Task

This is a method that requires specific goals, such as "35 percent more sales of high performance parts to college students," and a good estimate of the cost of reaching those objectives. The costs of achieving each goal are totaled to arrive at a proposed budget. Naturally, the jobber might not be able to afford all of the objectives, so the goals should be ranked according to importance and the budget modified as necessary.

Percentage of Sales

This is the easiest and most often used method of arriving at a budget. Based on what other businesses (especially those in the parts industry) are spending, the jobber commits a similar percentage of sales to advertising. The jobber might want to spend more to outrun the competition or less if there is little competition already.

According to figures published by the Automotive Service Industry, auto parts operations spend about 0.63 percent of net sales on advertising. This means that for every $1,000 in net sales $63 is spent on advertising. Chain operations and mass merchandisers spend several times that amount.

CO-OP FUNDS

Many manufacturers offer co-op advertising allowances. Some extra effort might be required to get in on the co-op money, but it could take care of a large percentage of the advertising bill.

A counterperson can check with the reps for each manufacturer to see if co-op funds are available and to get details about each company's program. Ask for literature and claim forms, then do some careful reading. No two programs are alike. Different manufacturers honor ads placed in different media. Payments can be cash or credits against future purchases. The amount available might be limited by past purchases, or it might be a percentage of an initial order.

AD DESIGN

After the market objectives are defined and a budget is established, the ads must be designed. Whether the counterperson is directly involved in the layout or copywriting or is responsible for hiring a professional, he or she must understand the basics of creating an effective ad. The first thing to consider is the ad's purpose, which is always to sell something.

Advertising experts have an acronym for the goals of a good ad: AIDCA. AIDCA means get **A**TTENTION, develop **I**NTEREST, **D**ESCRIBE the product, **C**ONVINCE the reader, and get **A**CTION. With these objectives in mind, the counterperson should study some ads, identify how they achieved each of these goals, and borrow ideas freely.

Choose the right items to advertise. Advertising is not the way to get rid of hard-to-sell items. Choose products that are in demand, in good supply, and representative of the store.

The layout must be simple. The design should lead the reader's eyes naturally from headline and art through copy, price, and signature. The signature is the store's name, address, and phone number, or any other similar details.

Tips on copywriting include:

- As the copy is read aloud, listen for awkward phrases, difficult to understand meanings, or unbelievable promises.
- Be stingy with words. Make each one count.
- Avoid too many qualifiers. Saying "Superb" and "Fantastic" too often makes customers suspicious.
- Always include prices. Whenever consumers see an ad without a price, they assume it is too high to print.
- Above all, get right to the point. Long introductions make people suspect they are being set up.

Some of the work can be handled by the media. Newspaper art departments can help with the design and production of ads. A printer can probably lend a hand with the artwork for handbills or direct mail pieces.

Radio stations can help with script writing, special sound effects, and musical backgrounds—often producing the entire spot.

Many television stations can also handle the entire production of an ad (for a fee, of course). For more help, consider advertising agencies. They can

FIGURE 9–13 The telephone can be an important sales tool when used in a systematic telemarketing program.

bring fresh thinking, outside objectivity, and a creative approach.

PROGRAMMED DISTRIBUTION GROUP

One of the most popular alternatives to personally developing an advertising campaign or hiring an agency is to affiliate with a programmed distribution group. Like any approach to advertising, this one offers both advantages and disadvantages.

For a modest fee (often offset by a percentage of their purchases with the host distributor), members share the efforts of an advertising agency retained by the group. The agency produces the ads and places them with the local media.

The chief disadvantage is that the ads are uniform for all of the participants. If the host warehouse carries three brands of spark plugs, the jobber might be involved in ads for all three, even if he or she does not particularly want to carry more than one.

TELEMARKETING

Telemarketing can be defined as any business use of the telephone. With relatively little expense, the phone can bring customer orders and inquiries to counter personnel, and it can take messages out to customers (Figure 9–13). In the first case, the phone is used in a reactive manner—the business responds to a customer—initiated call. In the second case, the business initiates the call, or proactive telemarketing.

BENEFITS OF PROACTIVE TELEMARKETING

The auto parts aftermarket is fertile ground for telemarketing. Like other wholesale/distribution businesses, traditional methods of going to market have become almost unbearably expensive, while profits are diminished due to intense competition.

Many parts suppliers use an outside sales force to maintain customer contact and promote products, but the effectiveness of an outside sales staff is limited by time and distance. Outside people might spend as much as 60 percent of their time traveling, waiting, attending meetings, and doing paperwork. Given these constraints, it is common for them to be limited to between four and ten face-to-face calls per day.

For every highly profitable or potentially profitable account, there will be many barely profitable customers—customers whose business is valuable, but whose trade simply will not justify the expense of regular face-to-face calls.

Contact with these customers might be more profitable by phone. While an outside salesperson can contact only a handful of customers a day, a telemarketing person can reach scores.

However, telemarketing will not replace outside sales. For example, a customer's inventory cannot be checked over the phone. But a telemarketing program can supplement an outside sales effort, control selling costs, and generate revenue. It can also

- Increase sales with existing customers by maintaining contact between face-to-face visits and by polling customers along delivery routes
- Introduce new products, programs, and special offers, freeing the outside sales staff to provide the services key accounts require
- Reactivate old accounts with much less expense than either face-to-face calls or repeatedly mailed letters
- Acquire new customers by qualifying prospects, identifying the purchasing decision maker, and making sales and/or appointments for outside salespeople
- Reduce the cost of selling
- Build and maintain customer loyalty. According to Pacific Bell, a recent study shows that 68 percent of the customers who quit doing business with a supplier did so because they felt ignored.
- Make marginal accounts profitable. By reducing the costs of selling and servicing

FIGURE 9-14 A counterperson should record every time an account is called, the date of the call, the presentation, and the outcome.

small and infrequent buyers, their business becomes much more valuable.

- Increase the effectiveness of the outside staff by relieving them of marginal and inactive accounts and by doing most of the legwork associated with prospecting.

SYSTEMIZING TELEMARKETING ACCOUNTS

Planning and commitment are central to the success of a telemarketing program. Since telemarketing will affect all operations—outside sales, the counter, inventory management, purchasing, and accounts receivable—management must plan accordingly and integrate telemarketing into the overall business structure.

In a large organization, telemarketing would be allocated to the manager, staff, and other resources. In a small parts store, the telemarketing responsibilities might be handled by a single person as part of his or her other duties.

Since telemarketing can improve market coverage in several ways and because each telemarketing objective requires different techniques, the counterperson can gain experience by starting with the goal of increasing sales to marginal accounts before expanding the program to cover other areas.

By analyzing the profitability of individual accounts, the counterperson or salesperson can determine which accounts can best be handled by traditional methods and which accounts should be assigned to telemarketing.

Outside sales people should continue to call on the most profitable customers, as well as those with a clear potential for greater future sales. Marginal accounts—those whose purchases do not cover the cost of selling—should be turned over to telemarketing.

The key to increasing sales with marginal accounts is to track their inventory usage and to time calls to coincide with their buying needs. If a counterperson calls too soon, the customer will not be ready to buy. If he or she calls too late, the customer might have already gone elsewhere.

To identify the purchasing cycles of individual accounts, a record-keeping and filing system is needed.

A card or folder is created for each account. Each time the account is called, the date of the contact, the offers made, and the results are recorded (Figure 9-14).

The account record is then filed according to call-back date. Since the account file is sorted by call-back date, the calling cycle becomes automatic and self sustaining.

During early contacts, the prospect's probable reorder date is best determined by simply asking the customer.

As the counterperson becomes familiar with each customer and his or her purchasing needs, earlier estimates of the optimum call-back times can be refined.

A similar strategy can be used to return inactive accounts to active status.

On the first call, the counterperson should determine why the customer is no longer doing business with the store. If the customer is still in the market for the store's products but has had service problems or feels ignored, he or she should be contacted on a regular basis until either a sale is made, or it is determined that further pursuit is pointless.

A call-back cycle is established for these inactive customers, and when a sale is finally made, they are upgraded to marginal account status and their buying cycles are tracked along with other customers.

PLANNING THE CALL

To maximize the benefits of telemarketing, each call must be carefully planned. Just as telemarketing involves a strategy for segmenting, assigning, and cycling the market, it also involves a carefully designed strategy for each call. This call strategy is the result of planning that takes into account the objectives of the call and the techniques required to achieve each objective.

There are three elements in planning each call.

1. Identify the goal(s) of the contact. Without objectives clearly outlined, it is difficult to plan a successful call.
2. Plan the opening statement. The first 10 to 15 seconds are cruical to the success of any telephone contact. A good opening accomplishes several things. It identifies the caller and company to the customer. It puts the customer at ease and establishes rapport. And it focuses the prospect's attention on the product or objective. In short, the opening "sets the customer up" for the sale message.
3. Design the sales message. The specific message will be different for each product or objective, just as the exact opening has to be adjusted for each customer.

DESIGNING THE SALES MESSAGE

A successful sales message stresses benefits over features. A well-planned sales message also recognizes telemarketing's main limitation: the prospect cannot see the benefits of the product. An effective telephone presentation paints an inviting picture using only words.

Selling words are the ones that make the difference between a dull, merely factual presentation and a sales message that evokes purchasing desires within the prospect. Selling words are the adjectives such as smooth, fresh, and clean that create agreeable impressions. They are also the "people" words that hook and hold the prospects: you, me, I, and the customer's name. And they are the powerful words: slash, cut, race, grip, and control.

FIGURE 9-15 Conducting a proactive telemarketing campaign usually requires a quiet place to talk.

THE CLOSE

After the customer decides to buy, the final step in telemarketing contact is the wrap-up. The counterperson must summarize the order, arrange for the next contact, thank the customer, and let him or her hang up first.

Summarizing the order avoids any confusion and gives the customer a chance to add items that might have been overlooked. Arranging for the next contact ensures that the customer will not be surprised and allows him or her to prepare the order ahead of time.

Thanking the customer is a mandatory courtesy. And letting the customer hang up first gives him or her the chance to air any last-minute concerns. After all, the caller was prepared for the call. The customer, perhaps, was not.

TELEPHONE LOCATION

Because of the need to concentrate on the prospect and the call strategy, the counter, with its incessant interruptions, is not the place to conduct a proactive telemarketing campaign (Figure 9-15). For the best results, the telemarketing staff should work in an environment that is relatively free of distractions.

If the counter is not the place to do telemarketing, it is usually the best source of telemarketing personnel. Because of product knowledge and experience, the counterperson is well prepared to handle customer objections.

When developing a telemarketing program, store personnel might talk to the professionals at the telephone company. They are in the business of selling telephone calls and have practically made a science of telemarketing. They can help plan the program and provide training for the staff.

REACTIVE TELEMARKETING

Outside sales, telemarketing, and advertising are important and necessary, but they are often directed at people who have shown no obvious interest in doing business.

More and more customers are shopping by telephone, and many sales are either made or lost according to telephone technique. When the phone rings, the counterperson is faced with the best of all possible prospects: a customer with a real need who wants the counterperson to sell him or her something.

Be polite and courteous. Answer the phone with a "smile." If at all possible, calls should be answered before the third ring. However, the need to answer quickly often conflicts with other important priorities, like making change for a sale, answering other lines, or pulling the last two items of an order before running back to the counter. These priorities have to be balanced and managed, not subjected to steadfast rules.

On Hold

A "hold" button allows the counterperson to answer quickly and still finish a current task before diverting attention to the new caller. If the counterperson must put a call on hold, it should be done politely. "Motor Distributors, please hold a moment," spoken in a cordial tone will do just fine. Never place a caller on hold without identifying the store and yourself. Always try to give an indication of how long the customer will be on hold, and poll lines on hold often so callers will not think they have been forgotten. If the system does not have a "hold" button, set the receiver down gently. Remember, it is an amplifier.

Calling Back

Inquiries that involve a large number of parts, such as engine and transmission overhauls or major brake or suspension work, are better handled by calling back rather than by having the customer hold.

Even for simple inquiries, calling back can buy the counterperson the time to finish a task that is best handled without interruptions, or it can free the counterperson for other more urgent requests, as long as the counterperson calls the customer back at the appointed time.

Calling back has another advantage. It buys the counterperson the time to upgrade a simple quote into an effective sales presentation with the technique called *message planning*.

Message Planning Some counterpeople answer inquiries in a random or customer-determined manner. A customer asks a question, and the counterperson tries to answer immediately. Or the prospect asks questions A, B, and C, and the counterperson answers in that order. These are automatic responses, normal reactions, but they are seldom the best way to proceed.

For example, if a customer calls for a price on an engine overhaul, perhaps the automatic response would be to simply give a quote. But to manage this opportunity, the counterperson should design a message that says, "This is the place to have your engine work done."

That message, and any other, is composed of several elements:

- All the information needed to motivate the customer
- A sequence that takes into account the impact of each element (that is, benefits first, price second)
- No elements that would divert the customer's attention from the sales objectives

Time Management

To give excellent service to many callers each day, it is necessary to exercise good time management.

On the phone, time management means taking as much time as necessary to establish, maintain, or enhance the relationship with the customer and exchange all the data needed to handle the request without backtracking—and not a moment more.

The exchange of information begins by identifying both the company and the counterperson. Likewise, the counterperson must get the name of the caller. If either person must call back, knowing the names can save time.

Getting all the information to fill an order can be time consuming, but it is essential, and there are ways to save time. Some counterpeople look up the first part, ask for information, look up the next one, ask for more information, and so on. Besides wasting time, this process can be very annoying to the customer.

Instead of just reading, the counterperson should picture the customer's car, think about what has to be done, and what options and accessories would affect the parts in question. With practice, the counterperson can get most of the information needed before opening the books. This technique also forms the basis for a solid related-sales strategy.

Exchanging information obviously requires speaking clearly and slowly enough to be understood. The telephone customer cannot see facial expressions, gestures, or body language. The counterperson has to communicate with just words, tone, timing, and inflection.

OUTSIDE CALLING

Once commonplace in the aftermarket, the outside salesperson can be found in less than one-third of the contemporary parts operations, according to some estimates. Many jobbers believe that one counterperson, behind the counter, with a phone, makes more economic sense than an outside salesperson who could possibly, on an exceptionally good day, make seven to ten sales calls.

While a full-time outside salesperson is outside of the range of many jobbers, a part-time counterperson, part-time outside rep is often an effective alternative that is economically feasible. The outside salesperson and the counterperson have much in common. However, some basic differences should be recognized.

The outside salesperson is more of a prospector. Some estimate that 80 percent of their time is spent listening to dealers talking about bad delivery. Even if the truck goes out twenty times a day, it is not enough. The outside person has to decide what is and what is not a founded complaint.

Part of the traditional outside salesperson's job entails visiting stocking dealers to keep their inventory up-to-date. The salesperson might be seen stocking parts in the ignition cabinet, putting away some hoses or belts, and taking orders at the same time. One call might take ten minutes, another a couple of hours.

But the major distinction to be made between the counterperson's responsibilities and those of the traditional outside salesperson is that while the counterperson sells products that the customer expects to buy, to a great extent, the outside salesperson should be pushing products the customer does not expect to purchase. For example, the outside salesperson could turn a dealer around to an emission control brand that he or she did not handle before. A salesperson should never leave the store without a line sheet. His or her job is to bring in orders that would not have come in before.

When filling in a part-time outside salesperson, the counterperson has to know how to overcome resistance when calling on new prospects. The traditional outside salesperson never leaves the store without a special or "deal" to offer. The deal is what will hook a new customer or at least keep that prospect's attention.

The knowledge, experience, and reputation of counter personnel play a large part in determining how successful an outside sales call actually is. The traditional outside salesperson pushes this point. The counterperson has to learn how to sell himself or herself when on the outside call. To get a new account or even to keep an established one, the counterperson has to talk quality products, efficient delivery, accurate invoicing, and assistance. And the counterperson must deliver on these promises.

MACHINE SHOP SERVICES

When promoting the jobber store and its products, the counterperson should not bypass the opportunity to promote machine shop services (Figure 9–16). As more people use a store's machine shop

FIGURE 9–16 Resurfacing rotors is a major source of machine shop work. Measuring with a micrometer will determine if there is sufficient thickness required for refinishing.

FIGURE 9–17 Point-of-sale aids describe a particular item and should be placed within customers' reach.

and are satisfied with the work, more parts will be sold at the counter. In effect, as the machine shop services are promoted, they will promote the rest of the store.

Unless the counterperson promotes these services, they will remain largely unknown because many customers never hear about them. Since many dealer customers are afforded delivery by the store, they never have a chance to stumble across these services. And many do-it-yourselfers will never ask about them. It is up to the counterperson to promote the shop's services as products that could help the customers.

The counterperson can talk to the manager about updating any signs to include a statement about the availability of quality machine shop services. He or she can look into changing ads in the phone book to incorporate information about all of the shop's capabilities: block boring, cylinder head reconditioning, rod rebuilding, crankshaft grinding. A counterperson can make a list of specials that would help bring in new machine shop business.

Every effort should be made to promote this shop because it will bring more business to the counter. Also, the more dependent a customer becomes on the shop service, the less likely he or she is to buy parts at "deal" prices from competitors. By keeping customers up-to-date on machine shop capabilities and by offering the best possible service, customers will come to depend on the store and machine shop for all their needs and parts.

MERCHANDISING

Good merchandising paves the way for increased sales. Manufacturers and suppliers often provide brochures, counter displays, and other point-of-sales aids. However, to increase sales the counterperson must use these aids effectively and continually think of new ways to help make merchandise move.

Point-of-sales aids are signs or displays that notify customers that certain items are in stock. These aids might describe features or benefits of a particular item. These displays should be placed where customers can reach them and handle them (Figure 9–17).

FIGURE 9–18 Because of its extensive use, motor oil attracts customers and can build profits.

Displays should not be placed where they block or clutter aisles or counters, but they should be set up along busy customer traffic areas. Displays should be clean, and a full supply of stock should be maintained in the display. Displays should be changed regularly. Properly displayed merchandise can increase sales, especially among walk-in customers, because it stimulates the desire to buy immediately, to ask for more information, or to plan to buy at a later date.

Different products require different marketing approaches because of varying uses or different types of customers that would normally purchase them. The following merchandising techniques illustrate ways to help sell various automotive products, such as oil, ignition parts, filters, belts and hoses, paint and body equipment, front-wheel drive parts, brake parts, and seasonal items.

MERCHANDISING OIL

Without oil, there would not be an automotive industry or an aftermarket. Yet very few jobber stores derive much profit from its sale. In fact, many merchants have come to view motor oil as a necessary evil because of the following:

- It is bulky.
- The containers often leak.
- It requires a lot of handling.
- There is pressure to carry a variety of brands and grades.
- Customers are very sensitive to its price.

In spite of these drawbacks, motor oil offers some real potential for drawing customers and building profits (Figure 9–18). Oil changes are the most frequently performed DIY task and the reason cars most often visit a garage. There simply is not another product that brings customers back to the store as reliably and frequently as motor oil.

Price Indexes

Everything done to promote motor oil sales has to be done with extensive consideration of the customer's sensitivity to its price. The counterperson has to survey competitors' prices in order to determine what price the customers consider "reasonable and customary." From the study of prevailing prices, the counterperson should determine two price indexes:

1.   The range of prices encountered in a variety of outlets on an everyday, nonadvertised basis.
2.   The range of advertising prices

The range of everyday prices is a good indication of what customers will accept without feeling as though they are being gouged. The second index will indicate the type of prices that must be offered in order to advertise. Advertising even a little too high is inviting customers to stay away. Purchasing and inventory decisions, as well as pricing and advertising strategies, must be based on the knowledge of prevailing selling prices.

Purchasing Philosophies

Unlike most parts that the jobber buys from either a warehouse distributor or directly from the manufacturer, oil can be bought from a variety of sources:

- WDs
- Specialty oil distributors (like WDs for oil)
- Oil jobbers (local distributors for the products of a particular refiner)
- Manufacturer (refiner)
- Packager
- Competitors such as drugstore chains and warehouse clubs

All of these sources have to be monitored because prices and purchasing opportunities change rapidly.

How oil is purchased depends on the jobber's philosophy about profits on motor oil and the importance attached to staying within the bounds of customer expectations. Some view motor oil as a traffic builder, or a loss leader, and give it away at cost (or less), counting on related sales and repeat business to make up the difference. Others treat motor oil as a distinct profit center, paying close attention to movement and margin, and demanding the same return in investment as any of the other inventory investments.

Some choose to buy in huge quantities to keep jobbers' costs to a minimum and such a choice can serve either extreme of the profit philosophy. However, bear in mind that the "lowest possible cost, at any cost," might not really be the lowest cost.

Capital tied up in large, slow-turning inventories is capital that is unavailable for brisker investments, nor is it available to meet (and discount) other payables. The formula for calculating return on investment (ROI = Gross profit/cost of inventory × 100) will not show the effects of stifling the cash flow or abusing the time value of money, but those dangers are very real.

At the other extreme of purchasing philosophies, the jobber can buy oil in small quantities for no investment at all. In this case, the jobber pays for it monthly, and it turns twelve times a year. Because

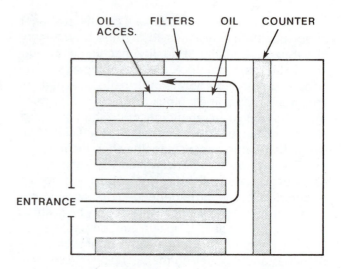

FIGURE 9-19   Oil displays should be placed in the back so customers have to pass other offerings on the way.

oil is sold in large enough volumes, fairly precise determinations of monthly demand can be calculated. The only drawback is the extra handling expense with frequent ordering, receiving, and stocking.

Another advantage of ordering in small quantities is it keeps the stock fresh and packaging is always up-to-date.

Many oil customers have strong brand preferences. By turning the oil rapidly, the jobber can carry a greater variety of brands and grades and increase customer satisfaction.

Finally, when the actual capital investment in oil is kept to nothing, the jobber can afford to sell it at a low margin and still come out with a decent ROI. Buying in quantities that allow twelve turns also allows the jobber to meet most everyday, nonadvertised prices. But, to occasionally advertise a real bargain price, the jobber will probably have to purchase a large quantity of a single product, perhaps in cooperation with a few other stores or as a member of a program group.

Location and Display

One of the basic tenets of effective merchandising is that staples should be located so as to cause the customer to see the store's other offerings on his or her way to the desired item (Figure 9-19). Supermarkets put the bread way in the back of the store, and convenience markets do the same with the refrigerator cases.

In an auto parts store, the customers that motor oil attracts should be routed through as much of the store as possible. Along the way, the gondolas and endcaps should be merchandised to grab the customer's eye and remind him or her of other automotive needs.

The display area for the oil should be kept fully stocked and clear of accumulated plastic case wrappers and boxes. It is helpful to stock the various oils in pairs of cases, one stack in the pair opened for single can sales and other sealed cases for case lot sales. This display method prevents customers from having to rip open cases to get a couple of quarts and makes it easy to restock the display as it sells down.

The area right around the oil will be the busiest, and this is where related items should be (Figure 9–20). Oil filter wrenches, drain pans, funnels, drip pans, waste oil disposal products, replacement drain plugs, and filters should be displayed near the oil.

Oil changes are where DIY involvement starts, and products that suggest deeper involvement can be displayed in the vicinity of the oil department. Service manuals, test equipment, jackstands, ramps, and specialty tools offer the DIY customer expanded horizons.

Advertising

Advertisements for oil should always feature related products. Another advertising technique is to make an exceptionally low price conditional on the purchase of some other selected product.

The oil on the shelves is an excellent place to slip a flier or coupon. In each case of oil, the counterperson can enclose coupons good for on-the-spot rebates on purchases of filters or any other merchandise.

One point about advertising low prices that is occasionally overlooked, with embarrassing results, is limiting the quantity available to each customer. Many a dealer has taken advantage of a competitor's generosity to stock his or her own shelves.

Selling Oil

Motor oil is a product that everyone needs, and most customers understand the importance of keeping their oil fresh and clean. Still, a few customers will not appreciate the importance of motor oil, and quite a few will be puzzled by all the brands, grades, and service classifications.

The two main characteristics of an oil are its viscosity and its service rating. Viscosity refers to the oil's thickness and its ability to flow at a given temperature. The Society of Automotive Engineers, in rating oils, measures viscosity at either or both of two temperatures: 0 degrees Fahrenheit and 210 degrees Fahrenheit. When viscosity is tested at 0 degrees, the viscosity number is followed by the letter W. When an oil has been tested at 210 degrees Fahrenheit, no letter is used.

The majority of oils sold today are multiviscosity oils containing additives that cause the oil to flow better at cold temperatures and lubricate better at high temperatures. Rating a multiweight or multigrade oil at 10W–30 indicates that at 0 degrees Fahrenheit it performs like a regular 10W oil. But at 210 degrees Fahrenheit it performs like an oil with a viscosity rating of 30.

Having a chart such as the one shown in Figure 9–21 displayed at the point of sale will help customers select the proper blend for their needs.

FIGURE 9–20 Related items should be displayed around the oil.

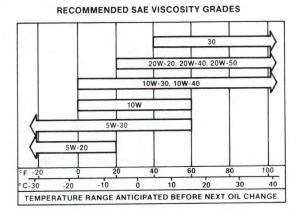

RECOMMENDED SAE VISCOSITY GRADES

TEMPERATURE RANGE ANTICIPATED BEFORE NEXT OIL CHANGE.

NOTE: DO NOT USE SAE 5W-20 OILS FOR CONTINUOUS HIGH-SPEED DRIVING.

FIGURE 9–21 A helpful aid for selling oil is a chart that tells the type and quantity of oil recommended for each car.

Oils used in gasoline engines are also rated according to service designation. SA and SB oils are intended for use in utility engines subjected to very mild operating conditions. SC lubricants are intended for use in cars and light trucks built before 1967. SD oils meet the protection requirements of vehicles built between 1967 and 1971. Finally, SE oils meet the special high-temperature requirements of cars and trucks built in 1971 and later.

Oils used in diesel engines use the prefix "C." As with gasoline engine lubricants the least efficient diesel lubricants are rated A. Only four ratings are used; CD is the most efficient.

Synthetic oils offer superior lubrication characteristics and service life when compared to petroleum-based lubricants. However, they are more expensive, so these advantages must be stressed. Also remind customers that synthetic oils cannot be mixed with one another or with petroleum-based oils.

The most important fact about motor oil that the counterperson can impress upon customers is that it becomes contaminated and its additives get used up.

The only assurance of adequate lubrication is changing the oil and filter at least as often as the vehicle manufacturer recommends. Heavy city driving requires more frequent changes.

MERCHANDISING IGNITION PARTS

Marketing tactics for ignition parts have to be designed for two very different types of customers—the professional mechanic and the do-it-yourselfer.

Serving Professional Customers

The traditional elements of marketing ignition products to the professional installer include the following:

- Place an inventory on customer location.
- Make regular sales calls to replenish inventory.
- Regularly adjust the inventory to meet the customer's changing needs.
- Provide hot-shot deliveries of the parts not stocked.
- Provide catalogs, buyer's guides, and price sheets.
- Hold clinics to keep the customer informed of changing ignition technology.
- Provide signs, posters, clocks, etc., to help the customer market those products.
- Offer a fair price.

The jobber can add to these elements in an effort to improve penetration into the dealer market.

FIGURE 9-22 Many factors must be taken into consideration when choosing an ignition line or lines.

However, none of these traditional marketing tools should be omitted.

In the ignition market, the jobber's toughest competitors are likely to be wagon peddlers offering lower prices. Beating them involves understanding their limitations. They must sacrifice service to offer their low prices. As a rule, wagon peddlers cannot provide clinics or hot-shot deliveries. Also, they cannot deliver a variety of products, such as brake pads, mufflers, and air conditioning parts.

Price obstacles can be overcome, at least partially, by taking some of the following steps:

- Choose ignition lines carefully.
- Work with the WD for better discounts.
- Participate in a program group.
- Trim normal profit margins on ignitions.
- Offer a "generic" line (or a mix of premium and generic).
- Give dealer accounts extended payment terms (dating) or even consignments on their initial stock.
- Offer to buy back the stock if the dealer is not satisfied with it.

When choosing an ignition line, a counterperson must remember that the manufacturers of the best ignition lines keep the jobber's dealer account updated with current catalog and price information through direct mailings, thus relieving the counterperson of that responsibility. This kind of service is invaluable for preserving customer satisfaction. The customer does not miss sales because of missing application information or lose profits by selling from an out-of-date price sheet.

A few other points should be considered when choosing the ignition line(s) (Figure 9–22). Since a large percentage of a counterperson's time is spent on ignition parts, any time saved on ignition research will be significant. Catalog quality, completeness of the line, and availability from multiple local sources are the most important factors in eliminating unnecessary research.

The line's reputation will have a lot to do with its acceptance and reputations can vary from area to area. A line's local reputation is affected by the quality of its factory rep, the quality of its local distributors, and the overall professionalism of the group of jobbers stocking the line.

Serving DIY Customers

Taking ignition parts to the DIY market requires a different approach, and the most effective vehicle is going to be advertising.

In a fairly large metropolitan area, direct mail and door-to-door fliers are often the most effective advertising media for ignition parts. With a reasonable budget, the jobber is able to reach the most important group of do-it-yourselfers—those in the immediate marketing area.

A chain of stores or a single store in a small town might do better with radio, television, or newspaper advertising.

Another effective way to bring ignition parts to the DIY's attention is to use reminders, or tip sheets, that emphasize the importance of ignition system maintenance and promote the products. These sheets can be dropped into the bag with the rest of each customer's order or even slipped into each case of oil.

Another possibility is to use reprints of articles that appear in the automotive section of the local papers or in consumer magazines.

Some spark plug manufacturers offer color posters showing plugs that have suffered various types of damage. These posters sometimes work as conversation pieces to get people thinking about their ignition systems. They can also help to solve problems customers are having with their cars. The counterperson can design displays to accomplish the same ends.

Related Sales

Most people buying ignition parts are performing regular or preventive maintenance. Rather than fixing just one thing, they are interested in their car's general welfare, and they are already predisposed to related sales. With these customers, the counterperson's task is to expand their awareness of all that is involved in a thorough tune-up.

Because the ignition system is not an isolated system, a thorough tune-up extends beyond that system. The starting and charging system, the carburetor, and the engine's internal condition are all interdependent upon the condition of the ignition system.

Therefore, in addition to simply selling the requested ignition parts, expanding the idea of a tune-up to a major service creates the opportunity to sell belts, batteries, alternators, regulators, cables, terminals, valve cover gaskets, and air, fuel, and oil filters.

Selling the idea of a complete service opens the door for sales of supplies such as battery cleaner/protector sprays, engine degreaser, carb cleaner, motor flush, water remover, fuel, and oil additives and engine oil.

And the counterperson might even sell a few tools, including test lights, voltmeters, battery terminal pullers, battery pliers, filter wrenches, compression testers, remote starter switches, feeler gauges, gasket scrapers, and parts cleaning brushes.

Merchandising these related items together will boost sales, both from impulse and the fact that the displays will already have made suggestions that the counterperson will repeat in selling encounters (Figure 9–23). Also, selling related items is much easier if the counterperson can accompany the customer to the ignition display area and show him or her what is needed.

For the customers who are asking for ignition parts in an attempt to correct (rather than prevent) a problem, the counterperson must ask questions and listen to the answers.

FIGURE 9–23 A display of ignition parts should be surrounded by related items, such as belts, batteries, alternators, filters, chemicals, and tools.

FIGURE 9-24 A counterperson can use cheat sheets for easy access to new parts numbers or other pertinent information.

FIGURE 9-25 Ideally, several filter brands should be available.

He or she must determine the customer's real problem, help correct it, and then encourage the customer to do a complete service to prevent future problems.

Cheat Sheets

Because of the frequency of ignition parts requests, many counterpeople memorize the more popular applications. But when the counterperson changes lines, he or she can use cheat sheets, which are nothing more than 3 × 5 cards with the new part numbers listed for the old applications memorized by the counterperson (Figure 9-24).

However, cheat sheets do not have to be limited to just the situations in which new lines are encountered. For example, when the counterperson can see that a change in the industry is going to result in a new set of popular numbers, he or she can make cheat sheets to prepare for the inevitable demand.

Counterpersons can also use a maxi sheet, in which all the part number information for a complete service is written on a single page. With really complete ignition lines, much of this work is already done in the catalogs. For instance, one ignition line catalog lists points, condenser, rotor, cap, and plug wire set all on the same line. However, a maxi sheet puts all that information and more on a single page. One of these for each of the most often requested tune-up applications can make a big difference.

Since the sheets are selling tools, the counterperson should include prices and reminders about related items such as tools, chemicals, supplies, and manuals. Felt-tipped pens can be used to highlight important information, and small adhesive-backed memo pads can be used to add temporary information such as items that are on sale.

MERCHANDISING FILTERS

Engines run on fluids—fuel, oil, coolant, and air—and fluids are excellent vehicles for carrying abrasives and other contaminants throughout an engine. Filters strip those contaminants from the fluids and thus enhance the vehicle's dependability and prolong its life.

Part of a counterperson's job is to supply those filters, to educate customers about their importance, and to be sure the customers' vehicles get the right filter along with the related items necessary to complete the service of which filters are a part.

Two of the most important considerations in the jobber's filter marketing plans are the line(s) to be carried and the rep that comes with them.

Choosing a Line

The major filter manufacturers have worked to create a high level of brand awareness among car owners. Offering customers a filter they are not already familiar with is a tough way to start a sale.

Handling a familiar major line also offers the advantage of completeness. The opportunities to sell filters are far from limited to just the frequent car and light truck applications. Heavy trucks, agricultural, marine, and stationary industrial equipment are all covered very well by the major manufacturers, but rarely by the generic short lines.

However, there often is a place for a private labeled price line in the jobber's marketing plans. It might be necessary to offer DIY customers an inexpensive alternative. To seriously pursue the filter market requires stocking one of the major brands and, ideally, having access to several of the others (Figure 9–25).

Importance of a Good Rep

When all other factors are close to being equal, the jobber should select the filter line by the rep that comes with it.

With service items like filters, a survey is a powerful marketing tool. The factory rep is the person the jobber will rely on to survey the equipment of agricultural and fleet accounts. A good rep will be eager to perform these surveys and will do a thorough and accurate job.

Location and Display

Merchandising filters is a great way to pick up their movement with walk-in customers.

Some stores keep filters in back because if they are stocking a fairly complete selection of filters, and particularly if stocks include a lot of filters for non-automotive applications, it is easier to deal with them if they are all in one place in numeric order. Also, many of the filters come in odd-sized boxes, and some of the larger ones are packaged in unappealing, brown cardboard boxes.

However, the most popular automotive air, oil, fuel, and transmisison filters are packaged, by most manufacturers, to be displayed and capture the customer's attention. Those bright packages are wasted potential if they are hidden in back.

It does not take many filters to build an attractive display with sufficient "mass" to create the impression of extensive coverage (Figure 9–26).

With the easy-to-read application listings on the filter packages, many customers will even do the research part of the counterperson's job, which buys the counterperson extra time for getting to know the customers a little better or selling related items.

However, a counterperson must beware of those applications on filter packages. In addition to saving time, they can also cost some sales, particularly on late-model cars because filters in the distribution pipe are still in boxes that do not reflect the latest applications.

Also, the applications on the boxes are not always as complete as those in the catalogs. So, a customer who does not find his or her car listed on the box will sometimes give up, figuring it can be purchased elsewhere or later. A counterperson should watch filter customers for signs that they are having trouble finding what they need.

The best place to display filters is next to, or on the way to, the motor oil. By stocking fast-moving oil filters on top of the oil display, customers will easily make the complete purchase. Filters can also be merchandised on end caps at each end of a bulk oil display or across the aisle from the oil or along a main corridor that leads from the oil to the counter or checkout area.

Other items in the store can be used to assist filter merchandising efforts. For instance, some filter manufacturers offer fender covers bearing a logo and sometimes a slogan. These items can be sold instead of plain covers.

A counterperson must keep the following points in mind when merchandising filters:

- Some older trucks and cars still use both oil and air filters, which are inadequate and a nuisance to maintain. Encourage customers to replace these filters with the paper element air filters.
- Many older cars are equipped with canister type oil filters. Check to see if a kit is available to replace it with a spin-on type.
- Avoid the practice of visually matching filters. Source oil filters have internal valves, while others, similar in appearance, do not. Also, it is extremely important that the filter fits its housing properly.
- Deformed or split oil filters are caused by a relief valve failure and not a defective filter.

FIGURE 9-26 Providing catalogs at displays makes it easy for DIY customers to look up their own applications.

FIGURE 9-27 Belts and hoses can be displayed where space is available—on pegboards, on gondolas, or on the walls above other items.

 SALES TALK _____

To prevent serious damage to the engine, the counterperson should advise a customer not to simply replace a damaged filter returned as defective without explaining about stuck relief valves (usually located in the oil pump).

- Every carburetor and kit sale should include a new fuel filter. The most common cause of a "defective" new rebuilt carburetor is dirt in the fuel supply bodying between the carburetor's needle and seat, causing the valve to stay partially open and resulting in excessively rich mixtures and gas running all over the engine.
- Do not replace fuel filters for injection systems with some intended for carburetion type vehicles.
- Water is especially destructive to diesel engines. Replacement fuel filters are available that have water-separating capabilities for applications that were originally just filtered. Check the catalog for these.

MERCHANDISING BELTS AND HOSES

Marketing and merchandising belts and hoses can yield amazing sales results for several reasons. First, any group of products that can do as well as belts and hoses, without any particular marketing effort, should be able to yield results with a good push. Second, belts and hoses are practically the opposite of the kinds of goods that are used as cost leaders by mass merchandisers. Consumers are very much aware of the prices of motor oil, filters, and spark plugs; but with rubber products, availability, service, and quality are much more important.

Location and Display

In many jobber stores, the belts and hoses are hung on racks in the back—usually way above everything else—where customers cannot see them and where it takes a long hook to reach them. More visible displays will undoubtedly increase sales.

If the space is available, these rubber products can be hung in rows on a pegboard wall or gondola. Some merchandisers have fingers that hold the belts by their sleeves, so they are displayed on edge, side by side. If space is a problem, the belts can still be moved up front and hung around the perimeter of the store above the other displays (Figure 9–27).

Perhaps a poster that encourages replacement of belts and hoses could be included with a display dedicated to cooling system products, such as cleaners, flush kits, and coolant. A burst hose, burnt valve, or shredded belt is always a better conversation piece than slick graphics.

Other rubber line products deserve merchandising, too. Some DIY customers might not know that they can buy heater, vacuum, fuel, and power steering hose by the foot. A counterperson should keep the hose dispenser in plain view and make it easy to find the hose clamps.

Related Sales

Merchandising is not the only way to increase sales of rubber goods. Every request for an alternator, regulator, or battery should be met with a question about the alternator belt's condition. Slipping belts account for more dead batteries than most people realize. Power steering pump, smog pump, and air conditioning compressor or clutch sales should also include new belts because often the unit being replaced has frozen up and scorched the belt.

Water pump replacements, coolant installations, radiator work, and cooling system flushes are all tasks that should include a careful inspection of hoses and belts. A counterperson should suggest to a customer with a car four years old or older that he or she check the belts and hoses. Also, the counterperson should point out that the heater hoses are as old as the radiator hoses (as is the bypass hose). If they are overlooked, they can cause just as serious a breakdown as a blown radiator hose. Finally, when a customer buys a new hose, always ask him or her about installing new clamps.

Inventory

Developing a good inventory for a store means knowing what sells in the area. A sales ranking report from a supplying WD will give a much better idea of local demand than a national classification.

Keeping track of lost sales is just as important. While the WD's sales ranking is a good starting point, its scope is still quite a bit broader than the jobber's specific market area. Lost sales let the counterperson know about activity in the market that is not highlighted in the WD's more comprehensive report.

Keeping a history on belt/hose movement is also important. These records will help the counterperson eliminate numbers that are not moving. Another reason for maintaining sales records is that a ranking of numbers for a particular store is even more specific than a WD's ranking. Thus, these records will be a significant aid for designing fast-moving inventories for customer acounts.

Servicing Fleets

The auto repair industry is not the only place to market rubber products. Fleets are a tremendous opportunity. Companies with anywhere from a half dozen cars to hundreds of trucks will be interested in an inventory of belts, hoses, and related products.

The tool to use in designing these inventories is the fleet survey. Rubber companies are eager to help in this area. They will usually provide survey forms and probably a knowledgeable person to complete them. Once the survey forms are completed and the results are tabulated, a suggested stock is proposed.

In addition to the belts and radiator hoses, the factory representative will suggest bulk heater, vacuum and fuel hoses, as well as clamps and other supplies offered in that company's line.

There is an added benefit to getting rubber products into a fleet: most fleets support some manufacturing or distribution effort, and those enterprises are full of opportunities for rubber goods. The jobber that supplies the shop's needs is likely to be called upon to supply something for the plant.

Servicing Industry

Industry is a big user of rubber products. Factories and plants are full of belts connecting electric motors to machinery and make heavy use of belts and hoses on forklifts, utility vehicles, and trucks.

Industrial belts are different than either automotive or fractional horsepower belts. They are available in letter sizes from A to E, with the letters indicating the width, and the numeric portion of the part number giving the effective length (outside circumferences are slightly longer).

Industrial belts have deeper profiles and different wall angles than automotive or light-duty belts. The differences are important, and if the incorrect belt is supplied for an industrial application, the jobber's product will soon have a reputation for early failure.

Hydraulics are another important component in industry. In just a few hours of instruction, the counterperson can learn to make hydraulic hoses. Industry also uses air hoses, water hoses, steam cleaner and pressure washer hoses, and welding hose.

Servicing Other Markets

Light-duty or fractional horsepower belts are another good opportunity to extend rubber product sales. Heaters and air conditioners use them, as do many light appliances. They are used in lawn and garden equipment, too.

Shop equipment is another application for light-duty belts. Anyone in carpentry or cabinetmaking has several tools with light-duty belt drives.

Golf courses, cemeteries, and park and recreation agencies use a lot of light-duty belts on their lawn care equipment, and the golf course might use recreational vehicle belts on their carts. Marinas use fractional horsepower belts in their fuel pumps and hoists. Some areas have a market for snowmobile belts.

Farm equipment uses plenty of belts, coolant hoses, and hydraulic hoses. A farm shop has as much need for air hoses, water hoses, steam cleaner hose, and welding hose as any fleet or industrial shop. Surveys are an excellent way to tap the farm market, too. Even if an inventory is not sold, the jobber can adjust stock to meet the farmer's needs.

MERCHANDISING PAINT AND BODY EQUIPMENT

The most effective way of merchandising paint and body equipment is to offer the best service possible. Services include fast delivery, technical information, problem-solving, and color matching.

Delivery Service

Good delivery service does not necessarily mean delivering every pint of paint the minute it comes off the shaker. Most painters will plan far enough ahead to allow at least a couple of hours before they really need the paint.

On the other hand, the jobber ought to gain the trust of customers through reliable delivery service.

FIGURE 9–28 A good delivery service need not respond immediately to a customer's request, but it should fulfill promises of arrival times.

If the counterperson knows he or she will not be able to get to the customer's shop in an hour, the reason (no excuses) it cannot be done should be explained, and a realistic delivery time offered (Figure 9–28).

If a good customer really needs the paint, perhaps the counterperson could offer to send over the paint mixer. If this is done, explain that this practice is not always possible and that a special trip is being made for the customer. And if the occasion requires it, ask the customer to plan a little better in the future.

TECHNICAL INFORMATION

Perhaps the most important service that the counterperson offers to customers is technical information, often based on the experience of others in the field. The counterperson not only gets the sales pitches from all the reps, but also has access to honest, unbiased opinions from other painters and auto body repairers.

When a customer uses a new product, the counterperson is usually one of the first to know about it. The counterperson should ask how he or she liked the new product and find out every good and bad point that the customer encountered.

If a customer likes a new product, the counterperson can use that opinion as a reference when suggesting the product to another customer. In this way, the counterperson can keep customers current and up-to-date using the best of the new products. And in doing so, the counterperson will keep the customer's trust.

To Lighten Color	To Darken Color
Increase air pressure.	Decrease air pressure.
Increase gun distance.	Decrease gun distance.
Increase fan size.	Decrease fan size.
Close fluid valve slightly.	Open fluid valve slightly.
Speed up stroke.	Slow down stroke.
Add more solvent.	Add more paint.
Use faster solvent.	Use slower solvent.
Allow longer flash time.	Allow shorter flash time.

FIGURE 9–29 Methods for making a color lighter and darker

Problem-Solving

Problem-solving and the manner in which it is conducted will have a significant effect on whether accounts are won or lost. Finding out what went wrong is sometimes difficult, but it can be done by asking the right questions. The counterperson must ask the customer, step by step, what procedures and products were used.

After assembling all the facts, the counterperson must review the causes of the problem by checking manuals or calling factory reps. Then the counterperson must give the customer a recommendation for repairing the job.

Color Matching

The counterperson's ability to help painters with color matching problems might be the best marketing tool at the disposal of the jobber.

No experienced painter should expect that any color will match right out of the can. Actually, this match happens fairly regularly, but most colors must be adjusted in some manner. The counterperson should have the knowledge and ability to give the customer help in making that adjustment. Service in this area will gain customer respect and loyalty. On the other hand, if the counterperson cannot offer this help, the customer is forced to go somewhere else.

Three procedures are available to help the customer achieve a proper color match. Gun and solvent techniques constitute one manner of handling such a problem. Special blending procedures can also be utilized to make a slight mismatch look perfect to the human eye. Tinting should be considered a last resort, because it involves adding tints to change the color.

When working with metallic colors, gun and solvent techniques can prove to be the fastest and easiest methods for achieving a match. The idea is to apply the color so that the metallic flake lies in the same position in the repair area as it does in the original finish. If the flake is allowed to sink to the bottom of the paint film, it will be covered by more of the other pigments in the color, making it appear darker. Metallic flakes lying near the surface will have the opposite effect; the color will appear lighter than the original finish.

Figure 9–29 describes methods for making a color both lighter and darker. Although a close color might need only an air pressure adjustment, more often a combination of a few of these procedures is needed to bring about a proper match. One easy way to communicate this information is to print it on small cards to be handed out to customers.

With a blending technique, a slightly off-shade color can be made to appear to be a perfect match when the color difference is gradually spread over a large area. It is easy to spot a mismatch between a door and fender or a fender and hood. If the color is gradually blended into the door or hood, however, the human eye cannot detect the mismatch.

The common blending procedure is easiest to perform with acrylic lacquer. The same process can be used with acrylic enamel or some urethanes but becomes more difficult in such applications.

An optional blending procedure is the two-gun technique, so called because the painter uses one gun filled with ready-to-spray paint and another with a blending solvent. This solution is usually a blend of slow-drying solvents, often with some clear solvent added. All steps imitate conventional blending techniques, except that the blending solvent is used to wet the edge of each coat of color applied beyond the repaired area. This will melt in the overspray, making the blend appear more even. The use of a blending solvent will also make it much easier to blend when using an acrylic enamel or urethane repair material.

If the painter finds that the mix is too far off for correction through gun techniques or blending, the counterperson might have to utilize tinting. A few simple rules must be followed to achieve a successful tint:

1. Keep the tint on the light side until the final tint is determined. Many colors change in lightness and in cast as they dry, so be sure to check color match dry-to-dry.
2. Many colors look different when sprayed than they do when dabbed, so always spray a panel to determine match.
3. Many colors vary in different lighting, so check all color matches in natural light.
4. Always tint with small amounts. More tint can be added, but when too much is added, it is difficult, if not impossible, to bring the color around again.
5. Be sure the tinted color is applied heavily enough to achieve full hiding so it is not affected by a color showing through.

The first step in determining proper tinting colors is to properly describe the color. Each color can be described in three dimensions:

1. Cast, also called hue, describes where the color lies in relation to others in the color wheel.
2. Lightness, or value, is the amount of black or white in a given color.
3. Saturation, or cleanliness, describes the richness of the color.

These characteristics can help to determine which tint to use. For example, when working on a blue, if the original color is greener, add green. If it is darker, add black, If it is brighter, add bright blue, and so on.

It is sometimes difficult to determine what needs to be added to the tint in terms of how the original finish looks. It is handy, therefore, to know which tinting color "kills" another color. Figure 9–30 shows the color to be tinted, followed by the two opposite colors that kill one another.

There are also a couple of special methods for tinting metallic colors. First, use an aluminum tint to lighten metallic colors whenever possible because a white can change the "flop" (the low-angle color). Second, when tinting with an aluminum tint, the coarse metallics will add sparkle to a color without greatly affecting its darkness, where a fine metallic will give the color a gray or dirty shade.

Color	Opposing Tints
Blue	Green—Red
Green	Yellow—Blue
Red	Yellow—Blue
Gold	Green—Red
Maroon	Yellow—Blue
Bronze	Yellow—Red
Orange	Yellow—Red
Yellow	Green—Red
White	Yellow—Blue
Beige	Green—Red
Purple	Green—Red
Aqua	Blue—Green
Gray	Blue—Yellow

FIGURE 9–30 Some tinting colors "kill" another color.

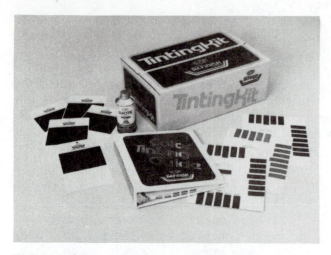

FIGURE 9-31 Tinting kits enable body shop personnel to custom tint paints. They should be stocked by any jobber with high-volume PBE sales.

As with any other problem-solving, it is important to educate the customer about how the problem is being solved. If the counterperson is good at tinting and blending, the task could become a full-time job just to help painters with color problems. If the counterperson is able to show how to match the color, the painter can learn how to solve his or her own problems. Special tinting kits are also available for PBE customers (Figure 9–31).

Additional Service

Body shops are finding it increasingly difficult to be profitable. Although painter/body repair people know how to repair a wrecked car and how to apply a beautiful finish, some have trouble controlling inventory and profit margins.

The counterperson, on the other hand, works every day with inventory and profit margins. When the relationship between the counterperson and the customer reaches the point where the customer feels he or she can confide in the counterperson, then the counterperson can add another dimension to the meaning of service. The counterperson can share experience-earned knowledge with the customer and help him or her become successful in business as well as in service. This service allows the customer to become more profitable and increase business and demand for products. More important, this extra service strengthens customer loyalty.

MERCHANDISING FRONT-WHEEL DRIVE PARTS

Front-wheel drive is rapidly replacing the more familiar rear driving axle designs, while proving to be a profitable innovation for the automotive aftermarket. The driving axles in a conventional domestic sedan would typically outlast the engine, transmission, and body. Axles, CV joints, and boots for front-wheel drive cars wear out much sooner.

Advantages of Front-Wheel Drive

Successful selling of front-wheel drive parts starts with understanding that these parts are expensive. The counterperson can help the customer accept the expense of a new axle or set of CVs by reinforcing the advantages of front-wheel drive.

Front-wheel drive designs eliminate the need for a driveline tunnel running down the center of the car, allowing more passengers to fit comfortably in a smaller compartment. Also, one of the most effective ways to reduce fuel consumption—and the cost of vehicles—is to make vehicles smaller.

Front-wheel drive also lends itself nicely to transverse engine mounting. A transverse engine requires a shorter engine compartment, allowing further reductions in overall vehicle size.

Another benefit of front-wheel drive is a shortened "power path." The shorter distance from engine to driving wheel means lower frictional and inertial forces to overcome. This fact, in turn, means that a smaller engine can accelerate the car briskly and that less fuel will be required to move it down the road. And, the smaller engine is less expensive to produce, keeping down the cost of the car.

Finally, front-wheel drive delivers good handling. The expense in front-wheel drive parts is due to the relative complexity of the CV (constant velocity) joints.

Encouraging Boot Inspection

The design of CV joints precludes the fairly trouble-free sealing that is done on U-joints. Since CV joints do not have little journals and bearing caps that can be sealed individually, the whole assembly and its lubricant has to be enclosed in a flexible boot (Figure 9–32).

These boots are subject to wear, puncture, aging, and cracking. Once they are permeated, the enclosed joint will not last long. Just a couple of weeks of normal driving with an open boot can easily ruin any CV joint. This breakdown could create a costly repair ticket for the consumer—one that could be easily avoided.

As DIY customers ask for other parts for a front-wheel drive vehicle, particularly items that require getting under the car to replace, the counterperson should suggest an inspection of the CV joint boots (Figure 9–33). If the counterperson does not encourage front-wheel drive owners to examine their boots, the joints are likely to fail on the road and be replaced by someone who, in all probability, gets the parts from another store.

Suggesting Repair Alternatives

For the customer who finds a damaged boot, there are several things the counterperson can suggest. The very best procedure is to remove the axle, disassemble the affected joint, clean it, lubricate it with fresh CV joint grease (different than ordinary chassis or wheel bearing greases), and install new boots.

If the customer is going to get that involved, he or she should also inspect the seals at the transaxle, wheel bearings, and brakes. Very few DIY customers are going to tackle that job. In this case, the counterperson should suggest the jobber's machine shop. The customer can bring in the axle and pick it up cleaned, inspected, repaired, and reassembled later in the day. If the jobber does not have a machine shop, the counterperson can make arrangements with one of the installer accounts to have the work performed.

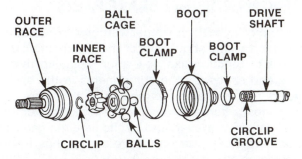

FIGURE 9–32 Typical components of an outboard CV joint

When making referrals, make sure the shop knows where each job is coming from. If possible, call the installer personally for each referral and make the appointment for the customer.

Some customers want nothing to do with removing an axle or taking their car to a shop. They want a quick, inexpensive fix. It is available, though it is no substitute for complete disassembly, cleaning, and lubrication.

Marketed under various names, there are CV joint boots that can be installed without taking the axle out of the car. Instead of slipping them over the axle and joint, these boots are open on one side and wrapped around the joint. The open side is then cemented shut, and the boot is clamped.

Split boots merchandise well. They are beginning to appear in attractive, see-through packaging. These catch the customer's eye and appeal to the average do-it-yourselfer's abilities. They are also valuable as conversation pieces from which to launch the sale of a more thorough repair.

Professional installers use split boots, too, particularly when their customers are on a tight budget and when the joint is still in good working condition.

Diagnosing Other Problems

A joint can wear out under a perfectly good boot, too. Without removing and disassembling an axle, diagnosing front-wheel drive problems is limited to hearing and feeling, which makes them difficult to diagnose (see Table 9–1).

If, during turns, the customer hears clicking or popping noises, the cause is probably the outer joint. If the noise gets louder when turning in reverse gear, it is almost certain that the outer joints are bad.

Vibrations during acceleration are most often attributable to wear in the inboard joint, although the outer joint is sometimes the cause. On cars with

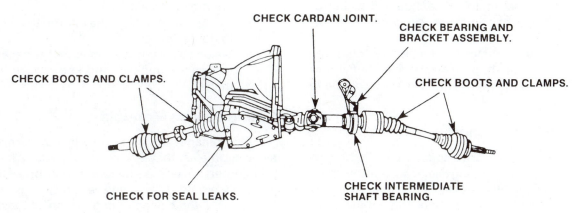

FIGURE 9–33 CV joint inspection points

TABLE 9-1: DIAGNOSING PROBLEMS IN FRONT-WHEEL DRIVE SYSTEMS

Symptoms	Check
Clicking or popping noise during turns	Outer joint
Noise increases when turning in reverse gear	Outer joint
Vibrations during acceleration	Inboard joint Outer joint Intermediate shaft bearing (on vehicles with equal length drive shafts)
Clunking noise when accelerating or putting car in gear	Inner joint Transaxle
Vibrations that increase with speed	Tire or rim Axles

equal length drive shafts, this problem can also be caused by a worn intermediate shaft bearing. Engine mounts are another possibility and, on transversely mounted engines, the upper mount, or torque strap, can be the problem.

A clunking noise when accelerating or putting the car into gear is usually caused by a worn inner joint or sometimes by wear in the transaxle itself.

Humming noises are almost always the result of wear in a bearing, not in a joint. Advise the customer to check the wheel bearings, the intermediate shaft bearings, and the bearings in the transaxle.

Vibrations that increase with speed are almost always the result of an imbalanced or out-of-round tire, or a bent rim, although some are the result of axles that have been bent in collisions or through improper towing.

Once a customer has pinpointed a bad joint, the options are more limited. The axle has to be removed, and the bad joint, or the whole axle, replaced.

Again, the customer can tackle the entire job, have the machine shop repair the axle, buy a new axle from the counterperson, or take the whole car to a mechanic.

Selling to Professionals

Most front-wheel drive parts will be sold to professional installers. Getting dealer accounts into the front-wheel drive business usually requires only information. With the possible exception of the pliers needed to tighten boot clamps, virtually every pro-

fessional mechanic already has all the tools needed. The only thing a few of them lack is information.

Most domestic U-joint companies in the front-wheel drive business offer clinics that will show dealers how to get the business and do the work. Hosting those clinics will give the counterperson an opportunity to show those dealers which store is the logical source for the parts.

Display

Since most retail customers are unfamiliar with half shafts, CV joints, and boots, displaying them will grab customers' attention and raise curiosity about these strange-looking parts.

A couple of rows of split boots at the top of an end cap, an axle on pegboard hooks midway down (a used one, cleaned up, with the boots partially cut away to reveal the joint), and packages of CV grease, and perhaps a couple of new joints on the bottom, make an effective display.

The counterperson might also want to display some of the posters available from the front-wheel drive parts supplier and a sign announcing that the machine shop handles front-wheel drive axle repairs.

MERCHANDISING BRAKE PARTS

Typically, a DIY customer will come to the counterperson asking about pads or shoes, expecting the brakes to function like new after replacing them. Pads and shoes are the components in the brake system that are easiest to replace, and they are the ones most often advertised. The brake system is much larger, however, and by the time the pads or shoes are worn out, the entire system has to be considered suspect (Figure 9-34).

An effective selling job includes the following:

- Determining what the customer really needs
- Getting the customer to think about the entire brake system
- Giving the customer the information needed to do the job right
- Making sure the customer has the necessary tools and supplies (and knows how to use them)

Selling Service Manuals

Good manuals will not only help customers to avoid problems, they will also sell parts. When the counterperson tells a customer disc brake hardware, shoe hold-downs, and return springs should be replaced, the customer might suspect the counterperson is only concerned about additional profit.

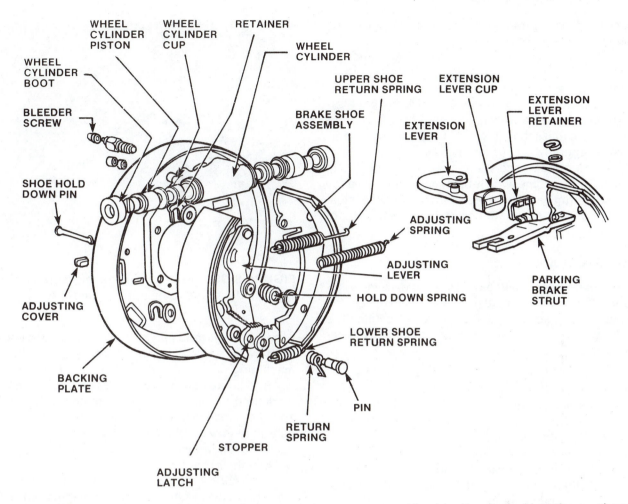

FIGURE 9-34 In a typical drum brake there are many items to wear out besides the shoes. Note the number of springs, clips, and hold-down hardware.

But when the customer sees in print that heat robs springs of tension, that calipers do not ride smoothly on rusted pins, that brake fluid does become contaminated and deteriorates, the counterperson will no longer be suspected of overselling. Manuals will confirm the counterperson's advice and do much of the selling (Figure 9-35).

Sales Opportunities

Marketing brakes means more than waiting for someone to ask for pads, shoes, or another item that is specifically part of the brake system. Other requests can tip off the counterperson to a possible need for brake products.

For example, any hint that a vehicle is about to be taken on a vacation or trip or pressed into service more severe than normal is an opportunity to inquire about brakes.

Tire chain purchases suggest the vehicle might be used in the mountains, where brakes can encoun-

FIGURE 9-35 Service manuals do a good job of stressing complete repair procedures to DIY customers.

ter extremes of prolonged high heat or be operated on slippery surfaces.

Products that enhance a car's performance afford the opportunity to suggest similar improvements in its stopping capabilities. Trailer hitches, wiring adapters, and connectors give the counterperson a clue to a vehicle that might soon be stopping more mass than it is used to. A braking performance upgrade is in order in this case, too.

Even appearance products can lead to brake sales. The customer who is buying a cleaner for alloy wheels might be very interested in the new, asbestos-free pads that eliminate unsightly black dusting of the wheels.

Related Items

Related item selling in brakes is extremely important, not only to build profits, but also to help the customer achieve the safest possible results. The following is a list of some related items for a complete brake job:

- Brake fluid
- Brake fluid dispensers
- Brake cleaners
- Assembly lube
- Disc pad quieting compounds
- Bench bleeders for master cylinders
- Pressure bleeders and adapters
- Wheel cylinder and caliper hones and replacement stones
- Caliper piston spreaders and piston removers
- Brake spring pliers and hold-down tools
- Hand tools, creepers, and jack stands
- Repair manuals
- Steel brake lines, unions, and adapters
- Aerosol bearing cleaner and bearing grease
- Wheel bearing packing fixtures
- Seal drives

Displays should be set up to begin the job of selling a complete service job. Brake pads, shoes, and hardware once came in fairly plain, dull boxes, but today, most manufacturers have upgraded packaging graphics to the point that these products can be used to build attractive, effective displays in the front of the store. Coupled with promotional posters, brightly packaged chemicals, and sharply carded brake specialty tools, there is no reason to leave brake items in the back.

A display could also feature fluid at a promotional price, literature explaining the importance of fresh fluid, carded master cylinder reservoir gaskets to protect the new fluid, a special price on pads,

FIGURE 9–36 Drums should be measured and remachined if found to be out of round.

shoes, and kits, and maybe even an offer for a can of fluid with each rotor or drum brought in for machining.

The brake lathe is another important element in marketing an assortment of brake items. The availability of brake machining services is a key factor in determining where some customers (particularly dealer, fleet, industrial, and agricultural accounts) will buy their brake parts.

Marketing brake parts and marketing machine work go hand in hand. When a customer calls for pads or shoes, the counterperson should always suggest that the rotor or drums be sent in for inspection and possible machining (Figure 9–36). Also, an invoice message offering dealer accounts free brake fluid with every pad sale that includes rotor inspection could be used.

A WORD OF CAUTION

As stressed in this chapter and throughout the entire text, competent counter personnel constantly acquire and expand their product and service knowledge. As you can see from the previous sections, many aspects of merchandising and selling require this type of knowledge.

But a good counterperson always knows his or her limitations in this area. Even the finest counter people are not always qualified to offer service or diagnostic advice on certain vehicle components or service procedures. Diagnosing and troubleshooting modern engines and computer-controlled systems require an organized approach and often special testing equipment. The process can be difficult for even the most experienced technician, let alone a well-meaning counterperson who has not inspected the vehicle, test-driven it, or even spoken to the actual owner in many cases.

In other words, there is always a limit to the assistance you can offer both professional and DIY customers. By offering incorrect or hit-and-miss advice, the customer might perform unneeded repairs and spend additional money until the problem is found and resolved. The obvious result is a very angry ex-customer.

And while professional repair technicians are always eager to learn about new products or possible servicing tips, they will quickly tire of hearing information they already know or have no use for. So choose the information and advice you offer very carefully.

MERCHANDISING SEASONAL ITEMS

For most jobbers, there have always been two seasons—fast and slow. Of course the two seasons might vary from region to region. For example, the fast season in New England might be the winter months and the fast season in the Plains might be the summer months—but nonetheless, the two seasons are very distinct.

What many jobbers have done to help offset this imbalance is to look for other lines that tend to sell in the slow months for that particular region (Figure 9-37). This method rounds out the ups and down in the selling cycle and makes the surges of both receivables and payables less severe. Jobbers might look for lines that do well in the winter for areas that

FIGURE 9-37 Attractive displays of seasonal items will boost sales. Stock seasonal items well in advance.

have traditionally strong summer sales and lines that do well in the summer for areas that have particularly strong winter sales. However, a careful analysis of the conditions in the area is necessary before spending money on acquiring an off-season line.

First of all, it must be determined exactly what months really are slow. To determine when the seasonal slowdown begins, a counterperson can refer to the company sales records. Another good way to spot the slow season is to examine the payables for the company—month by month. There is usually a spot where payables and purchases start to slide perceptibly. If a business is using payable information, the actual slide can be calculated as beginning two months before the payables began to descend. If one is using purchases as a guide, the slide actually started one month before purchases started declining, since the jobber is usually filling in inventory that was sold from one to three weeks after the fact.

After determining when the slowdown started, the jobber should assess what lines will help boost sales during those times. The following are a few seasonal suggestions:

Winter—Traditional
Engine heaters
Winter chemicals
Tire chains
Booster cables
Antifreeze
Battery chargers

Nontraditional
Chain saws
Snowmobiles and accessories
Space heaters
Snowblowers

Summer—Traditional
Polishes and waxes
Oil coolers
Radiators
Electric cooling kits
Appearance items
High-performance products

Nontraditional
Lawn and garden tools (power or manual)

Another alternative to lines that have definite seasonal cycles is to take on lines that require some selling rather than order taking. Some equipment lines are good examples of this type. Many counterpersons will not try to push a line of equipment until things slow down a little bit for them. They have found that when things are slow for them, things are also slow for their customers, and the customers have more time to listen to a sales pitch. They will wait until the slow period starts, then put together a promotion and actively solicit business from their customers.

Another way of using equipment to smooth out sales peaks is by looking at some of the lines of equipment offered by manufacturers today. Many manufacturers have done exactly the same thing to their product offering that the jobber is trying to do. The manufacturer realizes that seasonality in a line is not always good, so a manufacturer of battery chargers might also have a line of electric welding equipment. When one product is not selling, the other one is.

REVIEW QUESTIONS

1. Counterperson A considers the customer base when planning a marketing scheme. Counterperson B considers the product. Who is right?
   a. Counterperson A
   b. Counterperson B
   c. Both A and B
   d. Neither A nor B

2. Which of the following statements about price is not true?
   a. Generally, people know the prices of only about twenty items in any parts store.
   b. Counter personnel can affect price through purchasing.
   c. Price is the most important issue in marketing.
   d. To be competitive does not necessarily mean to be the cheapest.

3. Which of the following characteristics of the exterior or interior of a parts business can affect traffic flow, and, therefore, sales?
   a. parking availability   c. shelving
   b. lighting   d. all of the above

4. Counterperson A considers price memorization to be an important attribute of the job. Counterperson B does not think memorizing prices is important. Who is right?
   a. Counterperson A
   b. Counterperson B
   c. Both A and B
   d. Neither A nor B

5. Which type of delivery service promises a part to be delivered as soon as possible after the request is made?

   a. prorated delivery
   b. scheduled delivery
   c. hot-shot delivery
   d. all of the above

6. Which type of delivery service is the easiest to implement but the most expensive to maintain?
   a. scheduled delivery
   b. hot-shot delivery
   c. prorated delivery
   d. none of the above

7. What type of advertising builds an identity for a particular business?
   a. image advertising
   b. immediate response advertising
   c. television advertising
   d. all of the above

8. With what advertising medium does the advertiser have complete control over the market?
   a. television   c. radio
   b. direct mail   d. pennysavers

9. With what advertising medium is the message sought out by customers when they want to buy?
   a. direct mail   c. specialty items
   b. television   d. telephone directory ads

10. The business of making telephone calls to create or maintain business is called

    _____
    a. reactive telemarketing
    b. message planning
    c. proactive telemarketing
    d. all of the above

11. Which of the following statements is not true regarding machine shop services?
    a. Counter personnel should teach mechanics the methods of promoting the machine shop services.
    b. As more people use the machine shop and are satisfied by the work, more parts will be sold at the counter.
    c. The more dependent a customer becomes on the machine shop service, the less likely he or she is to buy parts from competitors.
    d. Machine shop services should be promoted in a similar manner that products or other services are promoted.

CHAPTER TEN

MAKING PARTS SALES

Objectives

After reading this chapter, you should be able to:
- List and explain the steps to a proper sale, including approaching the customer, qualifying the customer, identifying the customer's motives, and showing the customer product benefits.
- Explain how to handle objections, including understanding the reasons and usefulness of objections, anticipating objections, and knowing when to answer objections.
- Explain the seven techniques for answering objections, including indirect, boomerang, offset, another angle, reverse position, question, and direct denial.
- Explain how to handle real objections, such as price, product quality, and service dissatisfactions.
- Explain why it is important to respect competition and the counterperson's place of employment.
- Explain the importance of and procedure for selling related items.
- Describe telemarketing.
- Describe five common attitudes of difficult customers, including inattentiveness, silence, indecision, skepticism, and hostility.
- Explain how to keep customers by building existing accounts, handling customer complaints, understanding why customers leave, and catering to lost customers.
- Explain defensive selling.
- Describe ways to serve the professional customer, such as special counters and concessions.
- Explain how to handle stocking dealer inventory.

The relationship between counter personnel and the customer must be characterized by mutual loyalty, trust, and dependence. The counterperson must respect the customer's position, policies (if he or she is a professional mechanic), and time. In return, he or she hopes for the customer's confidence and good will. The professional customer must certainly be able to count on the counterperson to keep secrets and promises. Unethical treatment and undependable behavior will destroy this trust. Deceptive presentations, misleading statements, ambiguous phrases, overstatement, exaggeration, or careless handling of the literal truth will cause customers to shop elsewhere.

An ethical counterperson will not sell anything to a customer unless he or she truly thinks that the customer will benefit adequately from the purchase. Here is a good test: If the counterperson can put himself in the customer's shoes, assume the role of the customer, and buy what the counterperson is offering from across the counter, then he or she can recommend the purchase sincerely and in good faith.

Because the typical counterperson is an optimist, he tends to have a bias that must be discounted somewhat by prospects. There are, however, practical as well as ethical reasons for being frank and fair. Hard, aggressive selling is sometimes necessary because some products and some customers require it, but to carry this attitude so far that the result is a disgruntled or disappointed customer is not wise. Hard selling and satisfied customers, however, can coexist.

Hard, aggressive selling is not to be confused with high-pressure selling, a type of selling that violates fair treatment. High-pressure selling takes place when a counterperson offends by letting his or her personality become too aggressive or offensive, or by stooping to tactics that are unethical, deceitful, or dishonest. High-pressure selling is flagrant when a counterperson uses extravagant means to sell something to a customer that the customer neither wants nor needs.

Sometimes fair treatment to the customer competes with fair treatment to the boss, as in the case of a complaint or a request for an adjustment. Clearly, the counterperson is in the middle of a conflict of interests, owing something to the customers and to the boss. No rule is universally applicable. Whatever actions are taken, the counterperson must be honest and fair to both.

Because each customer is an assortment of desires with only limited purchasing power, he or she welcomes any assistance that will enable him or her more nearly to realize his or her desires and to provide for any needs.

The counterperson must sincerely want to serve and help prospects and customers. His or her assistance is more than permissible or desirable—it is mandatory.

In fulfilling the obligation to help prospects and customers, the counterperson must render whatever assistance possible, regardless of whether the prospect buys or not. Prospects and customers have problems, and they need help in coping with them on nonbuying days as well as on days when purchases are made. Whether there is a large order or no order at all, the counterperson should be a friendly counselor with valuable data and advice.

STEPS TO A PROPER SALE

Working behind the counter, the counterperson sees many prospects with many sales opportunities, but taking the prospect from his or her opening inquiry to the cash register involves an understanding of several selling techniques (Figure 10–1).

The job is fairly easy when all the customer wants is a set of points or plugs. But as the items become more costly, customers become more deliberate and the counterperson has to be more sophisticated.

The art of closing big-ticket sales requires an understanding of several steps:

1. Approaching the customer.
2. Qualifying the customer.
3. Identifying the customer's motives.
4. Showing the customer how your product's benefits will satisfy his or her motivations.
5. Dealing with each objection.
6. Laying a foundation for agreement.
7. Watching for the best moment.
8. Asking for the sale.

After mastering these steps, a counterperson will know how to close a sale professionally.

APPROACHING THE CUSTOMER

Approaching a prospect for the first time can cause tension because a person's natural fear of a stranger puts him on guard. Uneasiness in the presence of a stranger might make him suspicious, particularly if he has been the victim of some unethical counterperson in the past.

The goal of the first step of a professional sales presentation is to put the prospect at ease. The manner in which the counterperson approaches the customer will determine the success in achieving that goal.

Either consciously or unconsciously, customers will notice the counterperson's appearance. They are looking for a down-to-earth individual who can relate to their present state of affairs and help them make practical purchase decisions (Figure 10–2). Their assessment of the counterperson's character is being calculated, before words are ever spoken, based on appearance.

The length or style of a counterperson's hair means a lot less than whether it is neat or unkept. It is less important whether a counterperson wears jeans or slacks than if they are clean and sharp looking. On a first encounter, customers know little about a counterperson. Looking for some indication of competence, they assume (rightly or wrongly) that the physical appearance expresses the person's self-image. If a counterperson looks sloppy, cus-

FIGURE 10–1 Taking the prospect from his opening inquiry to the cash register involves an understanding of several selling techniques.

FIGURE 10-2 Customers want a counterperson who looks as though he or she is capable of helping them.

tomers might question his or her apparent lack of self-respect.

Upon approaching a prospect for the first time, a counterperson should be optimistically confident. This attitude is reflected by being relaxed and feeling on a level equal of the prospect. A good counterperson has reasons for feeling confident. He or she knows what the store has done and is doing. He or she knows how the store's products have helped other buyers in the past. He or she knows more about the store's products than the prospect does, and he or she has a sincere belief in the ability of the store, its products, and himself to help the prospect.

The counterperson should assume that his or her presentation will be heard, that the prospect will realize the desirability of the product(s), and that a purchase will be made. These assumptions prevent feeling defeated or overwhelmed.

By inviting and inspiring the prospect's confidence, the counterperson puts the customer in the mood to discuss and explain his or her problems. The prospect must have confidence in the counterperson and his or her propositions before the prospect will accept them. The customer must feel that the counterperson is honest and well informed.

 SALES TALK _____

When a counterperson sees a customer enter the store, he or she should stop any non-customer-related activities and recognize the customer's presence. Taking care of customers is more important than anything else to be done. Greeting the customer is the first step to any sale.

If a customer comes into the store while the counterperson is busy with another customer, he or she should still acknowledge the new customer and say he or she will be with the customer in a few minutes. Without rushing the first customer, the counterperson should not waste any time taking care of the first so that he can attend to the second as quickly as possible.

QUALIFYING THE CUSTOMER

Every prospect is different, and virtually every selling situation presents a unique set of opportunities and obstacles. To weave the most effective sales strategy the counterperson needs information. This gathering of information is known as qualifying the customer. Qualifying the customer simply means getting enough information about the prospect's motivations, needs, resources, and purchasing authority to intelligently recommend products or services.

A counterperson's opening questions should be designed to put the prospect at ease and elicit as much data as possible. The question, "May I help you?" is the worst possible beginning because it invites the prospect to say no. Instead, the counterperson should ask, "How can I help you?"

Another possible opening is "Can I show you anything in particular or would you like to browse around first?"

These questions restrict the prospect's answers to the positive and give him or her a way out without having to say no. His or her response will indicate whether he or she has a definite objective, is just looking, or is seriously in the market.

If the customer just wants to look around, the counterperson should let him or her. People under pressure tend not to buy.

If the customer hovers around one particular item or display, or looks at several models of the same item, the counterperson has an opportunity to probe further (Figure 10-3). He or she can offer to answer questions, provide literature, or give a demonstration.

Once the prospect is involved in a dialogue, the process of qualifying the customer begins. A request for something to get the grease off an engine might lead to the sale of a can of spray degreaser, a scraper, wire brush, solvent gun, pressure washer, or steam cleaner. A good job of qualifying gets the customer together with the product that best meets his or her needs.

To get the information needed, the counterperson has to ask questions. There are two types of

FIGURE 10-3 If a customer hovers around one particular item or display, the counterperson has an opportunity to open a dialogue.

FIGURE 10-4 Items that appeal to a customer's motive of self-preservation include lights, mirrors, flares, and even a CB radio.

questions: the closed question and the open-ended question.

A closed question is designed to limit the prospect's answer to a specific area. It controls the dialog and keeps the sales interview on track. A counterperson uses closed questions to lead the prospect toward the objective: "Do you think you'd rather have a pressure washer, a steam cleaner, or a combination machine?"

The open question leaves the prospect a lot of options for formulating his or her answer. It is used when the counterperson wants a better understanding of the prospect's needs: "What can you tell me about your present cleaning methods?"

In the process of qualifying the customer, it is important to determine if the prospect is the real purchaser. In some cases, especially when the purchase is sizable, a counterperson might be approached by someone who is interested in a product, but who is not authorized to make the purchase. This case often occurs with large fleets, utility accounts, and municipalities. The mechanics charged with maintaining the fleet might want the latest computerized tune-up scope, but to close the sale, the counterperson will have to work on the real buyer. However, the counterperson should remember that although the user does not have purchasing power, he or she can direct the counterperson to the actual buyer, maybe schedule an appointment, or give a favorable introduction.

IDENTIFYING THE CUSTOMER'S MOTIVES

A counterperson should listen for clues to the prospect's buying motivations. People buy products and services for a number of reasons. Understanding those reasons is a key to closing sales. Later in the presentation, a counterperson will want to relate the selling points and benefits of the product to the prospect's motivations.

Advertisers have spent a lot of time and money researching the motivations of consumers. A counterperson should remember the following most common purchase motivators.

Self-Preservation

This instinct is every buyer's first concern. Examples of purchases that meet this need are food, shelter, and medical care. A counterperson has products that satisfy this motivator, too.

An appeal to self-preservation might sell a prospect "up" from a cheap brake lining to a premium set. Opportunities for related sales include seat selts, brake, steering, and suspension parts, lights, mirrors, flares, even a CB radio (Figure 10-4).

Striving for Excellence

Being the best and having the best of everything is a strong motivator for some people. The professional mechanic who is determined to be the best needs the best tools and parts available to meet his or her needs.

Being Liked

Most people want to be liked and accepted by others. In the case of a garage operator, the purchase of a labor-saving piece of equipment might not only make his business more profitable, it might also result in employees perceiving him or her as a more likable boss.

Self-Improvement

People want to better themselves, to get ahead. To sell a good service manual or attendance at a clinic, a counterperson can bring this motivator into his or her presentation.

Being in Style

The garment business flourishes because of the style motivator. However, many auto buffs are also fashion conscious. In any of the automotive subcultures there is a fashionable wheel and an "in" set of speakers.

Making Money

The desire for profit is the ostensible motivator for any prospects who are also the owners of businesses. When selling to these people, a counterperson should emphasize the product's features that will cut expenses or increase profits.

Fun

Not all sales have to be justified in purely utilitarian terms. People want to have a good time. Everybody wants a treat once in a while, and everyone wants to feel loved and pampered. Products that can improve a customer's enjoyment of his or her car can be sold because they are fun to own and use (Figure 10–5).

Getting a Good Deal

Aside from impulse (fun) and self-preservation purchases, virtually all purchases involve a sense of value for the customer. Pointing out savings and making sure the customer is aware of any discounts are great ways to utilize this motivator (Figure 10–6). If the product delivers good value for customers, the counterperson will build a strong base of repeat business.

SHOWING THE CUSTOMER PRODUCT BENEFITS

After a counterperson has qualified a prospect and identified his or her motivations, he or she must build the case. Building a case for the purchase of a product consists of three parts:

1. Relating benefits of the product to the customer's motivations

FIGURE 10-5 Some customers make purchases just for fun or to be in style.

FIGURE 10-6 The opportunity to save money can convince some customers to buy an item now.

2. Meeting his or her objections
3. Laying a foundation of agreement

Relating selling points to motivations means showing the prospect how the product will satisfy his or her needs. If the prospect has expressed an interest in a product to help do a job with less time and effort, the counterperson should show him or her which features of the product deliver that benefit.

Successful counterpersons remember this key point: customers do not buy selling points or features; they buy the benefits of those features. For example, few people care if the car stereo uses field-effect transistors in its tuner section. The transistors are merely a feature. What customers are interested in is the improved reception, a benefit derived from the feature. Counterpersons who forget this distinction often get so involved in features (they usually sound quite hi-tech and impressive) that they actually confuse customers.

The proper use of features or selling points is a support for the benefits. Counterpersons should understand the features, but stress the benefits.

DEALING WITH OBJECTIONS

A counterperson will almost always encounter objections from the prospect. A customer who has no objections is usually a customer who has no buying interest.

Many counterpeople are intimidated by customer objections. The first thing to realize is that an objection is not necessarily a block to the sale; in fact, it can actually be an asset to the presentation.

REASONS FOR OBJECTIONS

Much resistance is the result of the customer's not knowing enough about his or her needs and about products that might answer those needs (Figure 10-7). Often, an objection is the prospect's method of directing the counterperson toward areas about which he or she would like more information. He or she needs more understanding of his or her requirements and of the benefits the counterperson promises before he or she can appreciate the product for what it really is and does. Sometimes the information a customer has is inaccurate or vague.

When a customer does not understand a sales explanation, that customer will probably offer opposition to buying. Such difficulty is the counterperson's fault because he or she must hold the customer's attention and tell such a clear story that the customer easily understands it.

Customers are reluctant to change habit patterns. They adopt a way of life and stick to it, repeating the same actions time after time. For some customers, the voicing of objections has itself become a habit. These customers give all counterpersons at least some token opposition whether or not they are going to buy.

Some opposition is presented in order to check out the counterperson's knowledge of a product and ability to answer objections. Sometimes the customer is really asking for justification of the purchase. For example, a dealer customer will want satisfactory answers to hurl at those who criticize the purchase or who demand an accounting for it. Furthermore, he or she wants assurance that what he or she is doing is right. The counterperson must dispel his or her doubts with facts and suggestions.

Personal preference and prejudice can cause customer resistance. Differences of opinion exist in customers' minds about the quality and suitability of various products.

Finally, buyers are often afraid to buy. They fear the product might not perform as the counterperson claims or that the counterperson might benefit more from the transaction than they will. They fear that soon after the purchase they will see a new product and wish they had waited to buy.

USEFULNESS OF OBJECTIONS

Objections can be turned into both assets and opportunities by the skillful counterperson.

First of all, selling is made easier when the customer objects, because the customer who talks is easier to deal with than the silent customer. When a customer refuses to take part in the conversation, a counterperson is unsure about the impression he or she is making, about points that are obscure and need amplification, about facts to stress, and about

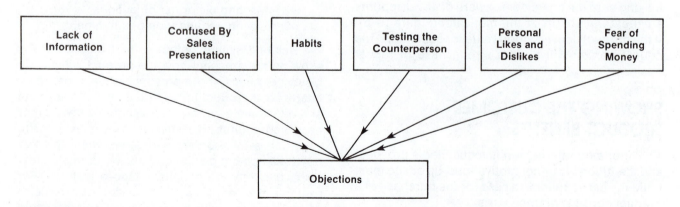

FIGURE 10-7 Several factors can lead to customer's objections, all of which must be dispelled by the counterperson.

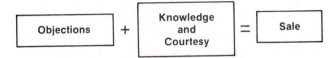

FIGURE 10-8 With product knowledge and courtesy, a counterperson can turn most objections into a sale.

buying motives to excite. The dangerous objections are those not disclosed. Second, objections indicate that the counterperson is making some progress. A sound generalization is that the objecting customer is the interested prospect, that he or she is beginning to experience desire and is giving thought to the sales explanation. A third reason that objections should be welcome is that their successful handling increases a new counterperson's confidence. Fourth, each objection gives the counterperson one more peg on which to hang reasons for buying. Finally, objections throw light on the customer's thinking. By underlining the buying decisions that are missing, they indicate what is needed before the counterperson can make the sale.

A counterperson need not fear objections. Most sales are made after the customer has voiced an objection (Figure 10-8).

Not only are questions and objections important signs of how to proceed, they are also evidence of the customer's attention and interest. If a customer asks questions, his or her mind must be focused upon the proposition. The questions indicate thought about what is being said; the customer is mentally agreeing or disagreeing with some point of the sales presentation.

A good counterperson encourages the customer to say what is on his or her mind. Furthermore, it is the customer's privilege to ask questions. Sometimes the most reluctant customer becomes a strong ally once objections have been answered.

FAVORABLE TIME TO ANSWER OBJECTIONS

Pinpointing the most favorable time to answer objections is an important factor in overcoming questions and objections. Experience indicates that four different times for answering objections are suitable depending on the situation. A counterperson might choose to

1. Anticipate the objection before it is raised
2. Answer the objection immediately when raised
3. Delay answering the objection
4. Ignore the objection if possible

Anticipating Objections

Perhaps the finest time to answer objections is before they are raised at all. This strategy keeps the prospect from thinking negatively or unfavorably in the first place. Also, the voicing of an objection at an awkward or regrettable moment is avoided. Thus, the counterperson robs the customer of any defensive objection he or she might be planning to conceal until the conclusion of the presentation, something the customer could use as a justification for ending the presentation without buying.

Also, prevention makes more time available for the counterperson to present a full, complete story. A person appears completely fair in cases where he or she raises and answers an objection or just weaves an answer into the presentation.

And most important of all, the prospect is prevented from taking a position. Once an individual states what he or she thinks, the person feels pressure to stick to the commitment and defend it.

A counterperson should note and list the most frequent and troublesome objections he or she encounters (Figure 10-9). Then, the counterperson should include in the presentation the most effective answers to these most common questions, complaints, and criticisms in the form of positive selling points.

The clearer, more logical, and more complete the story is, the less important a prospect's objections seem to him or her. If a prospect's attention is monopolized by numerous plus points and very few minus points, he or she will be inclined to consider the objections once planned to make far too insignificant to mention.

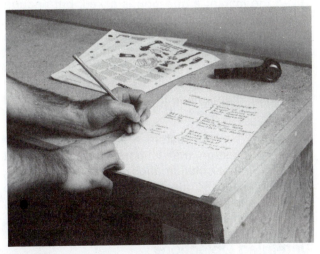

FIGURE 10-9 Listing the most frequent complaints about a product can assist a counterperson in dispelling these objections before they are made.

If a counterperson is not meeting many objections from customer after customer, then he or she can assume the presentation is reasonable, clear, and complete. But if more objections than usual are encountered, then more attention should be given to the early anticipation of them.

Deliberately Raising Objections

An exception to this recommendation is in those cases where the counterperson has an especially effective answer to a particular objection, an answer so powerful that he or she purposely waits for the prospect to raise the objection rather than answer it by anticipation. The answer he or she gives can move the sale to a point where the close itself is at hand. If the customer fails to raise the question, the counterperson should bring it up so that no strong sales point is overlooked.

Answering Objections Immediately

Customers will raise additional unanticipated objections, which the counterperson should address immediately. Not to do so, except in situations to be discussed later, is to have the customer question the counterperson's ability to answer his or her objections.

If the point is relatively insignificant, the counterperson can divert attention from it by pointing out a new feature, by asking a question that requires thinking along a different line, or by assuring the customer that he or she will answer the question a bit later. However, if a question is raised a second time, the counterperson has no alternative but to answer at once.

A thorough explanation does not have to be given the instant a question is asked, but enough should be said to satisfy the customer. Later, the answer can be expanded to make the most favorable impression. Never permit a situation to develop in which the customer gets the notion that questions are being disregarded.

Two exceptions to the immediate answering of an objection are

1. When the customer raises an objection or question that is unrelated to the point under consideration
2. When the question of price is raised before the counterperson has had a chance to build an adequate value story

Unrelated Objections

Where the customer interjects a question or objection not related to the point under consideration,

the counterperson can delay an answer until the orderly sequence of the presentation requires it. The counterperson can simply say, "That question is an interesting one. It is my intention to discuss it carefully with you in connection with another important feature. At this point, if it is entirely satisfactory to you, I would like to make sure you are thoroughly familiar with the features designed for your protection that we have just discussed." Then, without any pause, the counterperson should continue the sales talk.

Questioning Price

Many times the question of price is raised before the value story has been developed. In some instances it will be an excuse or a stall; in others, it will be an honest objection, as discussed later in this chapter. In either case, the counterperson should make every effort to delay the answer to this question until he or she has had the opportunity to convey some of the essential information that will contribute to the customer's appreciation of the product or service.

To an early objection to price, the counterperson might say, "There is certainly no question that price is an important consideration in every purchase and sale. It would be hard to find an individual who would question that fact. However, I'm also certain that you will not hesitate to agree with me when I say that unless this machine I'm preparing to demonstrate to you could provide the savings in time and the economy of operation that you want—well, you wouldn't be interested at any price. Before we even think about price, let's take a look at what the machine can do for your business" (Figure 10–10). With that statement the counterperson

FIGURE 10–10 Demonstrating a product or tool can often show its usefulness to a customer who then decides to buy it.

should immediately proceed to the demonstration and the building of the value story for the product. When these steps have been accomplished, then price can be mentioned.

Ignoring Objections

On some occasions certain objections can be ignored. Sometimes they are so casual that no real benefit can be gained by answering them. In such circumstance, the counterperson should continue with the presentations. If the point is serious, the customer is likely to repeat the objection. Then the counterperson is compelled to answer.

On the other hand, if the objection raised is a sincere one, it should under no circumstances be ignored. This technique of ignoring objections is recommended only in those instances where the thinnest type of excuse is offered by the customer. The other three suggestions for timing answers to objections are the ones that a counterperson will use most frequently.

TECHNIQUES FOR ANSWERING OBJECTIONS

The techniques that can be used to answer objections include the following:

- Indirect
- Boomerang
- Offset
- Another angle
- Reverse position
- Question
- Direct denial

There is a time for each one of these to be used effectively. Whatever the method of answering, the counterperson should be sure to listen carefully and courteously to the prospect. Although the counterperson might have heard the same objection from others, he or she should make the prospect feel by an answer that is considered and not hurried that his or her opinion is respected.

Above all, once a counterperson has answered the objection, he or she should move on to the next point.

Indirect ("Yes-But") Technique

This method dulls the sharpness of an objection or postpones it. The indirect method is most frequently used in combination with one of the other recommended techniques. It is a means to avoid flatly contradicting the prospect. For example, a counterperson using this method to answer an ob-

jection might say, "Yes, I can see why you feel that way. I felt the same way when I thought about it at first, but your friend, Bob Smith, who has used this machine in his business for many years, showed me how it saves him money."

Boomerang Technique

This method is sometimes called the *turnabout-response* or *translation* method. The counterperson who uses this method throws the prospect's objection back at him or her as the reason for buying. He or she in effect says to the prospect, "That's really the most powerful reason for you to choose this product!"

Some counterpeople plan some of their product pitches to make occasion for an objection that can be "rifled" back and followed with a close effort. For example, if the prospect says to a counterperson who has called to explain the diagnostic service plan of his garage, "...but I can't afford to buy a diagnostic tester," then the counterperson replies, "That's the best thing you could have said, because I know that if you couldn't afford to invest so nominal a sum in a piece of diagnostic test equipment you could hardly afford to lose the business you would gain with this piece of equipment. That loss would be many more times the cost of the equipment." Or when a customer objects, saying, "I'm too busy to come down to see you," the counterperson answers, "I want to see you because you are busy. If you weren't, this new product wouldn't interest you."

"Offset" Technique

The offset or superior point/fact technique admits the validity of the prospect's objection and then stresses a superior point that more than compensates for it. The counterperson capitalizes on the fact that no single product can be all things to all people.

Compensation is commonly used as a method for answering objections to price. When the prospect objects to what appears to be an excessive price, the counterperson points out the superior features incorproated in the product. He or she stresses the added conveniences that increase its value.

Sometimes, a counterperson might choose to write on a piece of paper the plus values he or she is pointing out and also a value that the prospect places on each of these. The dollar values are totaled and then the counterperson asks the prospect to add this figure to the price of ordinary products. The comparison between the new sum and the price quoted by the counterperson is certain to show the product in a favorable light.

Conversely, the prospect who objects because a product lacks certain features might accept the product when it is pointed out that the low price still makes it a good buy. The average prospect can quickly comprehend the fact that additional features cost more money. Consequently, if the basic product with a minimum of gadgets will serve his or her purposes, the lower-priced offering will be advantageous. Furthermore, the investment will be lower.

Finally, stress placed on a superior feature frequently will distract the prospect from earlier objections. This is especially true if the unusual feature has particular appeal. For that matter, many customers continue to buy from certain stores because the products they sell have one or more particularly unusual feature they desire.

Another Angle

Sometimes during a sales presentation, the counterperson evokes an objection that can be countered by suggesting another interpretation. He or she asks the prospect to re-examine the facts from another point of view that will show off the proposition in an entirely different light.

For example, imagine a counterperson trying to sell a conservative line of wheels to a service station owner who has a heavy inventory of flashy merchandise designed primarily to catch the eye. The dealer objects to putting in the offered line of conservative but quality wheels because he or she claims heavy sales in eye-catching merchandise indicate customers want that kind and not the other.

The counterperson replies, "Customers buy only what the dealer offers. If they do not see other lines, they will tend to buy from the range of stock carried. Other prospects will not even come into your garage if they are interested in well-designed, conservative merchandise when the display windows indicate that the garage doesn't carry it. Your concentration of sales in eye-catching merchandise might be of your own making and perhaps does not reflect the full range of taste of possible prospects. Why not put on a representative conservative line and see what happens?" The counterperson can bolster the suggestion if he or she can point out how this idea was tried successfully in other situations.

Reverse Position

An interesting technique for handling objections is a turnabout method in which the counterperson reverses positions with the customer.

When the objection is raised, usually in a general manner, the counterperson can say that he or she does not quite understand and will the prospect please help by pointing out or explaining what the difficulty is. Positions are actually reversed. The burden is now upon the prospect to show what is wrong while the counterperson listens, hoping that, as the prospect goes over the proposition, discussing its various features, he will help to sell himself.

The purpose of this method is to induce the prospect to talk. Often the prospect likes to build up his or her ego and will raise questions and objections that have little merit. The counterperson might profess interest, saying, "I never heard about that, Joe. Why do you think that is so?"

If the objection is weak and the prospect has only a poor idea of what he or she is talking about, the person will generalize and probably end up by saying, "The point wasn't important anyhow."

"Why" (Question) Technique

A supplementary technique for handling objections that many counterpersons use is to reverse positions by asking "Why?" Then the prospect must find a reason for his objection. Again, the counterperson asks, "Why?" and continues to ask "Why?" Thus the prospect does most of the talking.

The advantages of this method are many. First, it tends to clarify the objection. Sometimes, the prospect does not mean exactly what he or she has said or has not been quite clear. When the counterperson asks "Why?" the customer must rephrase the objection and perhaps add details.

Another advantage claimed for this technique is that it gives the counterperson a chance to collect his or her thoughts. As the prospect repeats the point, the counterperson has a chance to decide how to answer most effectively.

The most important reason for using this method is that the prospect might answer the objection. This technique is a modification of the "reverse position" technique described earlier. It places the prospect in the position of giving reasons for the thought. As long as the counterperson asks "Why" in an honest manner to clarify the situation and to start the prospect talking, the method is effective.

Direct Denial

By far the most dangerous technique to use is a direct denial. The counterperson denies the objection without equivocation and refuses to consider it. A denial of the statement made by the prospect is almost sure to put the person in a fighting mood. However, sometimes a denial is in order, such as when the reputation of the store or the counterperson is unjustifiably attacked. Failure to deny might only make the situation worse.

For example, the situation might occur in which a customer claims that lower prices have been quoted to a competitor. The person might have heard that this store has been engaging in some improper activity, although there is no proof. The customer might accuse the counterperson of some improper action. When these questions arise, quibbling is not the answer. The only answer is direct denial, accompanied, if possible, with proof of the truth.

In most cases where a counterperson wishes to disagree completely, he or she should tackle the problem obliquely, saying, "I'm sorry that story has reached you. I have no doubt you are sincere, and the person who told you probably was sincere or you wouldn't believe it. But it is completely false. There are no facts to justify it." This method avoids putting the prospect on the defensive, giving the counterperson an opportunity to explain further without having first irritated the customer by calling him a liar.

Another approach is to say, "I'm surprised that you believe such a story. You're a good businessperson and know it would be impossible to hide such a deal from our regular customers. Let us give you the full story."

At such times when the counterperson senses that a direct denial would only offend or embarrass, he or she could simply answer "Oh" to a customer's accusation. Generally, the customer will start talking again and the counterperson can use the "why" technique instead.

THE REAL OBJECTION

There are actually two types of objections: the excuse or "stall" objection used to help the prospect avoid either the hearing of the proposition or the need for a purchasing decision, and the sincere objection that is evidence of real interest in the proposition.

The distinction between these two main groups of objections is not always clear. For example, the statement, "I'm too busy now," might be made by two different prospects. For Prospect A it might represent an excuse to avoid hearing the proposition; for Prospect B the statement might be an honest objection. Prospect B might actually have other obligations that prevent him or her from talking. The counterperson must use his or her judgment to distinguish between the two.

Generally speaking, the techniques already mentioned can be used to meet both types of objections. However, the manner in which sincere and real objections can be met can be enlarged upon.

Price Level

No one would question that price at some time and in some manner is a part of every business transaction, but it is nearly all there is to selling.

The counterperson should be convinced in his or her mind that the price is right and should convey that conviction to the prospect with a tone of assurance. Only then can the counterperson expect that the prospect's attitude will also be a positive one. There are, however, some methods of approach to the price objection that can be most helpful, including suggestions on when to quote price; what to do if the buyer says the price is too high; dealing with the "cut your price" buyer; and meeting the comparison-price problem.

When to Quote Price

The most advantageous time to quote price is after the counterperson has had a full opportunity to build up the value story for the products or services (Figure 10–11). By the time he or she has secured the prospect's interest, stimulated his or her desire, and secured his or her confidence, value has been made to outweigh price.

There is, on the other hand, one time when it might be advantageous to quote the price in the sale, right after attention and interest have been secured. This procedure is recommended when the counterperson knows the price of the product, or service is considerably higher than what the prospect had in mind. In such a case, the counterperson can say, "Yes, the price of this battery charger is $150. It might seem a substantial sum to pay, but I can show you conclusively, and without any question, that it is worth what is asked, and more."

Then, without any hesitation, the counterperson should develop the presentation. Freed of the price fear in the prospect's mind, the counterperson can build a value story that makes the cost recede and accents the desire or want for the product.

Price Objection

Many counterpersons believe that their prices are too high. Such a negative attitude is often conveyed to the prospect. Then if the prospect ques-

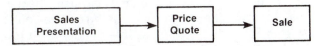

FIGURE 10–11 In most cases, the best time to mention price is after a clear sales presentation is completed.

tions the price, in most cases the fault lies in the counterperson's sales procedure, not with the price.

To the prospect who says, "Your price is too high," the counterperson should demonstrate in a positive manner the many benefits and features of the product. He or she should emphasize to the customer that one gets what one pays for, and that in the long run far more value for the money will be received by buying the higher priced item.

Another method that is helpful when this objection is offered is to break the price down into small units. To tell a garage prospect that the cost of a top quality battery charger is $150 sounds like a tidy sum. But a prospect might reconsider if the counterperson says, "This quality battery charger will serve you at least five maintenance-free years. At a yearly cost of $30, less than six cents per week, you'll be able to charge an unlimited amount of batteries. Even if you charge only one battery a week at $5 a charge, you would make $260 gross. Subtract the yearly cost of $30 and you have netted $230. So, all things considered, your initial investment is extremely minimal."

Price Reductions

Every counterperson will face at one time or another the prospect who haggles over a quoted price in an effort to have it lowered. If the counterperson has quoted the price with confidence based on a value story and has not indicated any possibility of a lower one, much of the force of the customer's attempt to bargain has been removed.

An appeal to the customer's sense of fair play can also have an effect as the counterperson tells him or her, "We have one price on this item for every one of our customers. I'm certain that you would not want to feel when you left here that someone could get a better price than you. You will never have any occasion to think this." And then you should reiterate the features of his or her offering.

Comparative Price Objections

Prospects will sometimes object that "your competitor sells the same product for less." If the prospect's statement is not true although he or she might honestly believe it is, the counterperson must uncover the facts that are responsible for this attitude. The counterperson should ask comparative questions, such as, "Did you see the same size wheel in the other grinder and was it mounted on five points to eliminate any vibration as this grinder is? It frequently happens that two grinders look the same but their performance is quite different." This procedure is recommended rather than issuing a direct denial that can at best only antagonize the customer.

If the comparative price statement is true, the counterperson can also point out certain additional advantages such as terms, service, delivery, or reputation that justify the decision to buy from this store since these factors should far outweigh any slight price difference that might exist.

Product Quality

To avoid the fear of price competition and product comparison, always sell quality. Quality is represented by the total sum of the characteristics that make a store's products or services good values.

If the item is lower priced, the counterperson should show that the quality is right in terms with the price paid. He or she can say, "Yes, there is no single product that can be all things to all people, but this one with its low price offers more value benefits than any other on the market. Let me show you."

On the other hand, if the product or service is higher priced, the counterperson should demonstrate superior quality that more than compensates for the extra dollars spent, in both the short and the long run. In this case, he or she might say, "There's an old saying and I'm sure that from your experience you know it to be true, 'You get what you pay for.' In the long run and even in the short run, you're getting much more, dollar for dollar, by selecting this better quality—it's superior value."

Service Dissatisfactions

Service dissatisfactions include criticisms of credit policy, deliveries, adjustments, maintenance, or any other service rendered by a store. Some of these objections will be volunteered by the prospect, but others will require a counterperson to use skillful questioning to get a clear picture of what is disturbing the customer.

If the prospect claims that credit terms are not long enough, a counterperson can prove by actual cases and testimonials that the goods sold will be installed on customers' cars so readily that long-term credit is not needed, or demonstrate that the goods could not be sold at the lower price if credit terms were longer, or outline some means whereby local banking facilities can be used.

When faced with an adjustment problem, a counterperson should let the customer unburden him- or herself and then strive to settle the matter in an equitable manner that is fair to both store and customer.

Reciprocal Relationships

Every counterperson will at some time be faced with reciprocity competition (Figure 10–12). This situation is typified by the buyer who says, "I buy from them because they refer do-it-yourself customers to us." To answer this objection, the counterperson must convince the prospect that instead of harming other reciprocal relationships the garage customer might have, this store will become another source for some lucrative referrals. And if this store proves to be a better source for referrals, the garage customer can decide which store deserves more business.

RESPECTING THE COMPETITION

A counterperson should respect competitors and their products. In many lines, one brand of product is just about the same as other brands, and many prospects know this. Furthermore, competition involves stores, counter personnel, and products that have good points and loyal supporters. A prospect might have a deep admiration for a competing store or competing counterperson, or might think so highly of a competing product of a competing store that he or she buys it from that store. In view of this, it is a mistake for a counterperson to show lack of respect for competition in front of the prospect. When the prospect or customer brings up the question of competition, the counterperson should say, "There is no question that store carries fine products. To be truthful, there isn't a store in this area that doesn't. However, I firmly believe that we carry the best products. Line by line, feature by feature, I can prove this to you."

A proper attitude toward competitors also includes fairness. A counterperson lowers him- or herself in the prospect's estimation if practicing unfair treatment or unfair criticism of competition. Resentment might be particularly strong when no defenders of competing stores are present or when a direct comparison requested by the prospect is not made fairly. Fairness brings the good feeling that results from operating ethically and also creates more enthusiastic customers.

Confidence is also an aspect of a proper attitude toward competition. It is the comforting knowledge that a counterperson has nothing to fear from any fair product or service comparison.

The successful counterperson collects and uses information about competition in a purposeful manner.

What to Avoid

Selling against strong competition takes tact, judgment, and control. Every product, counterperson, and store has its good points. Each good counterperson will stress those points. The prospect buys from the counterperson who convinces him or her that the store's products will provide the greatest amount of satisfaction. If the following errors are avoided, the handling of competition will be less risky and costly:

1. Do not include any reference to competition in constructing the sales talk.
2. Never initiate the subject of competition. If it is to be mentioned, let the prospect make the reference. Furthermore, a counterperson should not compare his or her store's products with another unless the prospect demands it.
3. Do not stray or be maneuvered away from the primary task, which is to explain what the store's products will do for the prospect. Do not be dragged into a discussion of competitive topics.
4. Have no ambitions to win mudslinging contests. Most prospects detect and resent any disparaging remarks made by competing counterpersons.
5. Never make a statement about competition before checking its accuracy.

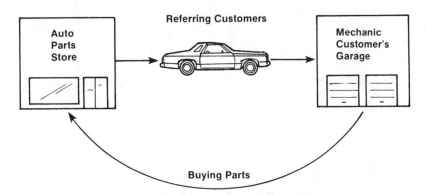

FIGURE 10-12 A reciprocal relationship involves two related businesses, such as a parts store and a garage, who give business to each other.

6. Do not expose flaws in competition in the hope of making sales through this tactic.
7. Do not welcome gossip, and never repeat it.
8. Never criticize competition.

RESPECTING THE PLACE OF EMPLOYMENT

Obviously, if it is unwise to criticize the competition, a counterperson should not criticize his or her own place of employment. Resist the temptation, in an effort to smooth away an objection, to agree with the customer. It is certainly important that some common denominator of agreement be found, but every remark a counterperson makes against his or her store can only result in added loss of respect for the store, the counterperson, and the products and services.

THE FOUNDATION OF AGREEMENT

After handling objections, the counterperson can move on to the next important step for closing a sale. Before he or she can expect a successful close, a positive attitude must be secured from the prospect. The counterperson must lay a solid foundation of agreement.

The closed question is useful at this point: "You feel that it's important to reduce operating expenses?" "Do you think a product like this could help you save time?" The point to remember is that anything the counterperson tells the prospect is open to question, but anything the prospect says is binding truth. The more "yes" answers the counterperson can get from the prospect, the more difficult it will be for him or her to say "no" to the close.

TIMING

Timing is another critical element in closing a sale. In most selling encounters, there is just one best moment to close the deal. Prior to that instant, the customer is not ready to make a purchase decision, and after that moment, interest and enthusiasm might be lost.

The ideal moment can occur anytime during the interview. The counterperson should keep in mind that the prospect might be ready long before he or she finishes telling why he or she should be ready. Although the following indicators are not guaranteed, they might signal that a prospect is ready to close the sale (Figure 10–13):

1. The prospect begins agreeing with the counterperson.
2. The counterpersn handles the "last objection" and senses relief or resignation in the prospect.
3. The prospect's body language shifts distinctly from defensive to receptive.
4. The prospect brings in the real buyer— wife, husband, or boss.
5. Questions about price are repeated or give way to inquiries about terms.

CLOSING THE SALE

Basically, closing simply means getting a commitment from the prospect. Successful salespersons usually share the following characteristics.

First of all, effective closers expect the sale. The very best closers seem to be certain of bringing each interview to a successful close. Their confidence might be justified, or it might be the product of an inflated ego, but whatever the source, that expectation of success often results in sales.

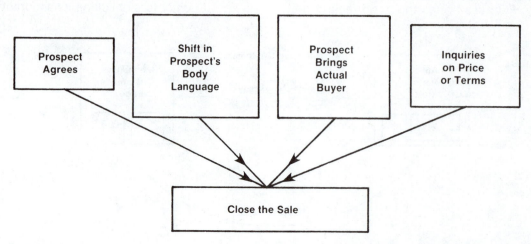

FIGURE 10–13 Any of several indicators can tell the counterperson that it is time to close the sale.

Second, good closers always let the prospect know that they expect the sale. In its mildest form, this tactic merely plays on the reluctance of most folks to deliberately disappoint others.

Finally, and most important, a good closer always asks for the sale. Many counterpeople who do a fantastic job of qualifying prospects, matching benefits to needs, laying a foundation of agreement, and handling objections, then stop and wait for the customer to start shelling out his or her money. And nine times out of ten, the customer does not. Why? Because purchasing is not totally rational. Few buyers ever feel they have enough facts to make a completely informed choice. But most people also recognize that they cannot afford the time to acquire all the facts, either. So, in each purchase, an element of doubt exists.

The counterperson's job, then, is to reduce that final reservation to the point where it can be overcome with a subtle nudge. Asking for the sale provides that nudge. It gives the prospect the encouragement to ignore the last mild uncertainty.

A counterperson should ask for the sale up to three times before accepting a no sale.

The best way to ask for the order depends on each selling situation. Here are a few examples:

- "Can we deliver this for you?"
- "Would you like me to carry this out to your car?"
- "We can have it in your shop Friday. Will that be okay?"
- "How about if we put this on next month's billing?"

The counterperson should make it as easy as possible for the prospect to give a commitment.

AVOIDING OVERSELL

Overselling is selling the customer more than is needed to do the job. Whenever a counterperson is more interested in short-term profits than the long-term best interests of the customer, he or she is overselling. In the automotive industry, where abuses of a few have created a highly suspicious consumer, overselling should be avoided at all costs.

Word-of-mouth is the most powerful advertising channel, and overselling at one store is good advertising for the competition.

Another form of overselling means continuing with the sales pitch long after the customer is ready to buy.

Selling too much can irritate customers and damage their financial health, and too much selling pitch can ruin otherwise good deals, but in terms of its lasting negative effects on the jobber, selling what a counterperson cannot deliver might be the worst form of overselling.

Overselling is bad for customers and the store's reputation, but it chews up profits in yet another way. Most valid selling occurs early in the sales interview, and squeezing the last nickel out of a prospect is not only dangerous in terms of damaged relations, it is also an outright waste of time.

SELLING RELATED ITEMS

Selling related parts when a customer buys a particular item can boost profits by 30 percent or more. But selling related parts also makes sense when the counterperson remembers that his or her job is to solve customers' problems, not just to sell parts. A customer's problem might not be solved by replacing a faulty part if the related hardware or chemicals is also not up to peak performance.

For example, if a mechanic customer purchases a water pump for a Chevy and the counterperson sells only the pump, not only did the counterperson miss a great sales opportunity but the customer's problem might not be totally solved. When the mechanic begins to install the pump, he or she might realize that an aerosol gasket remover would make the job easier, the fan clutch bearings are loose, and the belts need replacing. After returning from the store for these items, the mechanic realizes that the hose clamps are useless. So he or she hurries back to the store.

If the counterperson had suggested these related items in the mechanic's first visit, both the mechanic and the counterperson could have saved a lot of time (Figure 10–14). In addition, the mechanic

FIGURE 10-14 Selling related items is not only a great sales opportunity but also an effective way to ensure that a customer's problem is solved.

customer would have been very appreciative of the counterperson's initiative and, with this type of service, would more than likely return for parts again and again.

If the mechanic customer calls in the order on the telephone, the process is simplified. If the counterperson suggests a related item that the mechanic did not check yet, the counterperson can either hold or wait until the mechanic calls back to add it to the order.

Selling related parts is easier nowadays than in the past because of catalog formats and electronic cataloging. Many paper catalogs include information about related items on the page with a specific automotive part. Electronic cataloging with computers usually lists related items on the screen with the part. A counterperson's own technical knowledge will also assist him or her in selling related items. Understanding basic automotive components and how they function and are attached will greatly reduce the chance of missing the opportunity to sell related parts.

Most customers are willing to purchase related items because they are already sold on buying the main component. Most realize—perhaps with an additional explanation—that buying the main component will not prove profitable if they must run back and forth to the store for hardware or chemicals or if they install the new component without replacing adjoining items. Selling related items is usually one of the easiest aspects of a counterperson's job, but it is also one of the most important.

SELLING BY TELEPHONE

Many large companies, aware of the high cost of making personal sales calls in large territories, try to do as much selling as possible by telephone, which, as mentioned, is known as *telemarketing*. Even a counterperson working in a small jobbing store can use the telephone to increase personal selling potential (Figure 10–15).

Telemarketing can be divided into two categories: incoming calls and outgoing calls.

INCOMING CALLS

Unless the caller has dialed a wrong number, every time the telephone rings behind the counter the counterperson is already halfway to making a sale. The caller is already interested or he or she would not have called. A walk-in customer might just be interested in browsing, but a caller on the phone is already interested in a part, price, service, or other specific information.

The key difference between handling a customer on the telephone and handling a customer at the counter is speed. The counterperson should try to research the caller's request as quickly as possible. Because the caller is not looking at anything or talking to anyone else while on the phone, a two-minute wait on the phone can seem much longer to the customer than a two-minute wait at the counter. A pad and pen should always be available beside the telephone. The counterperson should write down all the necessary information and requests from the customer before searching for numbers to avoid backtracking and to save time.

If the counterperson knows he or she will require several minutes to handle the request, either ask the caller to call back or call the customer back personally. A "hold" button should be used only briefly for quick tasks such as answering another line or telling a counter customer he or she will be waited on in a minute.

OUTGOING CALLS

The counterperson can use the telephone to maximize selling potential. For example, when a new season approaches, the counterperson can utilize three minutes here and there to call regular customers and notify them of sales or discounts on seasonal items. Or, when a professional customer calls to make an order, the counterperson can utilize the opportunity to relay other information, which might lead to a larger order or another order in the future.

The counterperson can also use the telephone to personally notify customers of recent pertinent information such as manufacturers' defects and rebates or an upcoming sale. Customers will appre-

FIGURE 10–15 Using the telephone wisely can increase a counterperson's sales potential.

ciate the personal touch and the opportunity to save on items they had intended to purchase anyway.

When a counterperson decides to call regular customers, he or she should always have a definite motive, such as relaying information. Never call just to keep in touch, because to the customer, time is money. There should be a definite reason for each phone call, and the counterperson should start moving toward that goal from the first moment.

HANDLING DIFFICULT CUSTOMERS

Inevitably, prospects will be encountered who, for some reason, are difficult to handle. However, the following traits of difficult customers can be overcome and turned into a sale if they are handled properly.

INATTENTIVENESS

When the prospect's attention wanders away from a sales story, the counterperson should try at once to determine the cause (Figure 10–16). Perhaps the presentation is so vague that the prospect cannot or will not spend the effort required to follow it. Perhaps the method of presentation is failing to interest the prospect, or the prospect's mind might be on other matters.

One possible course of action is to mention the prospect's various wants, if the counterperson knows them, in hopes of finding a clue to the prospect's most powerful buying motives. A second and somewhat similar possibility is to summarize what the product delivers, all the while looking for clues.

FIGURE 10–16 A counterperson should try to involve an inattentive customer in the sales presentation.

Third, the counterperson might need to startle the prospect and recapture attention by enthusiastically presenting a new slant, a new case history, or a new advantage. Fourth, proof that is more meaningful than any given before might be offered.

Fifth, the counterperson might pause in the presentation until the prospect's attention returns. Sixth, in extreme cases, the counterperson might frankly ask for more attention.

SILENCE

Being met by silence is not as serious as it might seem because the prospect who listens can be sold, as long as the counterperson does not react adversely to the silence. An interesting aspect of a customer's silence is that it often causes a counterperson to speed up. Having covered one point and receiving no reaction, he or she hurries on to the next, and instead of developing at length the features that might enliven the presentation, it has hit only the high spots.

A good counterperson will ask for questions and bring the customer into the presentation. On occasion, he or she might even try to remain silent for so long that the customer, embarrassed by the silence, will break it. Then, after the customer has regained his or her voice, the counterperson will go on listening for objections that will lead to making a sale.

The proper procedure is to feed advantages and proof slowly to the prospect and to make sure he or she is listening. If questions such as, "That's true, isn't it?" are frequently interspersed in the conversation, the counterperson is taking positive action to draw out the prospect. An occasional question that cannot be answered by "yes" or "no" will help the counterperson check the progress being made.

INDIFFERENCE

Indifference is one of the most exasperating hurdles to overcome. The counterperson's job is to convince the prospect of a need, yet not make him or her aware that he is on the road to buying. The interview must remain informal and frictionless. The counterperson must engage in casual dialogue and, ignoring the prospect's apparent indifference, must continue to add advantage to advantage.

INDECISION

A counterperson confronts indecision when the prospect's uncertainty and hesitation cause postponement of the purchase. With this kind of difficulty, the counterperson must take the initiative by demonstrating friendly but firm guidance. By taking a positive, decisive attitude, he or she can help the prospect decide. Questions should be asked throughout the presentation to build up a series of commitments by the prospect. Relating pertinent case histories can be an extremely effective method of dealing with the indecisive prospect.

SKEPTICISM

Skeptical prospects are usually people who have been deceived by a salesperson or who know other people who have been deceived. Instead of making the skeptic back up claims, the counterperson should go ahead with the sales story and keep piling up facts backed by reliable proof. By being intentionally conservative, the counterperson makes it difficult for the prospect to voice any sort of disagreement. Above all, the counterperson must remain poised when dealing with the skeptical prospect and must refuse to argue.

HOSTILITY

When a counterperson encounters an antagonistic prospect (Figure 10–17), he or she must try to correct that attitude because it can have far-reaching harmful effects on the store and on the counterperson. Also, a hostile prospect tends to become a good customer once the erroneous opinion of the counterperson and the store has been corrected.

The worst course of action for the counterperson to take is to argue. The prospect will naturally become defensive, even if proved wrong, and will usually be too irritated to buy. Instead, the counterperson should listen sympathetically, thank the prospect for the information, and promise to investigate and try to remedy the cause of dissatisfaction. Then, he or she should get back to the customer with the remedy.

Of course, not all difficult customers can be satisfied. However, if a counterperson can convince these customers that their best interests are of concern, many will become good customers.

KEEPING CUSTOMERS

Frequently, a counterperson must spend a great deal of time with a particular dealer prospect before obtaining an order. Just as frequently, the first order is not large enough to cover the time spent in obtaining it. In some cases, three, four, or even more orders will be required to equal the development time of the account. A customer who then switches to a competitor must be replaced at the further expenditure of time. The cost of this time spent means that the

FIGURE 10–17 A counterperson should try to correct the attitude of the hostile customer.

counterperson cannot afford to let accounts slip away.

Because satisfied customers are hard to take away from their present source of supply and because the counterperson's personal standing with his or her customers is so important, a counterperson should be careful in handling all transactions and contacts with customers.

Customers return to and buy repeatedly from those stores that satisfy and please them. Customers want good values, accurate and adequate information about products, speedy service, and polite treatment. The customer especially wants to feel that each purchase and each return visit are appreciated by the counterperson (and his or her boss). Sound selling plus sincere appreciation for each sale build a loyal clientele whose esteem and goodwill are invaluable assets to the store.

As part of an annual analysis of personal performance, a counterperson should study each account acquired during the previous year and should make definite plans for holding it. Each account retained and each account lost during the period should be examined in detail. For each account lost, the counterperson should find a replacement, whether or not the lost customer can be regained. Landing a new account will reduce some of the pressure and restore confidence especially if the lost account was big.

BUILDING EXISTING ACCOUNTS

It is not enough for a counterperson to acquire and then hold an account: it must be built up to its full potential in sales volume. While trying to build the sales volume on a particular product line, the counterperson can be working simultaneously to sell other products. Among the more important reasons it is wise to work on increasing the volume of an account are:

1. Additional volume means additional income for employees and the store.
2. As long as a customer divides purchases among different stores, there is the risk of losing the account.
3. Less time is required to build up the volume of established customers than to obtain the same dollar amount by converting prospects into customers.

In obtaining the overall objective of increasing the volume of an account, the counterperson should persuade the customer as soon as possible after obtaining the first order to designate his or her store as the regular source of supply for the particular product he or she bought.

It is a mistake to be either too conservative or too aggressive in trying to increase the customer's volume of purchases. An overly conservative counterperson gives competitors an opportunity to pursue and obtain orders that he or she could have had. An overly aggressive counterperson can annoy customers.

Sometimes counterpeople are determined to write up a large order without proper consideration of the customer's needs. The best strategy is to follow a moderate course by holding on to current business while gradually taking over other items and other lines.

HANDLING CUSTOMER COMPLAINTS

More than likely, a counterperson will be required to handle complaints because he or she is the person who deals with the customer most frequently and knows the most about the customer and the buying circumstances in question. Four key points to remember when handling complaints are: fairness, sympathy, immediacy, and solutions (Figure 10–18).

Fairness

At the very start, the counterperson must assure the customer that he or she is determined to be completely fair and give the customer every courtesy and consideration. At the same time, the counterperson must reaffirm the customer's faith in the store, thus encouraging the customer to accept the store as one that does the right thing by its customers.

Sympathy

The counterperson should act as a sympathetic audience while the customer lets off steam and demands satisfaction. After all, the customer has a problem, no matter who is to blame. While the customer speaks, the counterperson should collect information and opinions and say nothing until all the facts are told.

Immediacy

Whether the counterperson perceives a problem to be urgent or not, all complaints should be handled immediately. The counterperson increases the customer's confidence in him or her by taking hold and handling the complaint right away. Referring the matter to the boss needlessly tends to magnify its significance.

Solution

The counterperson must first determine exactly what happened based on the facts, then he or she must rectify the problem. When the counterperson, the product, or the store is at fault, the fact should be admitted at once, followed by a promise of prompt and satisfactory adjustment. If the facts indicate that the customer is at fault due to improper usage, then the situation must be explained to the customer without ridicule or embarrassment.

For more information on handling complaints tactfully, see Chapter 11 on customer relations.

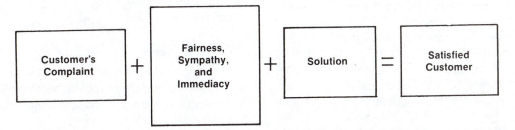

FIGURE 10–18 Handling a customer's complaint with fairness, sympathy, and immediacy can turn a disgruntled customer into a satisfied customer.

WHY CUSTOMERS LEAVE

Customers stop buying from a particular business for various reasons (Figure 10–19). The indifference of some counterpersons, high-pressure selling tactics, or high prices can drive customers away. Understanding problem areas is the first step in avoiding the loss of customers and in taking steps to win them back.

Service

Service is by far the most important issue. On one hand, service can be so good that unless the competitor's price is much better (15 percent or more), the customers would not consider switching. On the other hand, service can be so poor that no one would consider buying from a business at any price.

The word service expresses the sum of many facets. It includes timely delivery, adequate inventory, and skilled counter personnel who find the right parts the first time and pay attention to the customers' unmentioned needs. It also includes organization and discipline, never forgetting orders, and knowing where to get out-of-stock parts quickly and at the right price.

Service also means supplying information. A counterperson should be aware of clinics and training materials to pass on to professional customers, and he or she should schedule field work with factory representatives to keep dealers' inventories up-to-date.

Because counter personnel are usually the people to deal directly with customers, they have a great affect on how each customer perceives the service in a business. It takes only one ill-mannered counterperson or one with a haughty attitude to drive customers away.

Indifference of counterpersons or other employees is the number one reason for losing customers. Most customers will complain before leaving, providing an opportunity for a problem to be corrected and the customer retained. But if complaints are ignored, then the customer perceives the lack of reaction as indifference. The indifference itself becomes an additional complaint.

Sometimes a business may lose a customer over a credit policy. If the policy is sound, nothing should be changed. A customer who is denied excessive credit might go elsehwere, but unless the parts are being paid for, the store is really not losing any business.

Service is the total measure of a business's competence. If customers are lost because of poor service, they are being lost because of incompetence.

Price

Price is always an issue in keeping customers satisfied. With a few customers, it is the only issue.

At one time in the automotive industry, most products were sold for about the same price. Noticeable differences were usually due to different lines. And if the price difference was substantial, there was usually a big difference in quality.

Nowadays, manufacturers, in an effort to grab larger and larger market shares, have bypassed intermediate levels of distribution, selling directly to some customers who would otherwise have been supplied by WDs. In turn, these customers use their purchasing advantage to attract new business. Jobbers, feeling pressure by this competition, lean on their WDs for better prices, and the pressure is passed back to the manufacturers who respond with shorter, private label lines.

Sometimes explaining the marketing and distribution trends in the automotive industry will help a customer to better understand pricing. Other than explain, a counterperson can do little about a particular price.

Products

Products can lose customers for a business, too. For example, suppose a counterperson regularly sells starters to a local beverage distributor with a fleet of trucks. The counterperson notices that a lot of the starters are coming back as defects. The counterperson asks the WD to have the remanufacturer check into the problem and then take corrective action, but it does not happen and the customer begins to complain. Upon the counterperson's advisement, the manager or owner must choose between losing the customer or finding another supplier.

Even with a perfectly good line, a counterperson will still encounter customers who prefer a different brand. If several important accounts prefer the different brand, it might be wise for the business to change lines.

Warning Signs

Complaints are one of the most obvious warning signals of a problem in the buyer-seller relationship. Another signal is financial, such as a drop in a regular customer's total sales.

A shift in personal attitude or procedure can indicate a problem as well. For example, a customer might be increasingly influenced by a competitor, might pay less attention to business matters, or might allow associates to handle the business responsibilities he or she formerly handled.

CATERING TO LOST CUSTOMERS

Every business loses a customer once in a while, although losses should not happen often. Much can be done in an effort to regain a lost customer, but a counterperson must understand that no matter what efforts are made, a lost customer might not ever bring his or her business back. Taking care of problems immediately is one of the best ways to keep from losing customers and one of the best ways to get them back.

Many times a customer will end the relationship without telling anyone why. Because a counterperson should know why so other customers are not lost for the same reason, he or she should ask the lost customer. This inquiry also shows the counterperson's interest, although he or she has already lost the customer. The counterperson should be prepared to hear things he or she might not want to hear and should be able to remain calm because to deny problems or make excuses about them will not bring the customer back.

To convince the customer that problems will be solved, the counterperson must be able to admit problems exist. While the ex-customer is talking, the counterperson should take notes and then thank the customer for his or her input.

The counterperson should then discuss the complaint with bosses and co-workers to determine how the problems can be solved. He or she should be sure to get back to the customer with the information about how the problem will be handled in the future. The ex-customer might give the store another chance immediately, further in the future after thinking about it, or might not ever return. However, the counterperson's inquiries will probably prevent a similar problem from losing a customer in the future.

Timing

Timing is everything when attempting to regain a lost customer. The counterperson should not interfere with the "honeymoon" that the ex-customer and new supplier will be on for the first few months. During that period, both parties are excited about each other and are doing their best to keep each other happy. Also, the first few months might also be a punishment time for the customer when he believes he is giving the previous supplier what is due.

The counterperson should let the customer know that he or she appreciates any business the customer might give and would be happy to have all of his business again, but being too eager can create a situation where the customer plays one business against another. Then the counterperson should just wait. If the staff is competent, the products and service are as good or better than competitors, then time is the best ally. After the honeymoon phase is over, the customer might begin to see faults with the competitor and remember favorable aspects of the previous place of business.

After a few months have passed, some customers will return on their own. If not, the counterperson should increase efforts to make contact with the ex-customer by calling to see if anything is needed whenever the delivery truck is nearing his place, by giving referrals, or by inviting him or her to lunch.

The Ex-Lost Customer

If a customer does return, a counterperson should never remind the person that he or she made a mistake by leaving. An embarrassed customer will never come back. Instead, counter personnel should show their gladness and appreciation in the customer's return.

DEFENSIVE SELLING

Defensive selling implies a method of selling that protects the interests of the store. Because the do-it-yourselfer is not always clear and methodical in the diagnosis of a problem, he or she will sometimes purchase the wrong part, install it, find the problem has not been solved, then try to return the part as defective. Whenever a part is sold and returned in this manner, it is at the expense of the

HOW CUSTOMERS ARE LOST

Out of every 100 lost customers:
  1 dies
  3 move away
  5 form other friendships and contacts
10 are won over by the competition
15 are dissatisfied with your products
            and
66 LEAVE BECAUSE OF AN ATTITUDE OF INDIFFERENCE TOWARD THE CUSTOMER BY THE STORE OWNER OR EMPLOYEE.

FIGURE 10-19 Reasons customers leave a business.

store. For this reason, many stores have a policy of no returns on installed parts.

As a result, defensive selling is not only a protection for the store but for the do-it-yourselfer. It protects the customer from buying the wrong parts to solve the problem, and it saves time and energy as well. When the counterperson practices defensive selling, he or she is selling in such a manner that reduces the likelihood of returned parts.

DEFINING THE PROBLEM

The first principle involved in defensive selling is defining the nature of the customer's problem. For the counterperson to do this successfully, he or she must keep technical abilities sharp through personal study and technical clinics.

PROBLEM: A do-it-yourself customer wants to buy a starter because his car will not start.

ANSWER: Perhaps at a recent tech clinic, the counterperson learned about the starting/charging system and how each part is like a link in a chain. As he or she looks up the part number, he asks, "Does the starter click or make noise when the switch is turned?" A "yes" answer might prompt the counterperson to ask, "Will it start OK with a jump start?" Another "yes" answer might lead to the question, "Have you had the battery checked and, if you haven't, would you like us to test it for you?"

Like defensive driving in rush-hour traffic, defensive selling pays off. Instead of incorrectly selling a remanufactured starter, the counterperson might have correctly discovered an opportunity to sell a battery that will remedy the customer's "starter problem" (Figure 10–20).

TESTING TECHNIQUES

Technical suggestions will eliminate a large portion of DIY comeback complaints on part sales, but they will not always convince some do-it-yourselfers. Some DIYs will not take the advice of even the most qualified technician because they suspect that the only real motive is to sell them something they do not really need.

In the case of the previous example, if the customer did not believe the counterperson's technical explanation, and he or she refused to have the battery checked, then the counterperson can take another defensive step. The counterperson can test a starter in front of the customer to show that the starter operates correctly. The counterperson should warn the customer that the store might not necessarily accept returns on electrical parts (most

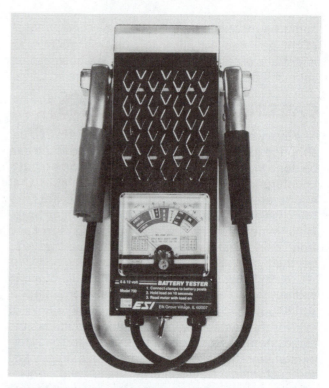

FIGURE 10–20 A variety of simple test equipment, such as this portable battery tester, can help diagnose customer problems and complaints.

do not) and that the new starter might self-destruct against a worn and somewhat toothless flywheel ring gear. The customer is then responsible for the consequences of the choice.

Most stores are equipped with testing devices to avert the dilemma of deciding whether the part was defective before it left the store or whether it was destroyed upon installation. In this way, the store is protected.

SELLING SHOP MANUALS

Another manner of defensive selling is the selling of shop manuals. They can help the customer avoid a false diagnosis.

The best place to sell books is at the counter (Figure 10–21). Books, after all, are not an impulse item. Nobody buys manuals unless reminded of their obvious money-saving virtues, especially when they are close at hand.

Manuals are not difficult to sell when the pitch is made at the opportune time. A good time to make the suggestion is when the customer is buying accessories for a new car; for example, a manual on how to maintain an expensive automotive investment. Another opportunity to sell a manual is when the DIY is obviously in a spending mood, having bought tools and other assorted items.

RETAILING THE REMANUFACTURED

The counterperson who practices defensive selling always sells parts not in the largest quantity possible but in the largest unit possible. The do-it-yourselfer might stride confidently up to the counter with a list of parts to rebuild the engine, but he or she might actually know little about fitting engine bearings or piston rings, driving the camshaft bearings, reaming the valve guides, grinding the valves, torquing the head bolts and bearing caps, or removing the crankshaft harmonic balancer. All of the above operations require special drivers, pullers, micrometers, or other tools for accurate, professional work.

The defensive suggestion is to sell remanufactured engines, transmissions, alternators, starters, carburetors, driveshafts, distributors, power-steering pumps, and anything that requires special tools and knowledge.

Price is the customer's primary objection, but the counterperson should always mention the supplier's warranty in the presentation of features and benefits. The counterperson should also explain that a few new parts installed in an old, worn engine or transmission do not make a newly remanufactured unit. Any type of parts-changing operation is far most cost-effective for both consumer and store alike.

REFERRING CUSTOMERS TO PROFESSIONALS

Not even the most astute technician can explain complex diagnostic procedures or pinpoint troubles by listening to a customer's fuzzy explanation of a

FIGURE 10-21 Defensive selling includes selling shop manuals to customers who might need more information.

problem. In addition, while the diagnostic skills of some DIYs might appear to be quite adept, delving into the complexities of modern computer controls and electronics, engine, and drivetrain requires some relatively sophisticated diagnostic procedures using expensive equipment backed up with comprehensive formal training.

However, some DIYs wrongly assume that they cannot complete a task. For example, the need for special tools often leads many customers to seek professional assistance. The do-it-yourselfer might think that a voltmeter or torque wrench is a special tool. The counterperson should explain that these, and many others, actually belong in a do-it-yourselfer's tool collection. They might not be used as often as a 3/8" ratchet, but they will be needed over and over in future years.

When the counterperson has explained the long-range economy of tool ownership, the customer will probably buy. If not, the counterperson might be able to rent it or refer the customer to a nearby rental firm.

Some customers simply lack confidence, not skill, and confidence can be boosted by providing information. The counterperson's greatest impact on the customers' confidence is his or her own attitude about their abilities. He or she should avoid speaking "down" to the customers and should converse with them according to their actual abilities. Some problems will be beyond the capabilities of a DIY. When a counterperson perceives that a problem is too complex for a DIY to handle, he or she should refer the customer to a local technician. However, there are a few defensive rules of professional referral a counterperson should know before innocently guiding the DIY customer and a professional account into possible conflict.

First, the counterperson should always ask the mechanic customer if he or she accepts DIY referrals.

Next, the customer cannot buy the counterperson's parts and then take them to the mechanic to install. Not only does the technician face the obvious replacement labor woes if the parts are defective, but he or she is losing the markup on the sale even before beginning the job.

Finally, the counterperson should never ask the mechanic for an estimate because DIY repair situations are not clear-cut for the technician. Not only must he or she unravel the aftereffects of unskilled workmanship, but the original primary problem must also be located. As a result, diagnostic times are often compounded considerably due to these mechanical malpractices.

The counterperson should always make discrete inquiries before referring a customer to a potentially valuable account. Some mechanic customers serve a very exclusive clientele. Some shops restrict their work to a list of preferred customers. Others specialize in fleet and commercial accounts. A few technicians cater only to people who drive BMWs or Peterbilt trucks, so beware of the unsolicited customer referral.

Too much business can often be a detriment to a busy shop whether it caters to a specific clientele or not. Each shop should be asked if they would like referred customers or if they have the time to look at a customer's problem. The counterperson should also give the mechanic customer a gracious way out of an unwanted commitment. Usually if a shop advertises heavily, then it is fair to refer customers there.

REFERRING CUSTOMERS TO COMPETITORS

Referring customers to competition can be risky. If a counterperson sends a customer to another jobber when the first store does not have the part, the customer might never return. On the other hand, the counterperson risks losing the customer by not offering a solution to the problem.

The counterperson has three options:

1. Tell the customer when the part can be obtained from the usual supplier, which usually takes at least a day.
2. Refer the customer to a competing jobber.
3. Purchase the part from another jobber and sell it to the customer.

Most customers do not want to wait long for parts and will go elsewhere with or without a referral from the counterperson. In this case, the counterperson is missing an opportunity to show an interest in the customer's problem and to show the excellent service of the store. If the counterperson chooses the second or third option—refer the customer to or acquire the parts from another jobber—the customer is likely to appreciate the prompt assistance, which will bring him or her back in the future.

In most cases, the third option—to acquire the parts from another jobber—is the best. The customer need not know where the part is coming from, only how soon it will be available. Usually, the part can be obtained in less than an hour. The customer might even purchase other items while waiting. Obviously, stores having several local branches should work through their chain before turning to truly outside sources.

SERVING THE PROFESSIONAL CUSTOMER

In most jobber stores, about 20 percent of the accounts are responsible for 80 percent of the business. Usually, these large accounts are professional mechanics. They form the ongoing financial foundation for the store and must be protected and appreciated for the key role they play in the store's success. Smart jobbers often tailor their services to better serve professional customers.

SPECIAL COUNTERS

One of the most common complaints from professional customers is that they are not waited on quickly enough. They complain that while they are being waited on, the counterperson stops to answer the phone or waits on two or three other customers first. Their impatience is understandable because any time they spend waiting is translated into lost profits for their business.

One of the ways jobbers have chosen to handle this situation is by building or designating a certain counter as the "Dealer Only" or "Wholesale Customers Only" counter. When shop accounts bring in $4,000 to $5,000 a month in business, they deserve special treatment. If a store loses one of these accounts, a great deal of advertising and promotional work will have to be done to get that amount of business back.

The dealer counter should include a special dealer telephone with an unlisted number to ensure that only professional customers use it. A number reserved for the dealer clientele makes them feel that the jobber store is definitely geared to take care of their business. Forcing the mechanic to dial a half dozen times to evade busy signals generated by the price-shopping customer will not expand a dealer base.

In many cases, a counterperson's order-taking skills will be sharpened by dealing with professionals because information is received, researched, and conveyed as quickly as possible.

Dealer counters can solve a variety of time-consuming situations if properly managed. For instance, the time spent picking up parts can be reduced greatly if the mechanic phones in the order. The counterperson can take the order, pull the parts, bill them, and have the order waiting at the dealer counter. All the mechanic has to do is sign the ticket, pick up the order, and move on. There is no need for this customer to wait behind the do-it-yourselfers.

To insure the customer of the fastest service, the counterperson should leave nothing to chance when taking an order. For example, suppose that only 10 percent of the 1975 Buicks with air had a different water pump. There is a 90 percent chance that the counterperson will be safe in choosing the standard pump. However, it is better to verify the choice with the customer. Otherwise the mechanic's bay might be tied up another day waiting for an overnight order.

The dealer counter is also the place where dozens of orders received each day from dealer accounts can be consolidated into a few convenient deliveries. The counterperson can designate a specific bin or space for each dealer account where various ASAP orders and stock replacement orders are assembled. This prevents the problem of only partially filled orders that results from a lack of organization. Efficient service is especially important to these businesses which are the major source of cash flow through the jobber store.

The dealer counter is the logical location to present bargains and promotions geared toward the professional. Often, the only time the mechanic stops moving is at the dealer counter. Here the mechanic can browse through pamphlets and catalogs. It is the ideal place to promote dealer tech clinics, put together all of the tech bulletins received from parts suppliers, post machine shop scheduling for the mechanic customer, or even install a tool rental department. All of these ideas will help promote a good relationship between the professional and the counterperson.

CONCESSIONS

Another way of recognizing good professional customers is by making special concessions to them that do not adversely affect the store's business.

For instance, in an effort to gain more control and limit expenses, a counterperson might organize scheduled deliveries. He or she sends a note to all the local customers that the store's truck will leave on an hourly basis from the store, making a regular route with deliveries.

At the same time, a letter is sent out to the store's consistent customers, mentioning the new schedule, but specifically pointing out that their business has always been important to the store and that special trips will still be made to supply them with parts needed in emergencies. Of course, the counterperson should ask these customers to be discreet about this special treatment.

CONVINCING DEALERS TO STOCK INVENTORY

There are a number of advantages for the dealer who stocks inventory. By being familiar with these advantages, the counterperson can better sell the idea to those mechanics who have not considered it.

Mechanic shops make their profits primarily by the hour rather than by the sale. A slight loss of chargeable time each day adds up to a considerable amount by the end of the month. Any time spent picking up parts, or even waiting for deliveries, inevitably steals away the mechanic's profitable time. Obviously, it is much less expensive to have a half dozen caps and rotors delivered to the stocking dealer than to deliver each one to a waiting mechanic.

PROFITABILITY

Profitability will also be increased if the stocking dealer receives a discount. Although the jobber might lose a few dollars discounting, the added volume of a loyal dealer will more than offset the loss. The dealer will especially appreciate the price difference when the per-hour profit on parts and labor is calculated. This advantage is lost when parts are gathered from a variety of sources.

CONVENIENCE

Convenience is another benefit that the dealer who stocks inventory receives. The mechanic who does not have to spend $10 of time chasing a $4.28 wheel seal will produce more chargeable hours of labor than the mechanic who must scrounge for the parts.

In addition, having the part is also profitable for the mechanic selling to the customer who might postpone the not-so-urgent repair until another day. A frayed fan belt or a soft radiator hose is a sale made for the dealer who "just happens to have one on the rack." Nobody ever sold from an empty shelf.

AVAILABILITY

Availability means being able to find the needed part at the flip of the catalog page. Availability also means being only a step away from the parts cabinet. If the mechanic customer needs a part not included in stock, he or she knows that the jobber can provide it because it is listed in the catalog. Availability is not affected by a missing part number because the counterperson will be there to deliver or to do the necessary restocking.

The mutual profitability of dealer inventories is obvious. The time/money factor, profitability, convenience, and availability increase sales at both ends of the jobber/dealer relationship.

SERVING DEALERS WHO STOCK INVENTORY

Serving the dealer and inventory requires familiarity with the dealer's types of business, selling technique, and replacement parts requirements.

The rule for servicing the dealer is simple: If counter personnel cannot take an active interest in serving the inventory, then the jobber should not expect the dealer to take an interest in stocking it. The counterperson's servicing (sales) technique is very important.

The jobber/dealer relationship will be cemented when the counterperson adopts the concept of becoming the mechanic customer's silent partner. The concept of the silent partner means that the counterperson should invest the time and effort necessary to create a mutually profitable jobber/dealer relationship. The counterperson has to visualize being the mechanic's personal counterperson. Becoming familiar with the mechanic customer's business will enable the counterperson to communicate with the customer in a way that better serves his or her needs.

To maintain a good working relationship, the basic duties of the visiting counterperson should include:

1. Keeping the inventory at a usable and profitable level.
2. Maintaining current and well-organized catalogs and price sheets.
3. Occasionally reorganizing the racks, shelves, and cabinets.
4. Keeping the dealer involved with new products, promotions, services, and training clinics.

 SALES TALK _____

Stocking dealers are an advantage to both themselves and their jobbers. Servicing stocking dealers is the primary task of the outside salesperson. Good service on a regular basis is the only method of tying jobbers and stocking dealers close together.

DESIGNING THE INVENTORY

Because no two mechanics sell parts in a similar pattern, the inventory must be customized to meet the dealer's particular needs. A good approach for the parts professional who is familiar with a dealer's past sales records would be to place in stock items the dealer regularly sells.

For example, if last month's tickets include an inordinate number of fan belts, the counterperson could place in stock most of the popular numbers recommended by the supplier. The counterperson can customize the model stock by adding or deleting until it meets the dealer's requirements. In any case, if the stock fills at least 60 percent of the customer's needs, then 60 percent less time is involved in chasing fan belts.

Customizing the inventory also includes suggesting a stock that is compatible with the clientele the dealer serves. A foreign car shop will use Volkswagen fan belts; a truck shop will use something that fits a Peterbilt. A four-wheel drive repair shop will use an above-average number of universal joints, bearings, seals, brake parts, shock absorbers, and steering parts. The local air conditioning specialist needs a convenient supply of freon, compressors, clutch bearings, hoses, seals, refrigerant oil, and miscellaneous parts to be efficient and productive. If the inventory does not fit, it is an unprofitable investment for the shop and an unpleasant advertisement for the store that installed it. Consequently, an analytical approach is much more effective than blind enthusiasm when selling inventory to the stocking dealer.

DETERMINING INVENTORY TURNOVER

Determining inventory turnover is essential to maintain profitability for the dealer and the jobber. Obsolescent or slow inventory ties up productive dollars which could be spent on items that would increase sales for the jobber and the dealer.

Tracking turnover is difficult on dealer inventories because sales tend to be more cyclical due to the demands of the clientele and the particular location and circumstances of the dealer's business. For example, the dealer's commerical accounts might schedule repairs according to seasonal budgetary intervals. A mechanic servicing fleet accounts might show sporadic inventory activity because of the uniform aging and service requirements of a group of identical vehicles. The first heavy snowfall deluges the automatic transmission mechanic with burned out transmissions, and vacation time resembles a gold rush for the air conditioning mechanic.

In most cases, seemingly erratic sales patterns can be tracked by routinely recording part numbers delivered in an inventory catalog. This method will provide an accurate accounting for the turnover of each part number. Small shops might not turn over each number four times per year or even once per year. Other part numbers might sell a dozen times each year. Consequently, dividing the actual or projected annual sales by the initial inventory investment will give an indication of the average turnover for that particular item. To illustrate, a $500 stock of belts might generate $3,000 in retail sales. A $1,000 profit represents doubling the original investment. Related labor income might represent another $1,000 profit if $10 labor is added to each $30 belt sale. This situation, in effect, quadruples the original investment even if each number is not turned over more than once each year.

Naturally, this method does not take into account the individual belts delivered from the jobber store or belts that would have been sold without making the investment, but it does reveal the basic profitability of the item. On the other hand, most sales and wasted time are rarely entered into the monthly profit and loss statement. Sales made and time saved do show up on the bottom line, which is the goal of the counterperson installing dealer inventory.

REVIEW QUESTIONS

1. Which of the following is not a step to a proper sale?
   a. qualifying the customer
   b. reversing position
   c. identifying the customer's motives
   d. demonstrating product benefits

2. Which of the following could be a customer's motives when buying?
   a. style
   b. getting a good deal
   c. self-preservation
   d. all of the above

3. A counterperson may delay answering an objection when the _____ .
   a. objection concerns price
   b. objection is unrelated to the present discussion
   c. customer does not seem to expect a reply
   d. all of the above

4. Which technique for answering objections turns the objection into a reason for buying?
   a. offset
   b. reverse position
   c. boomerang
   d. direct denial

5. Counterperson A quotes the price of a product that costs more than a customer has in mind right at the beginning of the sales presentation. Counterperson B never quotes the price but attempts to sell the product's merits and then write an invoice. Who is right?
   a. Counterperson A
   b. Counterperson B
   c. Both A and B
   d. Neither A nor B

6. A reciprocal relationship is between what people in the automotive aftermarket?
   a. counter personnel
   b. two businesses who give work to each other
   c. manager and counterperson
   d. all of the above

7. Counterperson A attempts to close a sale after 15 to 20 minutes. Counterperson B attempts to close a sale when the prospect begins agreeing. Who is right?
   a. Counterperson A
   b. Counterperson B
   c. Both A and B
   d. Neither A nor B

8. What is oversell?
   a. selling more than a customer needs
   b. continuing a sales presentation after the prospect is ready to buy
   c. both a and b
   d. neither a nor b

9. How can a counterperson sell to a silent prospect?
   a. Ask questions.
   b. Remain silent until the prospect responds.
   c. Maintain the proper speed of the presentation.
   d. All of the above

10. What should a counterperson do if a customer takes his or her business elsewhere?
    a. Find an account to take the place of the lost one.
    b. Call the competition and complain.
    c. Both a and b
    d. Neither a nor b

11. Defensive selling is selling _____ .
    a. to hostile customers
    b. that protects the store's interests
    c. that satisfies the customers' interests
    d. none of the above

12. Which of the following is not a feature of defensive selling?
    a. selling shop manuals
    b. pretesting parts
    c. defining the customer's problem
    d. reducing a price

13. What is the first thing a counterperson should do when considering referring a customer to a professional customer?
    a. Ask the mechanic for an estimate.
    b. Sell the part to the customer.
    c. Ask the mechanic if he accepts referrals.
    d. Ask the customer if he accepts referrals.

14. Why should a counterperson ever consider sending some customers to a competitor?
    a. to solve the customer's problem
    b. to instill good customer relations
    c. to show excellent service
    d. all of the above

15. Which of the following will improve customer relations with professional customers?
    a. a designated counter
    b. special telephone line
    c. special deliveries
    d. all of the above

CHAPTER ELEVEN

CUSTOMER RELATIONS

Objectives

After reading this chapter, you should be able to:

- List three ways to become more familiar with customers, including referring to them by name, remembering preferences and prejudices of regular customers, and keeping a record of all regular customers and their transactions.
- Explain how the customer's images of self, counter personnel, and store affect customer relations.
- Explain how responding to a customer's needs improves customer relations.
- Explain various aspects of counter courtesy, including eye contact, taking customers in turn, responding in a friendly manner, taking breaks out of sight, and remembering customers' needs.
- Explain the necessary differences in communicating with the professional customer as opposed to the do-it-yourselfer.
- Explain how to get pertinent information from any customer.
- Explain how to handle difficult customers and situations, including inattentive customers, delinquent accounts, and angry customers.
- Explain how to prevent and handle comebacks and complaints.
- Explain the importance of good customer appreciation.
- Explain how to handle telephone calls, including use of the "hold" button and time management.
- Describe several extra services that the counterperson can provide to improve customer relations, including offering bags or boxes, hanging a bulletin board, and sponsoring clinics.
- Prevent shoplifting.

Relating positively to customers is one of the most important skills a counterperson needs to develop. It does not matter how quickly and efficiently a counterperson can locate a part in a catalog if the customer is put off by the counterperson's attitude and chooses not to deal in that store. Likewise, if a counterperson is amiable, reliable, and sincerely attempting to fulfill a customer's needs, then the customer might ignore a fault of the counterperson or even be willing to pay a little more for parts. Some people patronize a particular store solely because they like the person behind the counter.

In any store that deals with the public, customers are the business. Although many jobbers might have several large accounts, no store can thrive with just a few customers. Part of the counterperson's job is to attract and keep as many customers as possible.

Customers talk to other customers, and customers talk to prospective customers. What they say about a counterperson and a particular store depends almost entirely on how they have been treated. That is, it depends on customer relations.

If customers experience good customer relations, they will be walking, talking advertisements for the business (Figure 11-1). Likewise, if a cus-

tomer experiences negative customer relations, he or she will probably not keep quiet about it.

Customer relations, then, is practically everything a counterperson does in the course of selling parts to a customer. The counterperson is the direct link between the store and its customers.

FIGURE 11-1 Good customer relations creates satisfied customers.

BECOMING FAMILIAR WITH CUSTOMERS

There are certain steps that a counterperson can take to work more successfully with customers. One of the simplest yet most rewarding is learning the names of as many customers as possible and addressing those individuals by their names.

Another courteous habit is that of remembering the preferences and prejudices of regular customers. Ideally, the counterperson should keep a record of all regular customers (Figure 11–2). Such a record might include the customer's name, the date of the last visit, the items last bought, and any other significant facts about the customer. These customer records can be reviewed quickly each morning, permitting the counterperson to give more tailored and personalized treatment to any of the recorded customers who drop in that day.

These records also serve as a mailing list or as a telephone call list. Where installment credit is available, the counterperson can keep posted on the customer's balances and recommend add-on purchases at the appropriate times. With these methods, a counterperson could convert counterperson-customer relations from a stranger-to-stranger type into an acquaintance-to-acquaintance relationship, and as many of these as possible to a friend-to-friend relationship.

IMAGES: CUSTOMER AND BUSINESS

Everyone has a self-image. Although that self-image is not always obvious to others, everyone gives little clues from time to time that reveal how he or she sees the self.

Counter personnel who pay attention to those clues improve customer relations. For example, if a customer is very proud of his or her skill in diagnosing electrical problems, a counterperson should not give obvious advice concerning electrical problems. On the other hand, such a customer will probably be glad to answer genuine questions, not insincere inquiries, about his or her area of expertise.

Most customers have a particular skill or interest, perhaps not even pertaining to the automotive industry at all but interesting nonetheless, which the counterperson might inquire about from time to time.

IMAGES OF COUNTER PERSONNEL

The customer also has an image of the counterperson, which dictates what he or she expects from

FIGURE 11–2   Ideally, the counterperson should keep a record of all regular customers.

FIGURE 11–3   The counterperson's appearance affects customer attitudes.

the encounter, and the counterperson's appearance has a lot to do with the customer's image. While appearance is not the most important aspect of a counterperson's personality, it does make a personal statement about the person's attitudes toward him- or herself and the job. Clean uniforms and careful grooming go a long way to instill confidence in customers (Figure 11–3).

IMAGES OF THE BUSINESS

The appearance of the store as well as the appearance of counter personnel influences the customer. A clean, neat counterperson establishes a positive image with customers, and a clean, neat store is equally important.

Although a counterperson is rarely in charge of decor or layout, he or she can maintain the appearance of the store in several ways:

1. Keep a rag handy for wiping dirt and grime from used parts off the counter.
2. Keep the counter free of clutter such as small parts, notes, paper clips, and other items.
3. Keep displays organized and well stocked.
4. Keep stock and supplies neatly shelved to improve appearance as well as to facilitate finding parts.
5. Help with basic cleanup by throwing away part wrappers and labels and by removing empty boxes from sight.

MEETING CUSTOMERS' NEEDS

People have different attitudes, needs, wants, and pet peeves along with images of self and counter personnel. The more a counterperson knows about and pays attention to these differences, the easier his or her job will be. A successful counterperson must review a customer's wants and needs to determine what can be done to meet them.

The customer's most obvious want is to fix or service his or her car. The customer also wants to find a person who can help select the best of all possible methods of installation as well as parts. Here is where the appearance and image of the counterperson come into play. It is important to the customer that the counterperson looks like a person who can help.

Every customer is different. In fact, even the same customer will not always behave exactly the same way every time he or she enters the store. Despite individual differences, however, customers have a number of things in common. For example:

1. Everyone is a potential buyer or actual source of income to the store. They have the power to buy.
2. Each one is an opportunity for the counterperson to provide true service.
3. No customer is ever an interruption of what the counterperson happens to be doing before he or she walked in or called.
4. Each expects and deserves to be treated with courtesy and respect.
5. All customers are affected by the counterperson's attitude. All respond favorably to genuine friendliness, interest, honesty, and sincerity; all respond unfavorably to aggressiveness and arrogance.
6. Every customer judges counter personnel and the store partly on apearance, whether fairly or not.

COUNTER COURTESY

Probably the best description of what a counterperson should do at the counter is this: provide prompt, efficient, and courteous service. No one likes to wait unnecessarily. Many customers have shops and customers of their own. And no one likes to be treated rudely.

From the time a customer walks through the door until the time he or she leaves, the person should be acknowledged and responded to favorably. Counter personnel should stop speaking to one another whenever a customer approaches the counter.

Customers usually do not want to feel like a number or just another face in the crowd. The first step toward making a customer feel noticed is to make eye contact. The counterperson should acknowledge the customer by looking at him or her and saying something such as "Good morning. I'll be with you in a moment." Eye contact should also be made when the counterperson is waiting on the customer by looking up from the parts catalog periodically while talking to him or her.

Another way the counterperson can make a customer feel important is to state his or her own name and to call the customer by name.

CUSTOMERS' TURNS

If the particular store has a take-a-number system for waiting on customers, then counter personnel should adhere to it exactly (Figure 11–4). If a counterperson recognizes one of the regular customers and can see he or she is in a hurry, then the counterperson can ask the person who is next if he or she would not mind letting the other customer go ahead. The person may or may not agree, but at least the regular customer will appreciate the effort.

If the store does not use a number system, the counterperson must be very sure to take each customer in turn. If the counterperson does not know who is next, he or she should ask.

In any case, the counterperson should again acknowledge the people waiting in line, if only to say, "Be with you in a minute."

COURTEOUS AND SINCERE SERVICE

When a customer approaches the counter, the counterperson should give a friendly greeting. Most

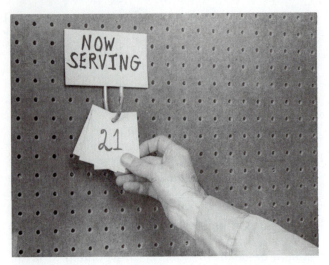

FIGURE 11-4 If a particular business has a take-a-number system, the counterperson should adhere to the numbers exactly when waiting on customers.

people have had the experience of approaching a person to buy something and being greeted with a blank stare, and the counterperson should make sure that experience does not occur in his or her store.

All customers expect fair and reasonable prices. Counter personnel rarely set the prices. However, the counterperson can do his or her part by making sure to use the right price sheets or discount rates. Also, every customer appreciates knowing about a special sale or discount.

Although a parts store or department cannot possibly have all parts in stock that a customer might want, a counterperson can try to make sure the customer does not suffer needless frustration over a part search. For example, a counterperson should make sure to check if the part is available before telling a customer that it is. Also, if the part is out of stock, the counterperson should always offer to order it. If it is an item that the store does not carry, then the counterperson should suggest possible sources that might have the item.

Everyone likes to be thanked. Customers are free to spend their money wherever they choose, so counter personnel should always thank customers who buy from them.

TAKING BREAKS

Breaks should generally be unscheduled and should be taken only when there are no more customers to be helped. Counter personnel should take all breaks and lunches in a separate area out of view of the parts counter (Figure 11-5).

COURTESY CHECKLIST

Most people respond favorably when someone recognizes and understands their needs and wants. Counter personnel should remember these four points concerning customer relations:

1. Everyone likes to believe other people are genuinely interested in them.
2. Everyone likes to be understood.
3. Everyone likes to be recognized and appreciated for his or her knowledge and accomplishments.
4. Everyone wants freedom of choice and tends to become annoyed when presented with only one choice or when pushed to make a certain choice.

COMMUNICATION

The heart of all customer relations is communication. Communication can be oral, such as what is said and the tone of voice used. And communication can also be nonspoken, or body language, such as facial expressions or gestures.

Both types of communication play a role in customer relations. A person's true thoughts and feelings will be evident in either words or actions or both. Confidence and sincerity will be obvious in both word and deed. Likewise, anger, mistrust, or dishonesty will be apparent in the communication of customer and counterperson.

But being knowledgeable, confident, and sincere is sometimes not enough. Knowing what type

FIGURE 11-5 Counter personnel should never take breaks where customers can see them.

of information must be communicated to different customers can be an important attribute of a counterperson. And knowing how to get pertinent information out of the customer can also be important.

The counterperson should remember that his or her job is not just selling parts but selling solutions to customers' problems. The counterperson should always give advice courteously, without talking down to the customer.

COMMUNICATING WITH PROFESSIONAL CUSTOMERS

Professional customers want the counterperson's help to satisfy their own customers and to avoid comebacks. However, professionals already know a great deal about vehicles and parts and might even have certain parts numbers memorized themselves. Specialists might know more about a particular automotive system than the counterperson will ever know. These technicians often know exactly what they need, along with an approximation of their total bill. Professional customers do not need, and most would resent, needless advice on basic mechanics.

However, professional customers might not be aware of recent information, such as manufacturer defects, discounts, or sales (Figure 11–6). They might not know where to obtain a part that is out of stock or not carried by a particular store. And they might be interested in knowing about customers that have shopped in the store and expressed an interest in services that the professional customer

FIGURE 11–6 Professional customers might not be aware of recent information, such as defects, discounts, or sales.

The counterperson should become familiar with the types of parts ordered repeatedly by regular professional customers to become acquainted with the type of work that customer does most often. The counterperson should try not to recommend a particular product line to a regular customer who has shown an aversion to it in the past. Recognizing and commenting on the professional customer's particular skills and talents will provide better customer relations than a lecture on automotive repair.

A few hindrances break down communication between the counterperson and mechanic customer. The most obvious failure is simply not to listen. For example, the counterperson might give the customer wheel bearings when he or she asked for wheel bearing seals.

Another problem in communication is when the speaker makes an assumption and, in doing so, fails to give all the information necessary to insure that the right product is selected. When a mechanic gives less information than needed to make a proper choice of parts, the counterperson should ask for the needed information instead of guessing.

When the counterperson communicates with the mechanic as an individual, communication is bound to be more productive. The more familiar the counterperson becomes with the mechanic customer and his oer her service business, the easier it will be to communicate with the customer and eliminate those misunderstandings that sometimes threaten a mutually profitable relationship.

COMMUNICATING WITH DO-IT-YOURSELFERS

Do-it-yourselfers usually need more information than professional automotive customers (Figure 11–7). These customers usually need and expect good advice on parts and methods. They view the counterperson as an expert, and he or she should utilize product knowledge and catalog skills to give the advice needed. Never should anything be sold to a customer that he or she does not actually need.

One way that a counterperson can assist these customers without taking too much time away from other customers is to have available printed how-to information. Once the counterperson determines what work the customer will have to do, the counterperson can give him or her something to read while assisting another customer or two. After the customer reads the information, the counterperson can clarify some important points and answer questions. Each customer's needs will be met without anyone waiting a long time.

FIGURE 11-7 DIY customers often benefit from the knowledge and experience of competent counterpersonnel.

Sometimes a do-it-yourselfer will attempt a job that cannot be finished either because of a lack of the skill or knowledge, the special tools, or the time.

The counterperson can help the do-it-yourself customer resolve the problem. Many times a counterperson can walk the customer through a repair that he or she is hesitant to undertake, selling special tools, test equipment, and books in the process. Offering the customer this kind of service builds the customer's loyalty to the store and the counterperson.

If the customer needs professional assistance, the counterperson can refer him or her to one of the store's professional customers who accepts referrals. This suggestion improves relations with both the do-it-yourselfer and the professional customer.

GETTING INFORMATION FROM THE CUSTOMER

Good communication with customers does not only depend on how the counterperson communicates with the customer, but also on how the counterperson gets the customer to communicate with him or her. The customer must communicate pertinent information for the counterperson to accurately assess the customer's needs. Many mistakes are caused by the customer omitting or misquoting important information and by the counterperson not asking for enough information before assuming what the customer needs. For example, what a customer refers to as a jeep might actually be a Scout. And what another customer calls a full-size Chevy might mean a truck, car, or even station wagon.

While the professional customer might possess accurate knowledge of the necessary information, occasionally this customer will assume that the counterperson, another professional, already knows what he or she is referring to. In this way, some facts might be omitted.

The do-it-yourselfer might not know all the information needed, resulting in omitted data or wrong answers to questions. The counterperson must verify the facts, perhaps by seeing the actual vehicle to be worked on and/or the old part, before selling or ordering expensive parts.

The counterperson can never assume anything. He or she must ask questions until the make, model, and year of the vehicle is determined, as well as a history of the problem.

HANDLING DIFFICULT CUSTOMERS AND SITUATIONS

Sometimes no matter how courteous or helpful a counterperson attempts to be, a customer will be argumentative or uncooperative anyway. The mildest form of difficult customer is one that is simply inattentive, not listening to the counterperson's explanations or repeatedly asking the same questions. With an inattentive customer, the counterperson should try to include the customer in the discussion as much as possible, asking questions all the while to be sure the customer is listening and understanding. And there will be those times and those customers when nothing the counterperson does can prevent a problem because the customer just simply does not pay attention.

DELINQUENT ACCOUNTS

The most apparent problem customers are delinquent accounts. For delinquent accounts, stores develop policies and procedures that make it easier to navigate around the tough decisions.

Some other examples of problem customers are the mechanic who calls for part numbers, then orders the parts from someone else at a lower cost; the customer whose every order is declared an "emergency;" the customer who will not give the right information until the wrong part has been delivered at least once; and the verbal abusers.

CUSTOMERS WHO ARGUE

The most unpleasant type of difficult customer is the one who argues, complains, or brings back parts whether the fault is his or hers or the counter-

FIGURE 11–8 The most unpleasant type of customer is the angry customer.

person's (Figure 11–8). The first fact a counterperson must remember when dealing with this type of customer is that various things having nothing to do with the counterperson or store can affect the way people behave. Remembering the following list can help the counterperson to be patient whenever a customer at the counter is being difficult.

1. Problems with physical condition or general health
2. Unfavorable experiences, whether in the immediate past or over a long period of time
3. Opinions, attitudes, and beliefs, especially when they appear to conflict with the counterperson's attitude
4. Emotions and feelings, often affected temporarily by stress or conflict
5. Differences in perspectives (seeing the same situation differently than the counterperson does, for example)

Although a counterperson is not responsible for a customer's problems, the counterperson can function more pleasantly if he or she remembers that in most cases the customer is not behaving in a difficult way because of what the counterperson did or who he or she is. Regardless of the customer's present outlook, the counterperson should make the experience in the store a pleasant one.

COMEBACKS AND COMPLAINTS

Comebacks and complaints are a part of any business, especially one that deals directly with people. While not a pleasant aspect of working in an automotive parts store or department, they need not ruin the entire day or more of a counterperson. Comebacks and complaints will affect the counterperson only as much as he or she lets them.

PREVENTING COMPLAINTS

Counter personnel should attempt to prevent complaints as much as possible. The first step in prevention calls for sufficient knowledge about the products and the customer's needs so that the counterperson can guide the customer in buying, recommending only those items required for the job to be completed and even refusing to sell if only disappointment and dissatisfaction would result.

A second step is to make certain that the customer and the counterperson think alike about the product to be purchased and what it will do. Product misunderstandings usually lead to strained relations.

As a third step, the counterperson should describe, in whatever detail necessary, the policies and procedures of the store. The customer should know in advance how inquiries are handled, the specific procedure for dealing with complaints, and what the store's position is on such matters as claims, allowances, damaged merchandise, credit terms, and returned goods.

Finally, much will be done toward holding complaints down to a minimum if the counterperson keeps all promises. If he or she promises to pick up merchandise that does not sell, it should be picked up with no hesitation. If delivery is promised by Wednesday, the goods should be there by Wednesday.

HANDLING COMPLAINTS

If a complaint cannot be prevented, then it must be handled. The first step in dealing with an irate customer with a complaint or an item to return is to encourage him or her to discuss the problem. The counterperson should respond with a simple, pleasant, no-commitment reply, such as "Oh, I'm sorry, Bill. What happened?" The counterperson should never appear angry. Maintaining a friendly response in the face of anger is almost entirely disarming to an angry person.

The customer will undoubtedly begin talking, and the counterperson should ask questions that help determine what steps were taken without hinting at any kind of blame one way or the other. After each angry sentence from the customer, the counterperson should wait a few seconds before replying. By doing so, this tactic cuts the rhythm of the customer's anger.

After talking for a while and discovering that the counterperson is not contradicting, the customer will usually begin to calm down. If he or she does not, the counterperson should try to make the customer realize that he or she cannot respond to both the customer's anger and needs. Because of this, the customer's anger will do him or her no good. In addition, the counterperson should begin speaking in a normal tone of voice, then gradually lower the voice. This method has a calming effect because more attention must be focused on listening.

The counterperson must not be concerned with placing blame because it is not evident that there is any blame to place. The customer is not always right, but the counterperson must be careful when proving he or she is wrong. An embarrassed customer will probably spend money elsewhere.

Often the easiest way for the counterperson to deal with a difficult situation is to put him- or herself in the customer's shoes. Most people, including counter personnel, are angry when something bought causes a problem, and usually the anger is directed toward the person who sold it. The counterperson should take whatever time necessary to understand the customer's viewpont (Figure 11–9).

When the counterperson understands the problem, he or she should state the problem to the customer as he or she understands it. If possible, it might be a good idea for the counterperson to have the customer write his or her viewpoint on paper because the customer is less likely to include exaggerated aspects of the problem in the written version.

If a counterperson discovers that the product is faulty, amends should be made immediately. If the counterperson is reasonably convinced that a product is to blame, it should be replaced. Of course, the counterperson must ask questions to make sure the product was used properly.

If the counterperson realizes that he or she caused the problem through a mistake or a wrong recommendation, then he or she should quickly admit the fault, apologize, and make a fast settlement. If store policy does not allow this, the counterperson should notify the boss so that the problem can be handled quickly and quietly.

If the counterperson determines that the problem was caused by customer misuse, then diplomacy must be used when stating so. Counter personnel can avoid pointing fingers through statements such as, "Sandscratch problems can often be avoided by using a better quality thinner in the primer. I think this could have prevented your problems." The act of referring to the problem as the subject rather than the customer's fault in causing it is called third per-

FIGURE 11–9 Careful listening is the key to understanding and helping an upset customer.

son, impersonal language. By using inoffensive techniques, the counterperson can get the point across without sounding accusing and without greatly embarrassing the customer.

Another way for the counterperson to explain a point or position is to show documentation such as the manufacturer's literature, repair manuals, basic textbooks, or other visual aids. With the use of documentation, the counterperson is better able to show the customer that he or she is not making up facts to suit his or her own needs.

Sometimes it will be impossible to determine fault because of a misunderstanding or counterperson's-word-against-customer's situation. Naturally, counter personnel cannot give in to every customer with a complaint. If the counterperson cannot in good conscience make the settlement that the customer demands, perhaps he or she can compromise within the bounds of store policy.

All problems should be solved as quickly, efficiently, and calmly as possible.

APPRECIATING GOOD CUSTOMERS

Customer relations include dealing with the good customers as well as the difficult ones. Many customers will pay their bills on time, recommend the store to friends, and usually do business without complaints. The counterperson must not only learn how to handle the difficult customers, but should also learn to show appreciation for the good ones.

A spoken thank-you should always be added to every business transaction. Bonuses or discounts for regular customers who buy a certain amount over a period of time are other ways of showing appreciation. Some businesses offer bonuses or discounts to customers who pay all bills on time.

Not only does the cooperative customer get the recognition he or she deserves, but the counterperson is regularly reminded of and regularly shows appreciation to the good customers. And most counter personnel will be better able to deal with the difficult customers by remembering that some customers are a pleasure to assist.

HANDLING TELEPHONE CALLS

Just as the appearance of the business and of the counterperson reflects an attitude to the customer entering the store, the way the counterperson answers and handles telephone calls reflects the attitude to the customer calling in (Figure 11–10). Most customers will assume that the counterperson's phone manners accurately reflect the store's competence. Courtesy and sincerity are just as important when dealing with customers on the phone as with customers in the store.

If at all possible, calls should be answered before the fourth ring. When answering the telephone, the counterperson should identify him- or herself. The caller cannot see the person answering the phone, and unless he or she knows the counterperson well, he or she may not recognize the voice. In addition, most people feel more comfortable talking to a voice with a name. Saying something like, "Good morning, Goodville Auto Parts. This is Charlie" gives customers a better image of the person behind the counter. Likewise, the counterperson should get the name of the caller.

Many times the phone rings while the counterperson is talking with a customer. If no one else is available to answer the phone, the counterperson should politely ask the customer to wait and then answer the phone (Figure 11–11). If the call will take a while, the counterperson should explain that he or she is assisting another customer at the moment. After taking pertinent information, the counterperson should request that he or she return the call as soon as possible or that the caller phone back in a few minutes. That way, neither the caller nor the counter customer is kept waiting very long.

The telephone customer cannot see the counterperson's facial expressions, gestures, or body language. All communications must be accomplished with words, tone, timing, and inflection. Counter personnel must remember to speak clearly and slowly enough to be understood and must request that the customer do so if he or she cannot be understood.

THE "HOLD" BUTTON

The "hold" button on most business telephones allows the counterperson to answer quickly and still finish the current task before diverting attention to the new caller. However, the "hold" button should never be used unless necessary. Putting a customer on hold while one counterperson tells another the rest of a joke is not good use of the device.

If a counterperson must put a customer on hold, he or she should identify the business and politely state something such as, "Please hold one moment." The counterperson should never pick up the phone

FIGURE 11-10 The way the counterperson answers the telephone affects the attitude of the caller.

FIGURE 11-11 Always acknowledge the presence of customers who approach the counter when you are on the phone.

and punch "hold" without saying a word, and he or she should never push the button before finishing the sentence or phrase. If the counterperson knows it is going to take a while before being able to deal with the caller, he or she should ask that the caller call back rather than put him or her on hold for a long period of time.

TIME MANAGEMENT

On the phone, time management means taking as much time as necessary to establish, maintain, or enhance the relationship with the customer, and exchange all the information needed to handle the request—but not a moment more. Some people habitually talk less while on the phone, and the counterperson might have to work harder to extract all necessary information from these customers.

Other people talk more while on the phone, and the counterperson might have to politely excuse himself. If that fails, the counterperson should consider raising the pitch (not volume) of the voice to convey a sense of urgency. Or the counterperson can simply stop speaking except for short replies to keep from blatantly ignoring the caller.

EXTRA SERVICES

Many small ideas have turned into big profits for businesses. Seemingly small tips that can improve customer relations include: offering customers a bag or box when they have an assortment of parts to carry, or hanging a bulletin board to inform professional mechanics and interested customers of seminars and other events (Figure 11-12).

Another way to build closer and better relationships with customers is to sponsor technical clinics either within the business or in another easily accessible location. Professional customers might be attracted to clinics discussing very technical and specific information such as ignition systems, air conditioning, or brakes. Novice car owners or users might be attracted to seminars on changing a tire, changing the oil, or jump starting a battery. An overt sales message should never be part of a clinic. Clinics are a service to regular customers and a way to attract new ones.

CATCHING THE SHOPLIFTER

Shoplifting is a big problem for retailers carrying any type of merchandise. According to one au-

thority, five to ten cents out of every dollar spent in a store goes toward the cost of stolen goods and the cost of preventing shoplifting and prosecuting the shoplifter.

Depending on a company's awareness of a problem and the measure it takes to prevent shoplifting, losses can range from between 1 and 15 percent of sales. These figures are especially damaging to auto parts businesses where a net profit of 4 or 5 percent is considered excellent.

In an average parts store, profits before taxes are 2.9 percent of net sales. If in one of these average stores shoplifting is only 1 percent of sales, eliminating that one percent would boost net profits by 34 percent. And nationally, 1 percent is a very low figure.

REASONS FOR SHOPLIFTING

A shoplifter could be anyone, from any walk of life and any age. The stereotypical criminals—drug addicts, kleptomaniacs, and professional thieves—actually make up the minority of shoplifters.

People shoplift for a variety of reasons. The kleptomaniac cannot help it. The addict needs to support a habit. The professional is pursuing a lucrative career. But these groups are so far in the minority that counter personnel will more often have to keep their eyes open for "ordinary" people.

Most people who steal do not need what they take and can afford to pay for it. Many shoplifters

FIGURE 11-12 A counterperson can improve customer relations by providing extra services such as hanging a bulletin board in the business to inform interested customers of seminars, clinics, or other events of interest.

FIGURE 11-13 Keep shelves stocked and organized to discourage shoplifting.

take luxury items to stretch their budgets. Juveniles may do it for kicks, on a dare, or for acceptance by their peers. Some people rationalize theft by citing high prices. Some simply want to see if they can "get away with it." Some take pleasure in "beating the system," taking a shot at the establishment.

Underneath all these "reasons" is the shoplifters' apparent belief that shoplifting does not really hurt anyone, that merchants can easily afford to part with some of their excessive profits.

Stopping shoplifters involves reducing the opportunities (and therefore the temptation), increasing the shoplifter's awareness of the risks, and increasing the risks as well.

REDUCING OPPORTUNITIES

Since counter personnel are the employees in a parts store or department most likely to be in contact with the public, it is usually up to them to catch or deter shoplifters. The first step is to reduce the opportunities for shoplifting. The possibility of stealing something undetected tempts people and, to a large extent, store owners and employees can keep everyone honest by withholding temptations.

Visibility is the first line of defense. Gondolas and endcaps should be kept low enough that counter personnel can see everyone in the display area from behind the counter. Counters and catalog racks should be low enough to provide good visibility, too. Counterpersons should be especially careful about the placement of double-tiered catalog racks. They are convenient and save counter space, but they can make it difficult to keep an eye on things out front.

Likewise, it is important that all the "tools of the trade" be within easy reach to prevent having to leave the area too often.

Counterpersons must be especially aware of tall floor merchandisers. Standing on the far side of a revolving tool display, a shoplifter can pocket whatever he or she pleases, covering the emptied slots with merchandise from adjoining hangers. The counterperson sees the rack turning, does not see any empty holes, and assumes the shopper is just looking.

Cluttered display areas are a shoplifter's paradise. Items taken from a shelf that is in disarray are seldom missed. If items are stocked in multiples of four, six, or eight, it is easy to spot missing merchandise (Figure 11-13).

Anything that slows down the thief or causes him to become conspicuous is a deterrent. For example, if electric drills and droplights are displayed out front, cords should be tied together. If display items are sold and used in pairs or sets—such as air shocks, helper springs, customer wheels—only one of the pair should be displayed.

Expensive items should be displayed behind the counter or in locked cases. Counterpersons should give a lot of thought to what gets displayed near entrances and exits. In the moments that a counterperson is busy researching parts in a catalog, a thief can easily slip something under his or her jacket, and glide through the door. It is far better to keep the areas around entrances clear, or to display only large, heavy, or awkward merchandise near the doors. Barrels of oil and chemicals and bags of floor sweep are safe choices for display near entrances.

Counter personnel should never have more doors open than can be reasonably watched. It is also important to restrict access behind the counter and to other areas of the store. It does not do a lot of good to keep expensive merchandise in back if it is easy for customers to step behind the counter.

Sound alarms on entrances and exits as well as at the entry points to restricted areas inside the store can help keep employees aware of customers' movements and discourage thieves.

Shoplifters can be made more visible by installing convex and one-way mirrors, peep holes, even TV scanners (Figure 11-14). If there are upstairs offices, a window overlooking the display area is an excellent deterrent. Signs warning shoplifters they will be prosecuted are also effective.

In addition to adjusting the physical layout to discourage crooks, counterpeople can further deter shoplifting by paying more attention to each customer. Counterpersons should note one big difference between legitimate customers and shoplifters: cus-

FIGURE 11-14 Strategically placed convex mirrors will allow store personnel to monitor all merchandise areas.

tomers want attention; shoplifters do not. If you make it a point to greet people as they walk in and periodically acknowledge their presence, the customers will be appreciative, but the thieves will become uncomfortable.

Shoplifters like to work when there are as few staff present as possible. Staggering lunch hours and breaks and not socializing in one area keeps "eye pressure" on the store.

INCREASING RISKS TO THE SHOPLIFTER

Apprehending shoplifters is an area where the counterperson must move with extreme caution. Several successful legal actions have been brought against merchants who accused people of shoplifting.

The laws that apply to the handling of shoplifters are different in each of the 50 states. The counterperson should be familiar with his or her state laws before attempting to apprehend or prosecute anyone. This can be accomplished by checking with local police, the district attorney, a lawyer, the chamber of commerce, and trade associations.

Generally, the laws say that a merchant, who has reasonable cause to believe a person has committed a theft, may detain that person in a reasonable manner in order to:

- Request identification
- Verify the identification
- Make a reasonable inquiry as to whether that person has in his or her possession unpurchased merchandise
- Investigate the ownership of that merchandise
- Call the police, and turn the person over to the police

These laws give the merchant the right to detain, investigate, and decide whether or not to prosecute, while protecting a counterperson from civil liability if, and only if, he or she has complied with the law.

The first point to consider is that the merchant must have reasonable cause to believe a theft has been committed. The counterperson should actually see the person take the merchandise and be able to testify that he or she saw the suspect take it. To strengthen the case, the counterperson should also see the person conceal the merchandise.

The counterperson must be sure the person does not get rid of the merchandise before stopped. If the shoplifter takes the contents of a box and throws away the box, the counterperson should be sure to recover the box.

The counterperson should also see that the person did not pay for the merchandise. If the counterperson loses sight of the person, it is possible that the individual put the merchandise back or paid for it. In either case, an apprehension may be deemed a false arrest.

Although it is not strictly necessary, it is usually best to approach the suspect outside the store. Allowing the person to leave the store further establishes an intent to steal, and it avoids disruptive scenes inside.

Another point the laws make is the store's responsibility to investigate the ownership of the merchandise before filing a complaint. One expensive court award resulted because a manager did not check out the suspect's claim that he had bought the merchandise elsewhere.

Counter personnel should understand that when a counterperson apprehends, detains, and files a complaint against a shoplifter, he or she has made an arrest. The fact that the counterperson turned the accused over to the police does not transfer responsibility for the arrest to the police. The counterperson made the arrest, and is liable if it is improper. Whether to prosecute or not is generally a decision made by store owners or managers, not counter personnel. No employee should attempt to deal with shoplifters without clearly understanding his or her company's written policy.

REVIEW QUESTIONS

1. Which of the following is not true concerning customer relations?
   a. Customer relations is practically everything a counterperson does in the course of selling parts to a customer.
   b. Relating positively to customers is one of the most important skills a counterperson has to develop.
   c. Many jobbers thrive on one or two customer accounts.
   d. Customers can be walking, talking advertisements for a business.

2. Which of the following is not an important step in becoming more familiar with customers?
   a. offering unlimited credit
   b. referring to customers by name
   c. remembering likes and dislikes of regular customers
   d. keeping a file of regular customers

3. Which of the following customer's images can affect where he or she chooses to do business?
   a. self-image
   b. images of counter personnel
   c. images of a particular business
   d. all of the above

4. In order to improve the image of the business, Counterperson A keeps the counter area clean. Counterperson B keeps displays organized and well stocked. Who is right?
   a. Counterperson A
   b. Counterperson B
   c. Both A and B
   d. Neither A nor B

5. What is the customer's most obvious want when he or she enters a parts store or department?
   a. to make a friend
   b. to fix or service his or her car
   c. to be recognized and liked
   d. to demean counter personnel

6. Which of the following is not an example of counter courtesy?
   a. taking breaks behind the counter
   b. assisting customers in turn
   c. making eye contact
   d. saying thank you

7. Which of the following is not true concerning counter courtesy?
   a. The counterperson should be sure to use the right price sheets or discount rates.
   b. The counterperson need not respond to customers until it is their turn at the counter.
   c. Most people respond favorably when someone recognizes and understands their needs and wants.
   d. Counter personnel should stop speaking to one another whenever a customer approaches the counter.

8. What is the heart of all customer relations?
   a. pricing
   b. appearance
   c. communication
   d. familiarity

9. The counterperson should remember that his or her job is not just selling but
   _____ .
   a. selling related items
   b. selling solutions to customers' problems
   c. memorizing parts numbers
   d. handling complaints

10. Which of the following does not have to be explained to professional customers?
    a. basic automotive mechanics
    b. manufacturer discounts or defects
    c. where to acquire out-of-stock parts
    d. sales prices

11. Which of the following can be used to assist the do-it-yourselfer?
    a. printed how-to information
    b. special tools or equipment
    c. referrals to professional customers
    d. all of the above

12. Which of the following is not information that the counterperson must know before selling any parts?
    a. make and model of the vehicle
    b. history of the problem
    c. name of the last person to work on the car
    d. year of the vehicle

13. Which of the following is not a typical difficult customer?
    a. those with delinquent accounts
    b. inattentive customers
    c. do-it-yourselfers
    d. angry customers

14. Counterperson A believes that keeping all promises will prevent customer complaints. Counterperson B believes that product knowledge and meeting the customer's needs will prevent customer complaints. Who is right?
    a. Counterperson A
    b. Counterperson B

    c. Both A and B
    d. Neither A nor B

15. Which of the following is not true when dealing with an angry customer?
    a. The counterperson must not be concerned with placing blame.
    b. If the counterperson cannot solve the problem in a few minutes, he or she should move on to the next customer.
    c. The counterperson should try to put himself or herself in the customer's shoes.
    d. The counterperson should use documentation when explaining a point to the customer.

CHAPTER TWELVE

THE MANAGER'S POSITION

Objectives

After reading this chapter, you should be able to:

- List several advantages of a management position, including salary, fringe benefits, greater autonomy, authority, prestige, credibility, receiving firsthand information, satisfaction in success, and career advancement.
- List several disadvantages to a management position, including long hours, commissions and bonuses beyond one's control, responsibilities and blame, and balancing interests.
- List five important qualifications for a management position: interest, competency, creativity, setting an example, and mental fortitude.
- Explain preparations necessary before looking for a management position.
- Explain the ways to find a management position, including answering advertisements and networking.
- Explain questions a prospect should ask an interviewer before accepting a management position.
- Define the phrase "management by objectives."
- Explain the importance and use of sales records.
- Explain the steps in building a counter team, including searching for, selecting, and interviewing a prospect.
- Explain the types of training necessary for a new employee, including technical and sales training, inventory control, pricing, cash flow, displays, and productive habits.
- Explain the importance of the following management tips: hire good people, understand power, cut costs, watch ROI, and prepare for changes.

A manager's chief function is to cut costs while increasing sales and profits. Moving into management is a natural step in a counterperson's career path, bringing more autonomy and better compensation.

However, like any career choice, management has its advantages and disadvantages. The counterperson will be giving up some benefits in exchange for others.

Preparation is necessary to qualify a counterperson for a management job. Unfortunately, not everyone who receives the required adequate preparation is guaranteed a management position because few management jobs are available and the competition for them is great.

ADVANTAGES OF A MANAGEMENT POSITION

Management offers several advantages over straight counterwork. The most obvious is money. In an industry that is not noted for high wages, the step up to management is quite an improvement. According to a recent industry survey, the average store manager earns about 65 percent more than the average counterman.

Managers not only earn more money, they often have more and better fringe benefits. Use of a company car, longer vacations, travel packages, use of a company vacation property, and flexible working

321

hours are just a few of the added benefits that are fairly common in some areas.

INTANGIBLE REWARDS

Management offers more than just material compensation. There are many less tangible rewards.

Greater autonomy is perhaps the most important. It is natural to want to direct one's own affairs and to operate independently. Rather than working within the limitations of a fairly narrowly defined job, positions with greater responsibility offer greater decision-making freedom.

Some people enjoy the additional authority that comes with management. Most people agree that it is more desirable to give the orders than to take them.

Prestige is another intangible reward for management. People have a higher regard for leaders than followers and hold management in higher esteem than staff.

Closely related to prestige is credibility. Salesmen (and customers with problems) seek out the manager because they expect him or her to have the authority to give them the results they want.

Being "in the know" is another advantage of management. Managers are usually the first to know about decisions that are being made by the owners and are more likely to participate in those decisions. Other important information—bulletins from WD's, literature from manufacturers and trade publications—usually goes to the manager before it arrives at the counter (Figure 12-1).

A lot of satisfaction can be derived from success, and managers are in a position to receive a bigger share of that satisfaction. When a store succeeds, it is due to a team effort, and everyone shares in the success. But the manager has the additional satisfaction of having put the team together.

Even when a winning idea originates with a delivery driver or stock clerk, the manager is entitled to a certain amount of satisfaction for having had the good sense to implement it.

Finally, management positions are a big plus in one's overall career plan. To future employers, management experience will indicate a valuable and desirable employee. And the experience will have prepared that individual for even greater responsibilities.

DISADVANTAGES OF A MANAGEMENT POSITION

Of course, management also has disadvantages (Figure 12-2). The money is usually earned with long hours and hard work. Most managers receive salaries rather than hourly wages, and there is no overtime pay. When the job requires it, the manager simply has to put in the time. In fact, a manager can put in so many hours (including work taken home) that his or her actual hourly earnings are below those of counter personnel.

FIGURE 12-1 Managers should see that counterpersons have access to bulletins and trade publications that will improve their performance and increase their knowledge.

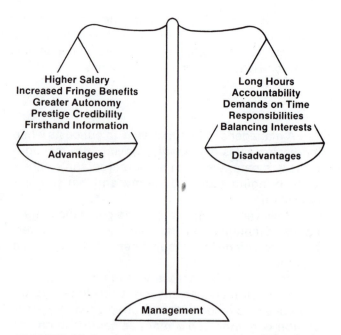

Higher Salary
Increased Fringe Benefits
Greater Autonomy
Prestige Credibility
Firsthand Information

Advantages

Long Hours
Accountability
Demands on Time
Responsibilities
Balancing Interests

Disadvantages

Management

FIGURE 12-2 A counterperson must weigh advantages against disadvantages when considering a move to a manager's position.

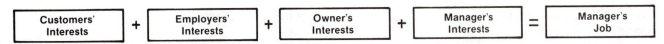

FIGURE 12-3 The manager's job consists of balancing and combining many different interests.

Also, a significant part of a manager's pay comes from commissions and bonuses that are tied to the store's performance which is never entirely within the manager's control. The economy can take a slide, a counterperson might offend and lose the store's best account, and the owners can make decisions that affect the store's ability to increase profit.

RESPONSIBILITIES

Autonomy is another mixed blessing, coming at the price of greater accountability. An employee following orders, company policy, or a prescribed procedure cannot be held responsible for the results, but a manager making fresh decisions about new problems is entirely accountable for the consequences.

With the privilege of giving the orders comes the responsibility for hiring, training, rewarding, disciplining, and terminating employees. Each of these tasks must be handled with thoughtfulness, insight, and an eye toward fairness. A manager's decisions in these areas will sometimes be perceived by employees (and owners) as careless, arbitrary, or unfair.

Being responsible for others can be a tremendous burden. Employee problems become the manager's problems—at least to the extent that the manager has to intervene if problems begin to affect the operation.

The prestige and credibility that come with management means people will be constantly making demands on one's time. Once the title of "manager" is appended to a person's name, insurance salespeople, financial planners, and people hawking office supplies will be vying for that person's time. Add to these the employees, customers, suppliers, and owners who have a legitimate need for that individual's time, and it is easy to see why one study concluded that executives are interrupted once every eight minutes.

As mentioned earlier, a manager shares in and derives considerable satisfaction from his or her store's success. When the store succeeds, the owner might perceive it to be because of the team efforts. When there is a failure, the owner might believe that the manager did not put together an adequate team, allowed the wrong conditions to develop, or formulated an inappropriate response to situations beyond his or her control.

Certain problems exceed the counterperson's authority, and these problems are passed on to the manager. The counterperson will rarely be expected to resolve conflicts in which the outcome might cost the company the customer. However, the manager does make those decisions.

The manager's job is a constant juggling act—balancing the equation in which the interests of customers, employees, owners, and him- or herself are often in conflict (Figure 12-3).

QUALIFICATIONS

The first step a counterperson must take in considering a managerial position is to weigh the advantages and disadvantages of taking that position (see Figure 12-2). If the counterperson decides that the advantages are worth the disadvantages, then the counterperson must decide whether or not he or she has what it takes to be a manager.

To make that decision, the counterperson must ask the following questions:

1. Do I have a real interest in the industry?
2. Am I competent?
3. Am I creative?
4. Am I willing to set an example of the kind of performance I am going to expect from my employees?
5. Do I have the mental fortitude needed?

INTEREST

Only a genuine interest in the automotive aftermarket is sure to keep a person tuned in to what is going on. The industry is changing rapidly. Not only is the automobile undergoing radical design changes but so is the market—with traditional suppliers losing sales volume as more and more retailing companies move into this sector. Successful managers are the interested ones.

COMPETENCY

Most counterpersons who want to move up to management are already competent. Not that a store manager has to be the most technically accomplished counterperson or a dynamic outside salesman, but it is going to be a little difficult for a person to tell others how to do a job that he or she does not already understand.

In recent years, a great deal has been written about effective Japanese management. For example, when Japanese workers are being groomed for management they are cross-trained in all aspects of their company's operations. They are generalists rather than specialists (Figure 12–4).

The counterperson who wants to become a manager is way ahead if he or she has worked on cars, delivered parts, and tried a hand at outside sales. Competence (though not necessarily excellence) in each of these areas will give that individual the credibility to lead drivers, counterpersons, and salespersons, and the insight to hire wisely.

CREATIVITY

The manager must be able to envision more than just the obvious or tried-and-true ways of dealing with a problem. He or she must often achieve large goals with limited resources, and it often requires a lot of imagination to trim costs and maximize resources. Example: What would a manager do to improve the effectiveness of advertising while cutting the ad budget in half?

SETTING AN EXAMPLE

There is a difference between leadership and dictatorship. The manager can tell counter personnel what kind of attitude and performance is expected or show them by example. Setting a good example always reaps the greatest long-term benefits.

MENTAL FORTITUDE

If everyone agreed that a thing needed doing, it would already be done. Instead, managers often have to champion an unpopular cause, such as convincing employees to go the extra mile or persuading the owners to allocate more resources.

At times, owners will want their manager to sell an idea to the employees, or the employees will expect the manager to advocate one of their views to the owners. Often these instances will involve disagreements and sometimes the manager must defend an uncomfortable position.

PREPARATION

Management is fairly easy to prepare for. Few high schools and colleges offer training in counter work, but virtually all of them offer valuable management courses, many in the evening.

A gold mine of literature is also available at a university library, business school, or large bookstore (Figure 12–5). People who have invested heavily in business educations and who have dedicated entire careers to studying management have distilled their experiences into some excellent books.

Finally, a counterperson interested in moving up should watch other managers. From the managers of successful organizations, he or she can learn a lot about effective management. And he or she can learn how not to do it from those who manage sloppy or lagging businesses.

Preparation does not include only studying books and studying managers. The counterperson

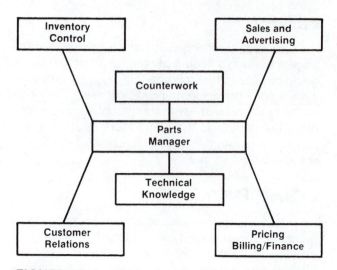

FIGURE 12-4 The best managers are cross-trained in all aspects of the business.

FIGURE 12-5 A counterperson should prepare for a management position by reading as much literature as possible about management.

FIGURE 12-6 A counterperson should check newspaper advertisements when searching for a management position, but most jobs are not advertised.

must have a track record as a worthwhile employee in his or her present position. Before an employer—even the present employer—will place someone in a management position, the individual must demonstrate perseverence, reliability, and integrity in his or her present capacity. He or she must also have a track record of being punctual and showing up for work consistently.

SEEKING EMPLOYMENT

Finding a job in management is difficult in some areas because of a shortage of management positions. Most stores and parts departments have only one manager, and if he or she is doing a good job, then employment at that location is doubtful. Chains with several outlets provide a better opportunity, especially if they are growing and opening new locations. Counterpersons who are serious about management should not rule out relocation or a long commute.

When looking for a management position, a person should remember that most good jobs are never advertised (Figure 12-6). Instead, promotions are made from within or through discreet inquiries with other people in the industry. Therefore, the interested counterperson should do two things:

1. Expand his or her job search beyond the known openings. Resumes should be sent to every interesting company, whether or not they are advertising for employees.
2. Tap the grapevine.

NETWORKING

WD salespeople, factory reps, customers, and other counterpersons usually know of the needs of other stores. Everybody who wants to work in management needs a group of people who can help find jobs, make referrals, provide good references, or just put in a good word here and there. This process of getting to know people throughout the industry is called networking.

Networking involves a whole underground sort of economy. A counterperson interested in a new position should attend events such as clinics, open houses, trade shows, and booster club meetings to meet as many people in the business as possible.

ASKING QUESTIONS

Before accepting a management job, the prospect should learn as much about the job and that particular business as possible. He or she should ask about or investigate the company's past performance, the kind of budget the manager will have, and how the manager's success or failure will be measured.

The prospect should ask a potential employer the following questions during the interview:

- Am I expected to be counterperson, outside salesperson, purchasing agent, and manager?
- How many times a year does the store's inventory turn?
- How many times would the owner like inventory to turn?
- What is the store's present net profit before taxes?
- Are bonuses tied to a sales quota or to profits?
- Will quotas be raised every time they are met?

MANAGEMENT BY OBJECTIVES

When a counterperson accepts a management job he or she will almost immediately be confronted with sales problems. Every sales problem constitutes an obstacle to a sales objective or goal. For example, a store's sales objective might be to increase sales to walk-in customers by 20 percent over a six-month period. Sales problems that could be hindering this goal might be that the store closes each day at 5 p.m., and most walk-in customers enter stores in evenings or weekends. Other sales problems could be that the entrance area contains

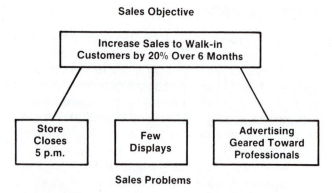

Sales Objective

Increase Sales to Walk-in
Customers by 20% Over 6 Months

| Store Closes 5 p.m. | Few Displays | Advertising Geared Toward Professionals |

Sales Problems

FIGURE 12-7 A manager must first solve several sales problems to achieve the sales objective.

very few displays, and that all the store's advertising has been geared toward professional mechanics.

Before the manager can realize the sale objective, he or she must devise ways to eliminate each sales problem (Figure 12-7). For example, perhaps employee's schedules can be rearranged to accommodate longer business hours, displays can be designed more attractively, and advertising can be altered. For each problem, the manager must weigh cost against results.

Objectives comprise the starting points for all productive business activity. Each objective must be precisely defined, such as the previously mentioned example. A manager should not simply state that he or she wants to increase walk-in sales. When objectives are not clearly defined, neither the manager nor employees know for sure when it has been met. For example, if walk-in sales are up one week, does that mean that the objective has been met?

The concept of management by objectives is utilized by most sales organizations, such as insurance companies. Their sales people are assigned objectives and are expected to work toward meeting them. What a new manager must realize is that a jobbing store is really no different than these sales organizations and that it is imperative for him or her to state sales objections to employees.

Three critical requirements exist in managing by objectives.

1. Every sales objective should be defined so clearly that its intent, scope, and limitations can be fully understood.
2. Every sales objective should be perceived by both the salespersons and the manager as being appropriate and attainable.
3. An orderly process should be used to solve problems that stand in the way of accomplishing sales objectives.

Even when this is done, the process is not finished. The manager has to evaluate the sales results in terms of the sales objectives. For example, using the previous example of increasing walk-in sales 20 percent, the manager might have chosen to rearrange employee's schedules to allow the store to remain open until 7 p.m. After examining sales records, the manager might realize that the store is actually losing money (or at least not gaining money) by staying open later on Mondays and Tuesdays because very few customers have walked in on those nights. The manager might then either rearrange the schedule again or offer special discounts on those nights to attract customers.

IMPORTANCE OF SALES RECORDS

The above-mentioned example shows how utilizing sales records can show a manager where a store's strengths and weaknesses lie. Thirty years ago the need for good record-keeping was not quite as great as it is today. At that time, the lines of distribution were clearly defined, so businesses usually sold to one class of trade and did not have to worry as much about covering all the distribution bases. Wholesalers sold only to retailers, who sold only to the ultimate consumer.

Now, however, with the advent of discount stores, discount warehouses, mass marketers, and huge service chains, the lines of distribution are not distinct. To make matters worse, the number of parts necessary to stock has almost quintupled in that same period of time.

Record-keeping is now more important than ever. Those people who have implemented an adequate system of record-keeping have a jump on those who have not, because accurate records of sales figures, collection figures and inventory can show the succeeding and failing aspects of a store. The following examples illustrate how records can be used to develop future business and cash flow.

Joe's Discount Auto

Joe's Discount Auto was founded back in 1979. The inventory sheets reveal that Joe's business is attracting the walk-in trade (Figure 12-8). The sheets show that the skin-packed merchandise he is stocking, such as his service line, light bulbs, oil change kits, and hand tools, are all moving well. He is getting a good number of turns on his inventory, and he is paying his bills on time. His overall business is good and his accounts receivables are low—only 20 percent of his total business. This means that 80 percent of his business is done in cash, which also points to a walk-in oriented business.

However, Joe would like to be able to take some more time off from work because he and his wife are the only two employees. If they are not there, no one is, and they are getting tired of working every Saturday and Sunday, not to mention the other five days of the week. The problem is that the volume is not high enough to justify hiring an additional person, and Joe feels that he has done just about all the advertising he can do. What can he do to increase his business profitably?

An analysis of Joe's sales records also reveals that almost 50 percent of his business is done on Friday, Saturday, and Sunday—again pointing to a high walk-in trade.

What if Joe took some time away from the counter on the other four days and made some sales calls to some dealers in his immediate area? The benefits would be two-fold. First, he might be able to increase his business on the slow days, because the days the garages and service stations are busy are the days that the walk-in trade business is not really very good. Secondly, Joe can get his volume up without any increase in overhead, since he has to be open on those days anyway. If the business does improve enough to justify hiring an extra person, he can add him on the days that he and his wife want to take off, and he will have moved into a better position without really risking anything in the way of extra operating overhead. There will be an increase in inventory, but it will result in fewer lost sales for the store, not fewer turns of the stock.

Sam's Auto Parts

Here is a store at the opposite end of the spectrum. Sam's Auto Parts has been in business for

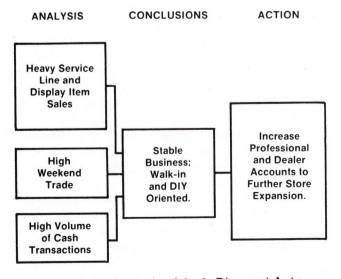

FIGURE 12-8 Analysis of Joe's Discount Auto

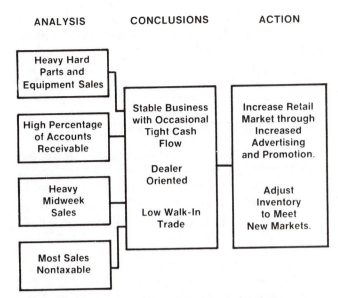

FIGURE 12-9 Analysis of Sam's Auto Parts

more than thirty years. Sam's records show that his busiest days are Tuesday and Wednesday and his sales of display-packaged merchandise are marginal at best. He is extremely strong in the area of hard parts, tools and equipment, and the more traditional kinds of parts, but when it comes to items like striping takes, headers and radios, he makes few sales (Figure 12-9). His volume runs about $40,000 per month, and his accounts receivable is 55 percent current, with the best being anywhere from 30 to 150 days past due. Nontaxable sales account for 87 percent of his business and he employs three people plus himself.

These records indicate a fairly well-established traditional parts store. The emphasis has been on the professionals and their needs. It is a high volume store, but one that is having collection problems and must round out the work schedule.

For example, a further delving into the records shows that Sam has people on hand Monday through Saturday, but only four of those days are really busy enough to justify the people that Sam has on board. If he goes to a shorter work week, he might lose some of his help because of the decrease in hours worked and, consequently, wages. He cannot increase his business much more with the help he has, because on four days they are just about as busy as they can adequately handle.

As a solution, Sam might try to increase his Saturday and Friday business by leaning a little bit more toward the walk-in trade. Most stores with a heavy walk-in trade are busier on the weekends than they are other times, so maybe by utilizing his existing help, he can kill two birds with one stone. The increase in walk-in business will help him out on the weekend, and since most retail business is done in cash, his cash flow will improve. Sam might have to purchase additional inventory to appeal to this group, but with careful monitoring of his inventory records, it should not be too much.

OTHER USES OF SALES RECORDS

These are just two small examples of the use of sales records in managing an auto parts business. Depending upon the amount of detail in the records, the manager might be able to go even further.

For example, if a store's invoices are printed with a point-of-sale computer, the system probably has an internal clock and the time of each purchase can be printed on the invoice (Figure 12–10). If a manager analyzed this information and found that the store is busiest between 9 and 10 a.m., and 2 and 3 p.m., maybe he or she could shift the hours around so that not everyone comes to work at the same time. Two counterpersons could come in at 7:30 a.m. and leave at 4:30 p.m. The other two could come in at 9 a.m. and work until 6 p.m. This would add the benefit of having a full crew in the store during the busiest time of the day, yet the store could stay open for longer periods of time without paying a huge amount of overtime.

Related Sales Record

Another sales record that is sometimes a little more difficult to pull out is the related sales record. The manager needs to look at a given category. For example, if 300 thermostats are sold annually, it

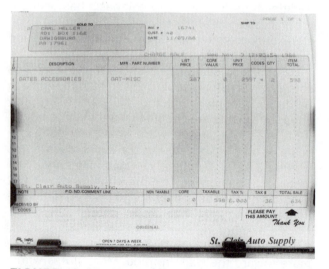

FIGURE 12–10 If a store's invoices are printed with a point-of-sales computer, the system probably prints the time of purchase on the invoice.

would make sense that about the same number of thermostat gaskets should be sold. If this is not the case, the manager should check to see if everyone is pushing a gasket with the thermostat sale. The manager could also check to see if other items such as water pump lube, gasket cement, and radiator caps were being sold.

Checking Advertising

Sales records can also be referred to in order to check the effectiveness of advertising. If a manager has always advertised heavily in the winter but not in the summer, he or she should examine the sales of the advertised items. If they are higher than would normally be expected, it is probable that the advertising has paid off. In that case, it would probably make sense to advertise in the summer months also. Unless the available sales information and records are used, the manager will never realize the store's potential.

If a manager runs an advertising campaign that does not work, then two months later runs one that does, he or she should not be too quick to say that the first one was poorly put together. Maybe the product mix was improper for the store and its image. If import parts have never been sold, the manager should not expect to change that product mix with just one ad campaign. He or she should look at the sales records to see if one or two import manufacturers are doing better than the rest.

Perhaps the way to start a campaign would be to target the vehicles that are already doing fairly well, or pick a line of imports already carried, such as spark plugs, and place special emphasis on that line. The sales records provide a starting point for the manager.

Management Philosophy

Probably the most revealing thing that sales records can tell a manager is his or her own management philosophy. Many managers have a philosophy in their head, and it might even be in writing, but a glance at their credits and warranties will reveal their actual company policies.

Sometimes a manager's philosophy will not be worked out in the actual policies of the company. For example, a manager might examine the records and find that warranty on rebuilts is running 15 percent of sales. The first reaction of the manager might be to jump on the manufacturer for quality control. If, however, the manufacturer has figures that show only a 5 percent warranty rate, the manager might have a policy of accepting all rebuilts back for credit, without question. Without checking records, a manager does not really know how well store policies are being enforced.

BUILDING THE TEAM

Another important aspect of the manager's job is the building of a professional counter team. To run a quality operation, a manager needs quality people.

TYPE OF PERSONS SOUGHT

Many parts managers or parts store managers were once successful counterpersons. They know what characteristics will produce efficient and profitable team work at the counter. A counterperson requires a certain amount of clerical skills, an ability to handle numbers skillfully, and an ability to do accurate research.

Some managers look mainly for a person with enthusiasm, optimism, and a willingness to learn. Others look for some type of automotive background. Some managers prefer to hire inexperienced counterpersons because they want to train them to do things their way without having to break them of habits acquired working somewhere else. Other managers prefer employees with counter experience because they understand the basic skills necessary and can potentially offer new ideas and techniques learned elsewhere.

Of all the qualities that are important in a counterperson, none are more important than an interest in automotive technology. Someone with interest will learn faster than someone who feels forced to learn, and the interested counterperson will continually update his or her knowledge because he or she enjoys the subject.

THE SEARCH

Finding the people for a team can be much more difficult than deciding what kind of people are desirable. The obvious routes are advertising in the newspaper, contacting employment agencies, and notifying the local unemployment office.

Advertising

Advertising through the media often attracts more people than the manager can handle. One newspaper ad can keep a manager buried in applications for a week and overwhelmed by phone calls from employment agencies for a month. The same can be said for using public agencies.

A manager can avoid time-consuming crowds and phone calls by advertising in a newspaper with-

out mentioning the name of the business. For a modest fee, most newspapers will provide a reply box at their address and the ad can request that resumes be sent there. The newspaper then forwards the replies to the manager, who can evaluate the replies at his or her own convenience and choose only to see the most promising candidates.

Agencies

If an employment agency is used, the manager can give very specific instructions for screening applicants. If they do not know how or fail to filter through all their candidates, the manager, once again, will be faced with a deluge of applicants. Government agencies might post all openings on a bulletin board, eliminating screening of applicants.

Aftermarket Grapevine

Employment experts say that the majority of good jobs are never advertised. Instead, notification about openings is served through the particular industry's grapevine, and the position is filled by someone already in the field.

This method of hiring saves dealing with the public and bureaucratic agencies. It tends to ensure that all the candidates will at least be familiar with—possibly even highly experienced in—the type of business in question. And it makes it easy to check the candidates' backgrounds and references. Instead of hiring someone off the street, the manager might be able to hire someone he or she already knows.

The aftermarket's grapevine is all around the manager. The manager's other counterpeople probably know several prospective employees, some of whom might want to make a change. The warehouse salespeople who call on the store see scores of counterpersons each week. They are in a position to know who is available and to pass the word about the manager's needs. The warehouse people who take the telephone orders also talk with hundreds of counterpeople each week and often become familiar with them.

But the best way to tap the grapevine is to become an integral part of it. Suppliers are always hosting clinics, open houses, or promotional dinners, which give the manager ideal opportunities to become acquainted with many other people in their industry. The meetings and social functions of the local trade associations afford the same opportunities.

Making oneself available as a speaker to the schools is a great way to meet the students and faculty. Instructors appreciate the input from real "experts" in the industry, and when they spot a particularly trained student, they are likely to refer him or her to the companies that have become familiar.

Raiding

Most managers understand the importances of knowing their competitors' strengths and weaknesses as they apply to specific products, prices, merchandising, and delivery service. But it is also important to know their people, because they can often be one of the manager's most valuable sources of talent.

Good counterpeople often develop a clientele that is more loyal to them than to their employer, and hiring a competitor's (former) employee cannot only bring a fairly high caliber of counterperson to the business but often their customers, too.

Businesses that are susceptible to raiding because of inadequate compensation, poor working conditions, abusive management, or a perceived lack of future, object that it is unsporting (even unethical) to steal talent.

However, the automotive industry is a free market in which all compete for workers as well as customers. A successful manager will keep this fact in mind when dealing with his or her own employees.

Raiding usually does strain relations because nobody likes to lose a good employee—particularly one they have trained.

However, occasionally a supplier, competitior, or customer might actually be happy to have another manager hire one of their people, such as when they are facing a cutback and would like to place some of their people in other jobs rather than lay them off. Mentioning a need at the appropriate moment can solve problems for all parties concerned.

THE SELECTION

The manager must select and hire the right person from among several candidates. Since the job involves a lot of work with words, numbers, organization, detail, and thoroughness, it makes sense to look for evidence of this in applications and resumes. Sloppy or incomplete applications and disorganized resumes might indicate that the person lacks enthusiasm or organization skills. References should be checked as well. Only after the manager has been favorably impressed by the candidate's application, resume, and references should the manager take the time to schedule an interview.

Interviews are the most time-consuming aspect of hiring and should be reserved for only a few of the most interesting candidates.

Appearance

The first thing to notice in the interview is the candidate's appearance. Like the appearance of the resume, it indicates something about the person's self-image. While there is considerably more latitude today about what kind of appearance is acceptable, a person's self-image is a useful indicator of potential performance. Also, the manager must be comfortable in presenting this person to customers.

Questioning

The purpose of the interview is to allow both parties to size up each other before entering into negotiations. The manager should want to learn specific things about the candidate's education, experience, and special skills, and also something about the applicant's values and aspirations.

Two kinds of questions are useful for this. Direct questions require immediate answers. For instance: "What was your grade point average?" "Why did you leave your last job?" Direct questions have a very limited range of answers and are useful for getting right at the information wanted. Unfortunately, direct questions can also be anticipated and answered with canned responses that tell very little more than what was specifically asked.

Open-ended questions can reveal much more. Some examples of open-ended questions include: "Why are you interested in a career in the aftermarket?" "Tell me about your most successful assignment." "What could your former employer have done to take better advantage of your abilities?"

Open-ended questions give the manager an opportunity to observe the candidate under pressure, to see how well that person can organize and articulate his or her views on several different issues.

Planning

Good interviewing takes planning. At best, poorly prepared interviews are a waste of time for both parties. The worst thing that can happen is that they result in an inappropriate hiring or rejection.

Not only does the manager need to be aware of what to ask, but also of what not to ask. Questions about race, religion, age, and criminal history might bring legal difficulties. The local automotive wholesaler's association can probably supply guidelines and even good application forms.

Cross-Questioning

Interviewing is a two-way street. Just as the manager has to learn a lot about the candidate, the candidate must learn a lot about the store before making a decision. The better the candidate, the more likely he or she is to be careful and selective about employment.

Quality people are concerned with more than just a paycheck. They care about where they work, who they work with, and where they are going. A manager should be concerned about an applicant that does not bring some pretty pointed and probing questions to the interview.

In the traditional aftermarket, some people state that their prices are justified by their quality and service. Some of this thinking has to apply to how managers shop for help, as well as how they sell parts.

A recently published survey on compensation in the industry indicates that the national average wage for counter personnel is just under $14,000. If this figure does not include overtime, it works out to $6.73 an hour. In some communities, grocery checkers earn twice that amount—with better benefits.

The same survey shows that the largest jobbers are paying their counterpersons more than 40 percent more than the smallest. One reason they became the most successful teams might be because they attracted and retained the best players.

TRAINING

The building of any team is incomplete without a training program. Succeeding can only be assured by staffing the most knowledgeable, skillful, and productive players.

Technical and Sales Training

Most training programs still concentrate on the two most obvious areas in which a counterperson must be well equipped: technical knowledge and sales ability. Technical and sales training are important, and a wealth of material is available from manufacturers, WDs, and programmed distribution groups. The trade press is another excellent resource for technical and sales information and provides a clearinghouse for information on what materials are available (Figure 12–11).

However, other areas need attention in a training program, too. Basic information about how a particular store fits into the distribution picture might help a new counterperson relate to customers and suppliers. To purchase intelligently and give customers credible explanations about prices a new counterperson might have to understand where the parts come from and what market forces make it

FIGURE 12–11 Suppling employees with up-to-date sales and technical data will build a stronger staff.

easy for the store to compete in some areas and difficult to compete in others.

Inventory Control

Inventory control is another area that every counterperson must understand. Questions a counterperson should be able to answer include:

- Under what circumstances are lost sales recorded?
- How well should a part be moving before it is stocked?
- How poorly should it sell before it is sent back?

Pricing

Pricing is based on an interplay between management's philosophy and market realities. Parts of varying quality are available at various prices, and for any parts that are sold, there must be a given markup to cover operating expenses. But competition puts a ceiling on the selling price.

Within those constraints, a manager wants a certain minimum profit or perhaps the maximum possible profit. To achieve the sales and profit goals, counterpeople have to understand the manager's philosophy. They have to be aware of what factors determine a store's return on investment (ROI) and the steps they can take to keep it in line with the manager's expectations.

Return on investment concerns all personnel. It is a measure of how well a store's resources are being employed and can be expressed as net profits as a percentage of assets. Each counterperson manages the ROI whenever he or she engages in a transaction that affects inventory—for instance, when the counterperson makes a decision to put a returned part into stock rather than sending it back to the vendor.

Cash Flow

Cash flow is a related concern that counterpeople can affect, but only when they understand its importance. A manager might not be too concerned with an account that charges a couple of hundred dollars a month and pays a little slowly. But when he or she orders a $2,000 engine, it is nice if the counterperson handling the deal insists that this particular invoice be paid promptly.

Displays

Some people seem to have a natural talent for developing good-looking and effective displays. However, for those without that ability, training might be the answer.

Productive Habits

One of today's buzzwords is "productivity" and it encompasses all the little attitudes and work habits that make the difference between working hard and working smart. The manager should take the time to teach productive work habits to counterpersons.

For instance, almost everyone can learn the business of organizing tasks and doing more than one job at a time. Several minutes can be saved if a counterperson organizes the items that must be researched in catalogs so that all items in a particular section are looked up before moving to another section or book.

Likewise, stocking shelves, either for retail sale or for storage in the stockroom, can be handled similarly. Stocking all items in one area before moving to another shelf can save several hours each month. Productive habits can save countless hours of work and frustration.

The powerful thing about time-saving techniques is that once a worker begins to use and appreciate one or two, he or she tends to develop more of them.

ADDITIONAL MANAGING TIPS

While the manager's job includes various tasks, it can be made easier by remembering the following tips:

- Hire good people.
- Understand power.
- Cut costs.
- Watch ROI.
- Prepare for changes.

HIRE GOOD PEOPLE

A good manager surrounds himself with good people. Some insecure managers feel threatened by excellent employees, so they tend to hire only people who have capabilities inferior to their own. They feel these people will never outshine them, make them look bad, or go after their jobs. This method is a self-defeating strategy, easily outdone by the competition.

UNDERSTAND POWER

A manager must understand power and use it wisely. Managers have two kinds of power: the authority that comes with the position and their own personal ability to motivate, persuade, or otherwise secure the cooperation of others.

Managers have the right to command company resources and are in a position to reward and punish. This kind of clout is indispensible but also incomplete.

The best managers also have a great degree of personal power. Their people perform—not because they have to—but because being on their team is exciting. Personal power is developed by gaining the respect and confidence of the workers by listening, paying attention, encouraging them, and treating them even-handedly.

Both kinds of management power are subject to erosion or enhancement. A manager can protect the authority of the management position by not abusing it and watching that it is not undermined by owners going around the manager to employees or employees bypassing management to get to the owners. This authority is also damaged when the manager fails to exercise it and thus allows a situation to determine the outcome or forces the owners to take action when management should have.

Likewise, a manager can lose personal power by abusing authority, treating workers unfairly, or breaking promises or confidences.

CUT COSTS

Cutting costs has a dramatic effect on profits and is therefore part of the manager's job. A good manager will periodically review the financial reports, treating each expense like an item on a checklist and looking for ways to function more productively and efficiently.

A manager should pay particular attention to the items headed "Fixed Expenses." That phrase is accounting jargon, and it does not really mean these numbers are fixed. Every manager's job is to contain costs, regardless of how the accountant might have classified them.

WATCH ROI

A manager can improve ROI by reducing inventory, increasing sales volume, increasing the margin of profit, and cutting costs. A manager can go too far in any of these directions—except increasing sales—and a balance has to be maintained. If a manager can show a healthy or improving ROI, the owners will not be able to find fault with his or her performance.

PREPARE FOR CHANGES

A manager must continually update his or her knowledge about the automotive industry and business management procedures in order to be prepared for changes in the industry or within the particular store. Managers should keep studying and networking. They need the input and insight of many other people in the industry to keep up with changes.

REVIEW QUESTIONS

1. A manager's chief function is to cut costs while _____ .
   a. pleasing employees
   b. pleasing customers
   c. increasing sales and profits
   d. receiving more pay

2. Which of the following is not an advantage of a mangement position?
   a. commissions from sales quotas
   b. fringe benefits
   c. autonomy
   d. prestige

3. Which of the following is a disadvantage to a management position?
   a. long hours
   b. accountability
   c. demands on one's time
   d. all of the above

4. Which of the following is not an important qualification for a management position?
   a. physical stature
   b. creativity

c. mental fortitude
d. interest

5. Which of the following statements concerning qualifications is false?
   a. A manager must sometimes defend an uncomfortable position.
   b. A manager should specialize in one certain task.
   c. Genuine interest in the automotive aftermarket will keep a person in tune with changes.
   d. Setting a good example always reaps the greatest long-term benefits.

6. Counterperson A earns a college degree in order to prepare for a management position. Counterperson B buys new clothes. Who is right?
   a. Counterperson A
   b. Counterperson B
   c. Both A and B
   d. Neither A nor B

7. Most stores describe a lost sale as one in which _____ .
   a. they do not get the order
   b. they must special order an item
   c. a customer buys only one of an item
   d. no customers request a particular item

8. When looking for a management position, Counterperson A applies to those companies advertising. Counterperson B applies to as many businesses as possible, even if they are not advertising. Who is right?
   a. Counterperson A
   b. Counterperson B
   c. Both A and B
   d. Neither A nor B

9. Which of the following is an important question for a management prospect to ask during an interview?
   a. Are bonuses tied to a sales quota?
   b. What are my expected duties?
   c. Will quotas be raised every time they are met?
   d. All of the above

10. Which of the following statements about objectives is not true?
    a. A sales objective is determined by the last month's profits.

b. Objectives comprise the starting points for all productive business activity.

c. An objective should be clearly defined.

d. Every sales problem constitutes an obstacle to a sales objective.

11. A manager can determine a store's strengths and weaknesses by examining _____ .

a. the counterpersons' behavior

b. sales records

c. the competition

d. advertising

12. Sales records can be used to examine _____ .

a. related item selling

b. advertising effectiveness

c. management philosophy

d. all of the above

13. What is ROI?

a. return on investment

b. a measure of how well a store's resources are being used

c. both A and B

d. neither A nor B

14. What is raiding?

a. hiring another business's employee

b. stealing the competitor's merchandise

c. using the competitor's advertising idea

d. firing an employee

15. Which of the following is not an important managing tip?

a. understand power

b. maintain the same level of success

c. watch ROI

d. prepare for changes

COUNTERMAN BUSINESS TERMS

The words and terms in this appendix are simplified definitions and, for the most part, are limited to business operations of an automobile aftermarket operation and counter personnel.

Accessories Parts that add to the appearance or performance of a vehicle.

Accounts Receivable Money due from customers.

Active Accounts Current customers who make frequent purchases.

Active Stock Merchandise in the store area readily available for sale to customers.

Automotive Aftermarket The distribution and sale of automotive replacement products.

Back Order Merchandise ordered but not shipped.

Bill of Lading A shipping document acknowledging receipt of goods and stating terms of delivery.

Break Point Where cost of shipping by a particular method changes significantly because of size or weight classifications. For example, parcel post shipments cannot exceed 40 pounds.

Cash Discount A discount given for the prompt payment of a bill.

Computer Terminal A keyboard system that permits a counterperson or operator to input invoices or obtain information from a computer usually located in warehouse distributor offices.

Core Items such as starters, alternators, carburetors, and brake shoes accepted in exchange for remanufactured items.

Counter Cards Advertising or display placards, usually with an easel back, placed on the counter.

Credit Memo The record of an amount paid to a customer usually for the return of an item or a core.

Dating Payments for merchandise extended to 30, 60, or 90 days without loss of cash discounts.

Dealers The jobbers' wholesale customers, such as service stations, garages, and car dealers, who install parts in consumers' vehicles.

Demand Items Items such as water pumps, bearings, clutches, and remanufactured parts that a customer needs for a specific automobile.

Deposit A specified sum of money a customer leaves with the jobber on special orders or to guarantee the return of a core.

Direct Mail Advertising literature sent to customers through the mail.

Discount A deduction, usually a percentge off the printed price sheet.

Display A special grouping of sale items or new product lines with point-of-sale materials designed to attract customers' attention.

Do-It-Yourselfers Retail customers who install the parts and accessories they buy.

Dual Account A classification of a customer according to the purposes for which he purchases. For example, a dual account customer can be one who both uses and sells parts, or a customer who uses parts for industrial and automotive purposes. The term is used in wholesaler policies that affect discounts, taxes, or credit.

Emergency Order An order placed for special requirements. It is not a regular stock or intermediate order.

End Cap A display fixture located at the end of gondolas.

Excise Tax Taxes on products paid by manufacturers or sellers usually included in the price of an item, but often added, as in the case of tires.

Extra Dating A discount is made available for purchased items that are delivered at a given date and marked payable in a given period, such as 30, 60, or 90 days. It is used to spread payment without loss of the cash discount.

Fixtures The furnishings in a store such as gondolas, end caps, display cases and racks, shelving, and counters.

Fleet A number of vehicles operated by one owner or company.

Franchise A specified sum of money an individual must pay for the privilege of owning and operating one of a chain of retail operations.

Freight Bill A bill that accompanies goods shipped describing the contents, weight, point of origin, shipper, and giving the transportation charges.

General Application Items Products such as oil, polish, antifreeze, and chemicals that apply to all makes of vehicles.

Gondolas Long, shelved display fixtures usually placed back to back to divide stores into aisled trafficways.

Gross Profit The selling price of an item less the jobber's cost.

Handling Charge The cost charged to a customer for returning an item to a supplier or manufacturer for repairs or adjustment.

Impulse Items Products customers buy on the spur of the moment to fill a want rather than a need.

Integrated Circuit A miniaturized electrical component consisting of diodes, transistors, resistors, and capacitors used in electronic circuits.

Interchange Lists Cross references for part numbers of identical items from different manufacturers.

Intermediate Order An order placed after a regular stock order to replenish stock until the regular stock order shipment is received.

Inventory Control A method of determining amounts of merchandise to order based on supplies on hand and past sales.

Invoice A sales slip used as a record of sales.

Invoice Register A system used to maintain a check on the issuing of invoices.

Jobber Owner or operator of an auto parts store usually wholesaling products to volume purchasers such as dealers, fleet owners, and industrial firms, and retailing to do-it-yourselfers.

Line The product list of a specific manufacturer.

List Price The suggested selling price to the final consumer.

Maintenance Free A battery that requires no water or charging under normal conditions for the length of the warranty.

Margin Same as gross profit: the jobber's cost subtracted from the selling price.

Marketing The total function of moving merchandise from the manufacturer into the hands of the final consumer including buying, selling, transporting, storing, and advertising.

Mark-Up The amount a jobber charges for merchandise above costs to earn a profit.

Media Vehicles for advertising such as newspapers, magazines, television, radio, and outdoor posters.

Merchandising The "second effort" of advertising such as building store displays of advertised items, posting newspaper advertisements and sales brochures, and promoting the event through public relations.

Modules Self-contained electronic circuits for computerized systems usually repaired by plugging in a new unit.

Microprocessor A very small computer-like device used to process signals from various circuits to achieve a control system that will adapt to operating changes.

Net Price The cost of an item to a particular purchaser such as dealer net, jobber net, or user net.

Net Profit The jobber's profit after deducting costs of merchandise and all expenses involved in operating the business.

Obsolescence When a part is no longer of use, either because of replacement by a superseding part, or due to lack of demand.

OE Original Equipment.

OEM Original Equipment Manufacturer.

Operating Capital The money a jobber needs in everyday operations to buy parts, pay salaries, and meet regular expenses.

OSHA Occupational Safety and Health Act. All companies with a specified minimum number of employees must comply with its safety regulations.

Overage More items received than ordered.

Overhead The costs of operating a business but not including purchases of merchandise.

Packing Slip A list of items and quantities accompanying a shipment.

"Paid-Out" A form or slip used to record such cash register transactions as returns or refunds on cores.

Percentage of Profit Profit accruing on the basis of the selling price of the product. (Gross Profit ÷ Selling Price = Percentage of Profit)

Perpetual Inventory A method of keeping a continuous record of stock on hand through sales receipts or invoices.

Physical Inventory Determining stock on hand through an actual count of items.

Pilferage Theft; shoplifting.

Point-of-Sale Materials (POS) Advertising materials such as window banners, placards, and counter cards used at a place of business. Also called point-of-purchase (POP).

Policy A guiding rule for the conduct of a business.

Price Leader   An item advertised at a very low price to attract customers into a store.

Price Sheets   Price lists usually accompanying catalogs from manufacturers or WDs; often in different colors for different types of customers.

Public Relations   Providing the media with stories of interest worthy of being reported in the news.

Purchase Order   A form from a buyer that specifies items desired, conditions of sale, and terms of delivery.

Rain check   A coupon that guarantees a customer the advertised price on a sold out item when it is available.

Rebuilt   A remanufactured part.

Related Items   The group of items and tools such as oil, filter, drain pan, filter wrench, and pour spout needed to perform a particular job on a car.

Reserve Stock   Merchandise usually kept in the storeroom to restock shelves and displays in the display area.

Retail   Selling merchandise to the walk-in trade, the do-it-yourselfers.

Returns   Merchandise returned by customers, usually for a refund or exchange.

Right-Hand Rule   Store layout designed to permit customers to move naturally to the right on entering.

SAE   Society of Automotive Engineers, which sets the standards for many products.

Salvage   That part or parts of an item retrieved from total loss or scrap, which is still suitable for restoration, use, or resale.

Selling Price   The price at which merchandise is sold, but might differ according to the type of customer: wholesale or retail.

Selling Up   Selling a customer a better quality item when a lower priced item obviously will not do the job he or she expects of it.

Shelf Talkers   Small placards or flags placed with the promoted items on display shelves to call attention to sale prices, seasonal items, or new products.

Spot   A radio or television commercial.

Stand-Up   A tall, self-supporting point-of-sale piece, usually flat and printed.

Staple Items   Items such as oil, coolant, additives, and car care products stocked in the display area that customers regularly purchase.

Statement   The bill, usually monthly, that a jobber receives from a WD for purchases, and also the bill a jobber sends to credit customers.

Stock Order   A purchase order issued by a jobber to replenish stock.

Supplements   Catalog and price sheet changes used to keep items and prices current until new catalogs are issued.

Tabloid Brochure   A colorful, printed advertising piece used in retail promotions as a newspaper insert, direct mail piece, and for door-to-door delivery.

Trade Discount   A discount from the regular selling price offered to high-volume purchasers such as dealers, fleet owners, and industrial companies.

Traffic   Customers.

Traffic Builders   All types of advertising, promotions, and merchandising materials designed to bring customers into a store.

Turnover   The number of times each year that a jobber buys, sells, and replaces a part number.

Vendor   The supplier or seller.

Walk-In   Do-it-yourself customers.

Warehouse Distributor (WD)   The jobber's supplier. The link between manufacturer and jobber.

Warranty   A document stating that a product will provide satisfactory service for a given period of time or the buyer will be entitled to a settlement according to the terms of the warranty.

Wats   Long-distance telephone service used by companies making a large number of long-distance calls daily.

Wholesale   The jobbers' price to large volume customers such as dealers, fleet owners, and industrial companies. Unlike a trade discount, it is usually a fixed price rather than a percentage off the retail price.

Will-Call   Merchandise held for a customer to pick up.

Window Banners   Large posters displayed in windows to attract passersby to items on sale or promotions in progress.

Wire Hangers   Point-of-sale advertising materials attached to or draped over a wire inside the store.

APPENDIX B

GROSS PROFIT TABLE

It is important to understand that the percentage of gross profit made on an item is not the same as the percentage of markup. For example, if an item costs the store $1.00 and is sold for $1.50, the percentage of markup is 50 percent ($0.50 divided by $1.00). However, the gross profit made on the sale is 33-1/3 percent ($0.50 divided by $1.50).

The formula for determining gross profit percentage is as follows:

$$\text{Gross Profit (percent)} = \frac{\text{markup percent} \times \text{cost}}{\text{cost} + (\text{markup percent} \times \text{cost})}$$

Example: If an item costs $6.67 and is marked up 20%, the gross profit equals:

$$\frac{.20 \times 6.67}{6.67 + (.20 \times 6.67)} = \frac{1.33}{6.67 + 1.33} = \frac{1.33 (\text{markup amount})}{8.00 \ (\text{selling price})} = 16.6\% \text{ gross profit}$$

As you can see, gross profit can be determined by simply dividing the amount charged above cost by the selling price.

The following table lists the markup percentages needed to produce the desired gross profit percentage:

To Make a Gross Profit of (percent)	Markup Cost (percent)	To Make a Gross Profit of (percent)	Markup Cost (percent)
9	10	31	45
10	11.1	32	47
11	12.35	33-1/3	50
12	13.75	35	54
13	15	36	56.25
14	16.35	37	58.75
15	17.65	38	61.25
16	19	39	64
17	20.5	40	66.75
18	22	41	69.75
19	23.5	42	72.5
20	25	43	75.25
21	26.5	44	78.5
22	28.25	45	82
23	30	46	85.25
24	31.5	47	89
25	33.33	48	92.25
26	35.1	49	96
27	37	50	100
28	39	60	150
29	41	72	333
30	43	80	400

WEATHERLY INDEX

The following is a sample listing of automotive code numbers that can be found in the Weatherly Index. Many catalogs list the Weatherly Index code number on the cover. The index is a compilation of three-digit codes that organize the parts catalogs into categories. For example, under code 000, all of the components related to the center drive would be found in catalogs with a 000 code. Complete information can be obtained from Master Products Manufacturing, 3481 East 14th St., Box 23985, Los Angeles, Calif. 90023.

000
CENTER DRIVE

002
Alloy Piston
Iron Piston
Piston

004
Piston Pin

006
Piston Lock Pin
Piston Set Screw

008
Piston Pin Bushing

010
Piston Pin Lock Ring

012
Piston Expander

020
Oil Control Ring
Piston Compression Ring
Piston Ring

022
Piston Ring Expander

030
Connecting Rod

032
Connecting Rod Bearing
Crank Shaft Bearing
Motor Bearing

034
Bearing Screw
Connecting Rod Bolt
Connecting Rod Clamp Screw
Piston Pin Lock Screw

040
Auto Crank
Crank Handle
Crank Hole Cap

042
Cam Shaft
Crank Case Heater
Crank Case, Valves
Crank Shaft
Crank Shaft Balance
Crank Shaft, Bearing Leak Detector
Crank Shaft Counterbalance

044
Cylinder Head

046
Cylinder Sleeve

048
Cylinder Block
Motor Block

050
Timing Chain

052
Timing Chain Sprocket
Timing Gear

056
Exhaust Manifold
Intake Manifold

060
Exhaust Valve
Intake Valve
Motor Valve
Valve
Valve Exhaust, Gas
Valve Hydraulic
Valve Lifter
Valve Rotator

062
Valve Spring

064
Valve Key
Valve Spring Retainer
Valve Spring Seat
Valve Stem Key

066
Rocker Arm Shaft
Valve Pushrod
Valve Roller and Lifter
Valve Silencer
Valve Stem Adjuster
Valve Tappet

068
Valve Grinding Bushing
Valve Guide
Valve Guide Felt
Valve Oil Seal
Valve Packing

072
Valve Insert Ring
Valve insert Seat

076
Flywheel
Flywheel Gear

078
Clutch
Clutch Parts
Clutch Pilot Bushing
Clutch Spring
Free Wheeling Attachments

080
Automatic Transmission Fluid
Clutch Disc
Clutch Facing
Clutch Lining
Clutch Plate
Fluid Drive Fluid

082
Flywheel Housing

084
Automatic Transmission
Automatic Transmission Cooler
Automatic Transmission Parts
Automatic Transmission Radiator
Selector Range, Transmission—Electric
Transfer Case
Transfer Case Gear
Transmission
Transmission Filter
Transmission Gear

086
Power Takeoff
Power Takeoff Parts
Transmission Auxiliary

088
Fabric Universal Joint
Fabric Universal Joint Bolt
Fabric Universal Joint Washer
Universal Joint Disc

090
Drive Shaft
Metal Universal Joint
Metal Universal Joint Parts

Power Takeoff Universal Joint
Propeller Shaft
Universal Joint Bell Housing
Universal Joint Bell Housing Support

092
Universal Joint Boot

094
Differential Adjusting Nut
Differential Gear
Differential Housing
Differential Spider
Pinion Cage
Pinion Lock Pin
Pinion Lock Screw
Pinion Shaft Set Screw
Ring Gear and Pinion
Worm Gear

096
Roller Chain
Roller Chain Binder Link
Roller Chain Sprocket

098
Iron Rivet
Ring Gear Rivet

100
FRAME AND AXLE

100
Axle
Axle Alignment
Axle Shaft
Axle Truss Rod

102
Axle Nut

104
Axle Flange Dowel Washer
Axle Key
Axle Shaft Shim
Propeller Shear Pin
Shaft Lock Key
Shim Sleeve Shaft (Speedi Repair)

108
Axle, Trailer Suspension
Axle, Truck Suspension
Front Axle
Multiple Axle Assembly
Rear Axle Assembly
Tandem Axle Assembly
Trailer Axle

110
Axle Bolt
Axlt Bolt Bushing
Control Arm
Control Arm Parts
Drag Link
Drag Link Spring
Front-End Coil Spring
Front-End Coil Spring Spacer
Front Wheel Suspension Parts
King Bolt
King Bolt Bushing
Knee Action Parts
Knee Action Spring
Spring, Coil
Steering Bolt
Steering Bushing
Suspension, Axle Trailer
Tie-Rod
Tie-Rod Ball Joint
Tie-Rod Bolt
Tie-Rod Bushing

112
King Bolt Dust Plug
King Bolt Lock Pin
King Bolt Set Screw
Steering Dust Plug

120
Brake Band
Brake Block
Brake Lining
Brake Lining Shim Stock
Brake Shoe
Shims, Caster and Camber

122
Aluminum Rivet
Brake Lining Rivet
Brass Rivet
Clutch Facing Rivet
Copper Rivet
Rivet, Blind

124
Brake Drum
Brake Drum Band
Steel Brake Drum Ring

126
Brake Adjusting Screw
Brake Cable
Brake Equalizer
Brake Ratchet

Brack Ratchet Pawl
Brake Rod
Brake Spring
Brake Yoke
Clevis Pin
Front-End Suspension Parts
Mechanical Brake Parts

130
Brake, Adjuster Slack
Brake Dressing
Brake Fluid
Hydraulic Brake Fluid

134
Air Brake
Air Brake Hose
Air Brake Parts
Booster Brake Hose
Brake Booster
Brake Booster Parts
Brake Disc
Electric Brake
Power Brake
Power Brake Parts
Vacuum Brake

135
Air Dryer

136
Anti-Jackknife
Anti-Skid Device
Hydraulic Brake
Hydraulic Brake Cylinder
Hydraulic Brake Hose
Hydraulic Brake Parts
Hydraulic Brake Spinning Wheel
  Stop Attachment
Reserve Brake Fluid Attachment
Reserve Brake Fluid Tank
Starting Motor, Hydraulic
Stuck Car Spinning Wheel Stop
  Attachment

140
Tire and Tube

142
Rubber Cement
Talcum Powder
Tire Boot
Tire Cement
Tire Dusting Powder
Tire Flap
Tire, Leak Sealant
Tire Patch Plug
Tire Patching
Tire Repair Patch
Vulcanizing Patch

144
Tire Valve
Tire Valve Core
Tire Valve Dust Cap
Tire Valve Stem
Tire Valve Stem Lock

148
Skid Control Device
Swerve Control Device
Tire Chain

INDEX

EXHIBIT A
KSS:Author DRIVER Element Program License Agreement

This is a legal agreement between you, the end user, and Comware Incorporated. You should carefully read the following terms and conditions before using the accompanying program. Your initial use of the application program indicates your acceptance of these terms and conditions. If you do not agree with these terms and conditions, you should promptly return the application program to the place where you obtained it.

The accompanying program contains, in part, copyrighted materials that are the property of Comware Incorported, to wit: the KSS:Author Driver Element program (DRIVER program). Comware licenses the use of this program by you only in conjunction with your authorized use of the accompanying program. You assume responsibility for the installation and use of, and results obtained from, the DRIVER program.

LICENSE

Unless otherwise prohibited, you MAY, in conjunction with your authorized use of the accompanying program:

   a) Copy the DRIVER program into any machine-readable form.
   b) Transfer the DRIVER program, along with the accompanying program, with a copy of this Agreement to another party only if the other party agrees to accept from Comware the terms and conditions of this Agreement. If you transfer the DRIVER program, you must at the same time either transfer all copies to the same party or destroy any copies not transferred. This includes all portions of the DRIVER program contained or merged into other programs. Comware grants a license to such other party under this Agreement and the other party will accept such license by its initial use of the accompanying program. If you transfer possession of any copy or merged portion of the DRIVER program, in whole or in part, to another party, your license is automatically terminated.

You may not reverse assemble or reverse compile the DRIVER program. You may not use, copy, modify or transfer the DRIVER program, or any copy or merged portion in whole or in part, except as expressly provided for in this Agreement. You may not sublicense, rent or lease the DRIVER program.

TERM

You may terminate this Agreement at any time by destroying the DRIVER program together with all copies. It will also terminate if you fail to comply with any term or conditions of this Agreement. You agree upon termination to destroy the DRIVER program.

DISCLAIMER OF WARRANTY

THE DRIVER PROGRAM IS PROVIDED "AS IS" WITHOUT WARRANTY OF ANY KIND, EITHER EXPRESS OR IMPLIED, INCLUDING, BUT NOT LIMITED TO, THE IMPLIED WARRANTIES OF MERCHANTABILITY AND FITNESS FOR A PARTICULAR PURPOSE. THE ENTIRE RISK AS TO THE QUALITY AND PERFORMANCE OF THE DRIVER PROGRAM IS WITH YOU. SHOULD THE DRIVER PROGRAM PROVE DEFECTIVE, YOU ASSUME THE ENTIRE COST OF ALL NECESSARY SERVICING, REPAIR, OR CORRECTION. SOME STATES DO NOT ALLOW THE EXCLUSION OF IMPLIED WARRANTIES, SO THE ABOVE EXCLUSION MAY NOT APPLY TO YOU. THIS WARRANTY GIVES YOU SPECIFIC LEGAL RIGHTS AND YOU MAY ALSO HAVE OTHER RIGHTS THAT VARY FROM STATE TO STATE.

LIMITATION OF REMEDIES

IN NO EVENT WILL COMWARE BE LIABLE TO YOU FOR ANY DAMAGES OR ANY LOST PROFITS, LOST SAVINGS, OR OTHER INCIDENTAL CONSEQUENTIAL DAMAGES ARISING OUT OF THE USE OF OR INABILITY TO USE THE DRIVER PROGRAM, EVEN IF COMWARE OR THE PLACE WHERE YOU OBTAINED THE DRIVER AND THE ACCOMPANYING PROGRAM HAS BEEN ADVISED OF THE POSSIBILITY OF SUCH DAMAGES, OR FOR ANY CLAIM BY ANY OTHER PARTY.

SOME STATES DO NOT ALLOW THE LIMITATION OR EXCLUSION OF LIABILITY FOR INCIDENTAL OR CONSEQUENTIAL DAMAGES SO THE ABOVE LIMITATION OR EXCLUSION MAY NOT APPLY TO YOU.

GENERAL

You agree that you will look only to the place where you obtained the program, and not to Comware, for any support, maintenance, assistance or the like with respect to the DRIVER program and the accompanying program, and that Comware shall have no liability to you in relation to these programs.

This agreement will be construed under the applicable laws of the state of Ohio.